Microfluidics for Advanced
Functional Polymeric Materials

Microfluidics for Advanced Functional Polymeric Materials

Liang-Yin Chu and Wei Wang

Authors

Professor Liang-Yin Chu
Sichuan University
School of Chemical Engineering
No. 24, Yihuan Road
First Southern Section
610065 Chengdu
China

Dr. Wei Wang
Sichuan University
School of Chemical Engineering
No. 24 Yihuan Road
First Southern Section
610065 Chengdu
China

Cover
Microchip – fotolia_©science photo;
Microfiber structure – fotolia_©nelik;
Nanostructure – fotolia_©chaoss

All books published by **Wiley-VCH** are carefully produced. Nevertheless, authors, editors, and publisher do not warrant the information contained in these books, including this book, to be free of errors. Readers are advised to keep in mind that statements, data, illustrations, procedural details or other items may inadvertently be inaccurate.

Library of Congress Card No.: applied for

British Library Cataloguing-in-Publication Data
A catalogue record for this book is available from the British Library.

Bibliographic information published by the Deutsche Nationalbibliothek
The Deutsche Nationalbibliothek lists this publication in the Deutsche Nationalbibliografie; detailed bibliographic data are available on the Internet at <http://dnb.d-nb.de>.

© 2017 Wiley-VCH Verlag GmbH & Co. KGaA, Boschstr. 12, 69469 Weinheim, Germany

All rights reserved (including those of translation into other languages). No part of this book may be reproduced in any form – by photoprinting, microfilm, or any other means – nor transmitted or translated into a machine language without written permission from the publishers. Registered names, trademarks, etc. used in this book, even when not specifically marked as such, are not to be considered unprotected by law.

Print ISBN: 978-3-527-34182-5
ePDF ISBN: 978-3-527-80366-8
ePub ISBN: 978-3-527-80365-1
Mobi ISBN: 978-3-527-80364-4
oBook ISBN: 978-3-527-80363-7

Cover Design Adam-Design, Weinheim, Germany
Typesetting SPi Global, Chennai, India
Printing and Binding betz-druck GmbH, Darmstadt, Deutschland

Printed on acid-free paper

Contents

Preface *xiii*

1 Introduction *1*
1.1 Microfluidics and Its Superiority in Controllable Fabrication of Functional Materials *1*
1.2 Microfluidic Fabrication of Microspheres and Microcapsules from Microscale Closed Liquid–Liquid Interfaces *3*
1.3 Microfluidic Fabrication of Membranes in Microchannels from Microscale Nonclosed Layered Laminar Interfaces *4*
1.4 Microfluidic Fabrication of Microfiber Materials from Microscale Nonclosed Annular Laminar Interfaces *5*
References *6*

2 Shear-Induced Generation of Controllable Multiple Emulsions in Microfluidic Devices *11*
2.1 Introduction *11*
2.2 Microfluidic Strategy for Shear-Induced Generation of Controllable Emulsion Droplets *12*
2.3 Shear-Induced Generation of Controllable Monodisperse Single Emulsions *14*
2.4 Shear-Induced Generation of Controllable Multiple Emulsions *16*
2.4.1 Shear-Induced Generation of Controllable Double Emulsions *16*
2.4.2 Shear-Induced Generation of Controllable Triple Emulsions *19*
2.5 Shear-Induced Generation of Controllable Multicomponent Multiple Emulsions *22*
2.5.1 Shear-Induced Generation of Controllable Quadruple-Component Double Emulsions *22*
2.5.2 Extended Microfluidic Device for Controllable Generation of More Complex Multicomponent Multiple Emulsions *27*
2.6 Summary *31*
References *31*

3	Wetting-Induced Generation of Controllable Multiple Emulsions in Microfluidic Devices 35
3.1	Introduction 35
3.2	Microfluidic Strategy for Wetting-Induced Production of Controllable Emulsions 36
3.2.1	Strategy for Wetting-Induced Production of Controllable Emulsions via Wetting-Induced Spreading 36
3.2.2	Strategy for Wetting-Induced Production of Controllable Emulsions via Wetting-Induced Coalescing 37
3.3	Generation of Controllable Multiple Emulsions via Wetting-Induced Spreading 38
3.3.1	Wetting-Induced Generation of Monodisperse Controllable Double Emulsions 38
3.3.2	Wetting-Induced Generation of Monodisperse Higher Order Multiple Emulsions 41
3.3.3	Wetting-Induced Generation of Monodisperse Multiple Emulsions via Droplet-Triggered Droplet Pairing 44
3.4	Generation of Controllable Multiple Emulsions via Wetting-Induced Droplet Coalescing 47
3.5	Summary 50
	References 52

4	Microfluidic Fabrication of Monodisperse Hydrogel Microparticles 55
4.1	Introduction 55
4.2	Microfluidic Fabrication of Monodisperse PNIPAM Hydrogel Microparticles for Sensing Tannic Acid (TA) 55
4.2.1	Microfluidic Fabrication of Monodisperse PNIPAM Hydrogel Microparticles 56
4.2.2	Volume-Phase Transition Behaviors of PNIPAM Microgels Induced by TA 57
4.3	Microfluidic Fabrication of Monodisperse Core–Shell PNIPAM Hydrogel Microparticles for Sensing Ethyl Gallate (EG) 62
4.3.1	Microfluidic Fabrication of Monodisperse Core–Shell PNIPAM Hydrogel Microparticles 62
4.3.2	Thermo-Responsive Phase Transition Behaviors of PNIPAM Microspheres in EG Solution 65
4.3.3	The Intact-to-Broken Transformation Behaviors of Core–Shell PNIPAM Microcapsules in Aqueous Solution with Varying EG Concentrations 65
4.4	Microfluidic Fabrication of Monodisperse Core–Shell Hydrogel Microparticles for the Adsorption and Separation of Pb^{2+} 67
4.4.1	Microfluidic Fabrication of Monodisperse Core–Shell Microparticles with Magnetic Core and Hydrogel Shell 68

4.4.2	Pb^{2+} Adsorption Behaviors of Magnetic PNB Core–Shell Microspheres *71*
4.5	Summary *75*
	References *76*

5	**Microfluidic Fabrication of Monodisperse Porous Microparticles** *79*
5.1	Introduction *79*
5.2	Microfluidic Fabrication of Monodisperse Porous Poly(HEMA-MMA) Microparticles *79*
5.2.1	Microfluidic Fabrication Strategy *80*
5.2.2	Structures of Poly(HEMA-MMA) Porous Microspheres *82*
5.3	Microfluidic Fabrication of Porous PNIPAM Microparticles with Tunable Response Behaviors *83*
5.3.1	Microfluidic Fabrication Strategy *86*
5.3.2	Tunable Response Behaviors of Porous PNIPAM Microparticles *87*
5.4	Microfluidic Fabrication of PNIPAM Microparticles with Open-Celled Porous Structure for Fast Response *90*
5.4.1	Microfluidic Fabrication Strategy *91*
5.4.2	Morphologies and Microstructures of Porous PNIPAM Microparticles *93*
5.4.3	Thermo-Responsive Volume Change Behaviors of PNIPAM Porous Microparticles *98*
5.5	Summary *103*
	References *103*

6	**Microfluidic Fabrication of Uniform Hierarchical Porous Microparticles** *105*
6.1	Introduction *105*
6.2	Microfluidic Strategy for Fabrication of Uniform Hierarchical Porous Microparticles *106*
6.3	Controllable Microfluidic Fabrication of Uniform Hierarchical Porous Microparticles *108*
6.3.1	Preparation of Hierarchical Porous Microparticles *108*
6.3.2	Hierarchical Porous Microparticles with Micrometer-Sized Pores from Deformed W/O/W Emulsions *108*
6.3.3	Integration of Nanometer- and Micrometer-Sized Pores for Creating Hierarchical Porous Microparticles *111*
6.4	Hierarchical Porous Microparticles for Oil Removal *114*
6.4.1	Concept of the Hierarchical Porous Microparticles for Oil Removal *114*
6.4.2	Hierarchical Porous Microparticles for Magnetic-Guided Oil Removal *115*
6.5	Hierarchical Porous Microparticles for Protein Adsorption *116*

viii | Contents

6.5.1	Concept of Hierarchical Porous Microparticles for Protein Adsorption *116*	
6.5.2	Hierarchical Porous Microparticles for Enhanced Protein Adsorption *117*	
6.6	Summary *118*	
	References *118*	

7   Microfluidic Fabrication of Monodisperse Hollow Microcapsules   *123*
7.1   Introduction   *123*
7.2   Microfluidic Fabrication of Monodisperse Ethyl Cellulose Hollow Microcapsules   *124*
7.2.1   Microfluidic Fabrication Strategy   *124*
7.2.2   Morphologies and Structures of Ethyl Cellulose Hollow Microcapsules   *125*
7.3   Microfluidic Fabrication of Monodisperse Calcium Alginate Hollow Microcapsules   *131*
7.3.1   Microfluidic Fabrication Strategy   *132*
7.3.2   Morphologies and Structures of Calcium Alginate Hollow Microcapsules   *133*
7.4   Microfluidic Fabrication of Monodisperse Glucose-Responsive Hollow Microcapsules   *136*
7.4.1   Microfluidic Fabrication Strategy   *136*
7.4.2   Glucose-Responsive Behaviors of Microcapsules   *140*
7.4.3   Glucose-Responsive Drug Release Behaviors of Microcapsules   *142*
7.5   Microfluidic Fabrication of Monodisperse Multi-Stimuli-Responsive Hollow Microcapsules   *144*
7.5.1   Microfluidic Fabrication Strategy   *145*
7.5.2   Stimuli-Responsive Behaviors of Microcapsules   *150*
7.5.3   Controlled-Release Characteristics of Multi-Stimuli-Responsive Microcapsules   *154*
7.6   Summary   *158*
        References   *158*

8   Microfluidic Fabrication of Monodisperse Core–Shell Microcapsules   *161*
8.1   Introduction   *161*
8.2   Microfluidic Strategy for Fabrication of Monodisperse Core–Shell Microcapsules   *162*
8.3   Smart Core–Shell Microcapsules for Thermo-Triggered Burst Release   *162*
8.3.1   Fabrication of Core–Shell Microcapsules for Thermo-Triggered Burst Release of Oil-Soluble Substances   *162*
8.3.2   Fabrication of Core–Shell Microcapsules for Thermo-Triggered Burst Release of Nanoparticles   *166*
8.3.3   Fabrication of Core–Shell Microcapsules for Direction-Specific Thermo-Responsive Burst Release   *168*

8.4	Smart Core–Shell Microcapsules for Alcohol-Responsive Burst Release *171*	
8.5	Smart Core–Shell Microcapsules for K$^+$-Responsive Burst Release *174*	
8.6	Smart Core–Shell Microcapsules for pH-Responsive Burst Release *176*	
8.6.1	Concept of the Core–Shell Microcapsules for pH-Responsive Burst Release *176*	
8.6.2	Fabrication of the Core–Shell Chitosan Microcapsules *177*	
8.6.3	Core–Shell Chitosan Microcapsules for pH-Responsive Burst Release *179*	
8.7	Summary *182*	
	References *183*	
9	**Microfluidic Fabrication of Monodisperse Hole–Shell Microparticles** *187*	
9.1	Introduction *187*	
9.2	Microfluidic Strategy for Fabrication of Monodisperse Hole–Shell Microparticles *188*	
9.3	Hole–Shell Microparticles for Thermo-Driven Crawling Movement *188*	
9.3.1	Concept of the Hole–Shell Microparticles for Thermo-Driven Crawling Movement *188*	
9.3.2	Fabrication of Hole–Shell Microparticles for Thermo-Driven Crawling Movement *190*	
9.3.3	Effect of Inner Cavity on the Thermo-Responsive Volume-Phase Transition Behaviors of Hole–Shell Microparticles *191*	
9.3.4	Hole–Shell Microparticles for Thermo-Driven Crawling Movement *193*	
9.4	Hole–Shell Microparticles for Pb^{2+} Sensing and Actuating *195*	
9.4.1	Fabrication of Hole–Shell Microparticles for Pb^{2+} Sensing and Actuating *195*	
9.4.2	Magnetic-Guided Targeting Behavior of Poly(NIPAM-*co*-B18C6Am) Hole–Shell Microparticles *195*	
9.4.3	Effects of Pb^{2+} on the Thermo-Responsive Volume Change Behaviors of Poly(NIPAM-*co*-B18C6Am) Hole–Shell Microparticles *196*	
9.4.4	Effects of Hollow Cavity on the Time-Dependent Volume Change Behaviors of Poly(NIPAM-*co*-B18C6Am) Hole–Shell Microparticles *199*	
9.4.5	Micromanipulation of Poly(NIPAM-*co*-B18C6Am) Hole–Shell Microparticles for Preventing Pb^{2+} Leakage from Microcapillary *200*	
9.5	Hole–Shell Microparticles for Controlled Capture and Confined Microreaction *201*	
9.5.1	Microfluidic Fabrication of Hole–Shell Microparticles *201*	
9.5.2	Precise Control over the Hole–Shell Structure of the Microparticles *203*	
9.5.3	Precise Control over the Functionality of Hollow Core Surface *205*	

9.5.4	Hole–Shell Microparticles for Controlled Capture and Confined Microreaction *206*
9.6	Summary *207*
	References *207*

10 Microfluidic Fabrication of Controllable Multicompartmental Microparticles *211*

10.1	Introduction *211*
10.2	Microfluidic Strategy for the Fabrication of Controllable Multicompartmental Microparticles *212*
10.3	Multi-core/Shell Microparticles for Co-encapsulation and Synergistic Release *212*
10.3.1	Microfluidic Fabrication of Multi-core/Shell Microparticles *212*
10.3.2	Multi-core/Shell Microparticles for Controllable Co-encapsulation *213*
10.3.3	Multi-core/Shell Microparticles for Synergistic Release *216*
10.4	Trojan-Horse-Like Microparticles for Co-delivery and Programmed Release *217*
10.4.1	Fabrication of Trojan-Horse-Like Microparticles from Triple Emulsions *217*
10.5	Summary *218*
	References *219*

11 Microfluidic Fabrication of Functional Microfibers with Controllable Internals *223*

11.1	Introduction *223*
11.2	Microfluidic Strategy for Fabrication of Functional Microfibers with Controllable Internals *224*
11.3	Core–Sheath Microfibers with Tubular Internals for Encapsulation of Phase Change Materials *224*
11.3.1	Fabrication of Core–Sheath Microfibers with Tubular Internals *224*
11.3.2	Morphological Characterization of the Core–Sheath Microfibers *227*
11.3.3	Thermal Property of the Core–Sheath Microfibers *227*
11.3.4	Core–Sheath Microfibers for Temperature Regulation *230*
11.4	Peapod Like Microfibers with Multicompartmental Internals for Synergistic Encapsulation *235*
11.4.1	Fabrication of Peapod-Like Microfibers with Multicompartmental Internals *236*
11.4.2	Effects of Flow Rates on the Structures of Peapod-Like Jet Templates and Chitosan Microfibers *236*
11.4.3	Peapod-Like Chitosan Microfibers with Multicompartment Internals for Synergistic Encapsulation *240*
11.5	Spider-Silk-Like Microfibers with Spindle-Knot Internals for 3D Assembly and Water Collection *241*
11.5.1	Fabrication of Spider-Silk-Like Microfibers with Spindle-Knot Internals *241*

11.5.2	Morphological Characterization of the Jet Templates and Spider-Silk-Like Microfibers *242*	
11.5.3	Magnetic-Guided Patterning and Assembling of the Ca-Alginate Microfibers *244*	
11.5.4	Water Collection of Dehydrated Ca-Alginate Microfibers with Magnetic Spindle-Knot Internals *246*	
11.6	Summary *248*	
	References *248*	
12	**Microfluidic Fabrication of Membrane-in-a-Chip with Self-Regulated Permeability** *253*	
12.1	Introduction *253*	
12.2	Microfluidic Strategy for Fabrication of Membrane-in-a-Chip *254*	
12.3	Temperature- and Ethanol-Responsive Smart Membrane in Microchip for Detection *255*	
12.3.1	Fabrication of Nanogel-Containing Smart Membrane in Microchip *255*	
12.3.2	Temperature-Responsive Self-Regulation of the Membrane Permeability *257*	
12.3.3	Ethanol-Responsive Self-Regulation of the Membrane Permeability *260*	
12.3.4	Reversible and Repeated Thermo/Ethanol-Responsive Self-Regulation of the Membrane Permeability *263*	
12.4	Summary *264*	
	References *264*	
13	**Microfluidic Fabrication of Microvalve-in-a-Chip** *267*	
13.1	Introduction *267*	
13.2	Microfluidic Strategy for Fabrication of Microvalve-in-a-Chip *268*	
13.2.1	Fabrication of Thermo-Responsive Hydrogel Microvalve within Microchip for Thermostatic Control *268*	
13.2.2	Fabrication of Pb^{2+}-Responsive Hydrogel Microvalve within Microchip for Pb^{2+} Detection *270*	
13.3	Smart Microvalve-in-a-Chip with Thermostatic Control for Cell Culture *272*	
13.3.1	Setup of Microvalve-Integrated Micro-heat-Exchanging System *273*	
13.3.2	Thermo-Responsive Switch Performance of Hydrogel Microvalve *274*	
13.3.3	Sealing Performance of Hydrogel Microvalve *276*	
13.3.4	Temperature Self-Regulation of Hydrogel Microvalve for Thermostatic Control *277*	
13.3.5	Temperature Self-Regulation with Hydrogel Microvalve for Cell Culture *279*	
13.4	Smart Microvalve-in-a-Chip with Ultrasensitivity for Real-Time Detection *281*	
13.4.1	Concept of the Microchip Incorporated with Pb^{2+}-Responsive Microgel for Real-Time Online Detection of Trace Pb^{2+} *281*	
13.4.2	Sensitivity of the Pb^{2+} Detection Platform *283*	

	13.4.3	Selectivity and Repeatability of the Pb^{2+} Detection Platform *284*
	13.4.4	Setup of Pb^{2+} Detection System for Real-Time Online Detection of Pb^{2+} in Tap Water for Pollution Warning *287*
	13.4.5	Setup of Pb^{2+} Detection System for Real-Time Online Detection of Pb^{2+} in Wastewater from a Model Industrial Factory for Pollution Warning and Terminating *289*
	13.5	Summary *289*
		References *289*

14 Summary and Perspective *295*
14.1   Summary *295*
14.2   Perspective *295*
       References *297*

Index *299*

Preface

Microfluidics, or the so-called lab-on-a-chip, has emerged as a distinct new technology since the beginning of the 1990s. The dimensions of the microfluidic channels and components are tens to hundreds of micrometers. The microfluidic devices can be used to flexibly manipulate the flow of microvolume fluids in microchannels, which are considered putting the lab on a chip. Due to the trend of miniaturization and integration of modern scientific and technological development, microfluidic technology has been widely concerned and valued by the international scientific and industrial communities. Since microfluidic technology can accurately manipulate small-volume fluids, it is rapidly extending from the original analytical chemistry platform for microanalysis and microdetection to high-throughput drug screening, micromixing, microreaction, microseparation, and so on. Due to its excellent ability to control fluid interfaces as well as excellent heat and mass transfer performances, microfluidic technology has become a novel and promising material preparation technology platform. Microfluidic technology has emerged in the construction of precisely controllable microstructured new functional materials with high performances and especially shows incomparable creativity and superiority compared with traditional technology in the design and preparation of some new functional materials with high added values.

This book, entitled *Microfluidics for Advanced Functional Polymeric Materials*, comprehensively and systematically treats modern understanding of the microfluidic technique and its great power in controllable fabrication of advanced functional polymeric materials. The contents range from the design and fabrication of microfluidic devices, the fundamentals and strategies for controllable microfluidic generation of multiphase liquid systems (e.g., discrete multiple emulsions and continuous laminar multiflow systems), and the use of these liquid systems with elaborate combination of their structures and compositions for controllable fabrication of advanced functional polymeric materials (e.g., solid microparticles, porous microparticles, hollow microcapsules, core–shell microcapsules, hole–shell microcapsules, multicompartmental microcapsules, microfibers, in-chip membranes, and microvalves). All the chapters together clearly describe the design concepts and fabrication strategies of advanced functional polymeric materials with microfluidics by combining the structures with the compositions of multiphase liquid systems to achieve advanced and novel functions. Vivid schematics and illustrations throughout the book enhance the accessibility to the relevant theory and technologies. This book aims to be a definitive reference book for a wide general readership including chemists, chemical engineers, materials researchers, pharmaceutical scientists, biomedical researchers, and students in the related fields.

The book is composed of 14 chapters. In Chapter 1, a brief introduction of the superiority and potential of microfluidics in the construction of microscale phase interfaces and preparation of novel functional materials are briefly introduced. In Chapters 2 and 3, microfluidic strategies for shear-induced and wetting-induced generations of controllable multiple emulsions are introduced, respectively. In Chapters 4–6, microfluidic strategies for controllable fabrication of monodisperse hydrogel microparticles, porous microparticles, and hierarchical porous microparticles are introduced, respectively. In Chapters 7 and 8, microfluidic strategies for controllable fabrication of monodisperse hollow microcapsules and core–shell microcapsules with an oil core and a stimuli-responsive hydrogel shell are introduced, respectively. In Chapter 9, the microfluidic strategies for fabrication of controllable hole–shell microparticles from double emulsions are introduced. In Chapter 10, a microfluidic strategy for template synthesis of multicompartmental microparticles, with accurate control over the structures of their inner compartments and the encapsulation characteristics of their loaded contents, is introduced. In Chapter 11, simple and versatile microfluidic strategies for controllable fabrication of functional microfibers with tubular, peapod-like, and spindle-knot-like internals are introduced. In Chapter 12, a microfluidic strategy for *in situ* fabrication of nanogel-containing smart membranes in the microchannel of microchips is introduced. In Chapter 13, fabrication and performance of microchips incorporated with smart hydrogel microvalves for thermostatic control and trace analytes detection are introduced. Finally, perspectives on the microfluidic fabrication of advanced functional polymeric materials are given in Chapter 14.

The authors' group at Sichuan University (group website: http://teacher.scu.edu.cn/ftp_teacher0/cly/) has been devoted to the microfluidic fabrications of polymeric functional materials since 2006. In the past decade, they have made significant contributions to the development of this field. Most of the contents in this book are the fresh achievements of the authors' group on advanced functional materials fabricated with microfluidics. Prof. Liang-Yin Chu wrote Chapters 1, 4, 5, 7, 12, and 14, and Prof. Wei Wang wrote Chapters 2, 3, 6, 8–11, and 13. The authors are very grateful to Prof. David A. Weitz at Harvard University who helped the authors a lot to carry out investigations in the field of microfluidics. The authors would like to thank all the current and former group members who contributed to the investigations on microfluidics, especially Prof. Rui Xie, Prof. Xiao-Jie Ju, Prof. Zhuang Liu, Dr Li Liu, Dr Nan-Nan Deng, Dr Mao-Jie Zhang, Dr Zhi-Jun Meng, Dr Ya-Lan Yu, Dr Jie Wei, Dr Lei Zhang, Dr Chuan-Lin Mou, Dr Li-Li Yue, Dr Gang Chen, Dr Ying-Mei Liu, Dr Hai-Rong Yu, Dr Xiao-Heng He, Dr Ming-Yue Jiang, Dr Shuo Lin, Dr Fang Wu, Dr Xiao-Yi Zou, Jian Sun, Hao Zhang, Ping-Wei Ren, Jian-Ping Yang, Shuo-Wei Pi, Xi Lin, Guo-Qing Wen, Yi-Meng Sun, Chao Yang, Wei-Chao Zheng, Mei Yuan, Xiu-Lan Yang, and Ming Li, for their creative researches on microfluidics. The authors gratefully acknowledge all the professors, friends, and colleagues who helped the authors' group to carry out investigations on microfluidics, and thank all the organizations who financially supported the authors' group for the continuous study in the field of microfluidics.

Finally, the authors would like to acknowledge the kind help of Dr Lifen Yang and the editorial staff at Wiley during the preparation and publication of this book.

September 2016

Liang-Yin Chu
Sichuan University Chengdu, China

1

Introduction

1.1 Microfluidics and Its Superiority in Controllable Fabrication of Functional Materials

Microfluidics, or the so-called lab-on-a-chip, has emerged as a distinct new technology since the beginning of 1990s [1]. The dimensions of the microfluidic channels and components are tens to hundreds of micrometers. The microfluidic devices can be used to flexibly manipulate the flow of microvolume fluids in microchannels, which are considered putting the lab on a chip (Figure 1.1). Due to the trend of miniaturization and integration of modern scientific and technological civilization development, microfluidic technology has been widely concerned and valued by the international scientific and industrial communities. In 2006, *Nature* magazine published a special issue on the topic of "*Insight: lab on a chip*," including seven related review papers [1, 2], where the editorial says that it might have the potential to become "a technology for this century." In 2010, *Chemical Society Reviews* published a special issue on the topic of "*From microfluidic applications to nanofluidic phenomena*," including 20 related review papers [3], which shows the promising momentum of development of microfluidic technologies. Since microfluidic technology can accurately manipulate small-volume fluids, it is rapidly extending from the original analytical chemistry platform for microanalysis and microdetection, to high-throughput drug screening, micromixing, microreaction, microseparation, and so on. Due to its excellent ability to control fluid interfaces as well as excellent heat and mass transfer performances, microfluidic technology has become a novel and promising material preparation technology platform (Figure 1.2). Microfluidic technology has emerged in the construction of precisely controllable microstructured new functional materials with high performances, such as microcapsules and microspheres, membranes in microchannels, and superfine fiber materials, and especially shows incomparable creativity and superiority compared with traditional technology in design and preparation of some new functional materials with high added values [4–35].

To sum up, stable phase interface structures of immiscible liquid phase systems that are constructed by the microfluidic technology can be mainly divided into two systems [4]: one is the emulsion droplet system with closed liquid–liquid interfaces, and the second is the laminar flow system with closed

Figure 1.1 Microfluidics: putting the lab on a chip.

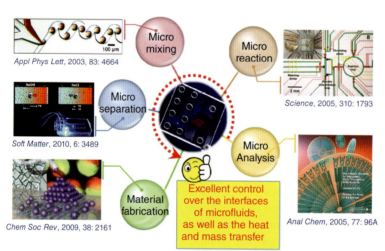

Figure 1.2 Microfluidic technology is becoming a novel technology platform for materials preparation because of its excellent control over the microfluid interfaces as well as the heat and mass transfer.

liquid–liquid interfaces (Figure 1.3). These two microfluidic-constructed stable phase interface structure systems can be used to prepare three categories of high-performance functional materials with accurate and controllable microstructures as follows [4–35]: (i) controllable fabrication of novel microspheres and microcapsules with precise microstructures by using emulsion droplet systems with closed phase interfaces as templates [4, 5, 7–32]; (ii) controllable fabrication of membranes in microchannels by using laminar flow systems with nonclosed layered phase interfaces [6, 33, 36, 37]; and (iii) controllable fabrication of novel microfiber materials by using laminar flow systems with nonclosed annular phase interfaces [4, 10, 34, 38]. As illustrated in Figure 1.3, microfluidic technology shows superior controllability and great potential in the construction of these three kinds of functional materials, and

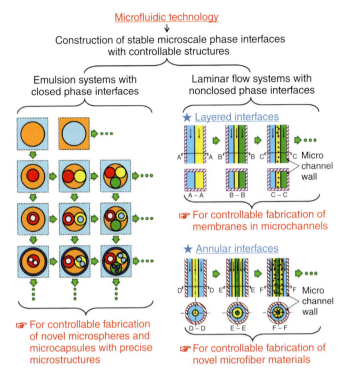

Figure 1.3 The system diagram of microfluidic method for the construction of stable microscale phase interfaces and for controllable preparation of novel functional materials.

can play its unique advantages in controllable construction of new functional materials with new structures, new functions, and high-performance features.

1.2 Microfluidic Fabrication of Microspheres and Microcapsules from Microscale Closed Liquid–Liquid Interfaces

Due to the small size and controllable internal structure, microspheres and microcapsules can be used as microcarriers, microreactors, microseparators, and microstructural units in drug delivery, substance encapsulation, chemical catalysis, biochemical separation, artificial cells, and enzyme immobilization, and have very broad application prospects. Microspheres and microcapsules are generally fabricated by using emulsion droplets with stable closed liquid–liquid interfaces (e.g., single water-in-oil (W/O) or oil-in-water (O/W) emulsions, W/O/W or O/W/O double emulsions, or even more complicated multiple emulsions) as templates, through subsequent polymerization, cross-linking, solvent evaporation, curing, and assembling in emulsion droplets or at interfaces. Traditional methods for the preparation of emulsion droplets are mainly achieved by mechanical stirring or fluid shear; thus, the sizes and the internal structures of the droplets and the resultant template-fabricated microspheres and microcapsules are difficult to be controlled precisely, which greatly affect the performances and applications of the microspheres and microcapsules.

Microfluidic technology, which can generate emulsion droplets by emulsifying disperse phase to continuous phase through microchannels with co-flow, flow-focusing, or T-junction geometries, can achieve continuous and precise control of the microstructures of emulsion droplets, exhibiting significant superiority in the fabrication of microspheres and microcapsules with controllable size distributions and microstructures.

Researchers from all over the world have made a lot of important progress in the use of microfluidics to construct microscale closed liquid–liquid interfaces and then fabricate monodisperse microspheres and microcapsules [4, 5, 7–32]. In the preparations of microspheres and microcapsules with microfluidic approaches, most of them are focused on the use of microfluidic-generated W/O or O/W single emulsions (as shown in the first row in the upper left corner of Figure 1.3) as templates for preparing monodisperse microspheres, or the use of W/O/W or O/W/O double emulsions (as shown in the second row in the first column of the upper left corner of Figure 1.3) as templates for preparing monodisperse core–shell microcapsules. Some studies have also attempted to prepare some materials with new structures such as multicore microspheres, Janus microspheres, and nonspherical particles by microfluidic technology.

The authors' group controllably constructed multiple emulsion systems with complex microscale multiphase multicomponent liquid–liquid interfaces by building series and parallel microchannels [31]. These emulsions are used as templates for controllably preparing multiphase multicomponent microspheres and microcapsules for the encapsulation of substances [29], as well as new multifunctional microspheres and microcapsules with complex structures [28, 30].

1.3 Microfluidic Fabrication of Membranes in Microchannels from Microscale Nonclosed Layered Laminar Interfaces

Because of the excellent performances in catalysis, separations, purifications, analysis and detection, controlled release, emulsification, and so on, functional membrane materials are considered as one of the important supporting technologies for sustainable development. If the combination of membrane materials and microfluidic technology is obtained, it will play the synergy of the two to achieve the integration of functional materials and components. In this way, it can not only promote the application of membrane materials in microseparation and microanalysis but also provide new catalysis- or reaction-separation coupling technologies for microchemical or microreaction processes, showing very broad application prospects [6]. Therefore, as a new technology platform, membrane-in-microchannel technology is increasingly subject to different disciplines of international attention [6].

In a co-flow microchannel, when immiscible multiphase fluids flows into the same microchannel, stable layered laminar flow patterns can be formed through microfluidic laminar flow technology [36] ("Layered interfaces" in Figure 1.3). In each phase, the fluid can maintain its flow pattern unchanged; chemical reactions such as polymerization and cross-linking only occur at the liquid–liquid interfaces, forming monolayer or multilayer parallel ultrathin membranes in the microchannels.

The microchannels can be divided into several independent channels by the membranes in microchannels. Due to the selective permeability or adsorption ability of functional membrane materials, selective separation, extraction, detection, and analysis can be realized with the membranes in the microchannels. Catalysts can also be effectively deposited on the membrane surfaces, thereby increasing the specific surface area of the catalytic material within a microchannel, to accelerate the rate of catalytic reaction in the microchannel. In addition, environmental stimuli-responsive smart membranes, which can regulate the effective membrane pore size and permeability in response to the change in physical or chemical signals in the environment, show incomparable superiority over traditional membranes [39]. If the smart membranes can be combined with microfluidics, it will undoubtedly provide efficient technology platform for the intensification of microseparation, microanalysis and detection, microreaction processed, and the enhancement of membrane performances.

Since the fabrication processes of membranes in microchannels are different from traditional membrane preparation processes, so far there are only a few reports on the fabrications of membranes with limited materials such as polyamide and chitosan in microchannels by using microfluidic laminar flow technology [6, 33, 36, 37, 40, 41].

1.4 Microfluidic Fabrication of Microfiber Materials from Microscale Nonclosed Annular Laminar Interfaces

Microfiber materials have a wide range of applications in optoelectronics, biomedicine, chemical industry, light industry, and other fields, wherein the hollow fiber membranes play an important role in the chemical separation processes. Currently, the preparations of microfiber materials are mainly achieved by using melt spinning method, electrospinning method, and other methods, while these methods are still difficult to achieve precise control of the microstructures of microfiber materials or impart multifunctional characteristics. Therefore, it is still necessary to seek new preparation processes and methods for the preparation of microfiber materials, and the microfluidic laminar flow technology is a very promising new method.

With microfluidic laminar flow technology, stable annular laminar flow patterns of immiscible multiphase fluids can be formed in the microchannels [4] ("Annular interfaces" in Figure 1.3). With these stable annular multiphase laminar interface systems as templates, microfiber materials including linear solid microfibers, hollow tubular microfibers, and core–shell composite microfibers can be fabricated by reaction or curing at the liquid–liquid interfaces or inside phases.

Since microfluidic technology enables continuous and accurate control over the annular liquid–liquid interfaces of laminar flows, it can provide optimal design of fibrous material synthesis systems. Therefore, compared with traditional spinning techniques, microfluidic technology has significant advantages in precise regulation and design of microfiber microstructures: it improves performances and imparts multifunctional characteristics of microfiber materials [4, 34, 36, 38, 42–47].

Microfluidic technology has been used to successfully prepare calcium alginate, polyvinyl alcohol, poly(lactic-co-glycolic acid), liposomes, chitosan, poly(ether sulfone), and polyacrylonitrile microfibers [34, 38, 42–47], which show excellent flexibility and extraordinary potential in the construction of microscale annular liquid–liquid interfaces and preparation of microfiber materials.

References

1 Whitesides, G.M. (2006) The origins and the future of microfluidics. *Nature*, **442**, 368–373.
2 Daw, R. and Finkelstein, J. (2006) Insight: lab on a chip. *Nature*, **442**, 367–418.
3 van den Berg, A., Craighead, H.G., and Yang, P. (2010) From microfluidic applications to nanofluidic phenomena. *Chem. Soc. Rev.*, **39**, 899–1217.
4 Atencia, J. and Beebe, D.J. (2005) Controlled microfluidic interfaces. *Nature*, **437**, 648–655.
5 Joanicot, M. and Ajdari, A. (2005) Applied physics – droplet control for microfluidics. *Science*, **309**, 887–888.
6 de Jong, J., Lammertink, R.G.H., and Wessling, M. (2006) Membranes and microfluidics: a review. *Lab Chip*, **6**, 1125–1139.
7 Utada, A.S., Chu, L.Y., Fernandez-Nieves, A., Link, D.R., Holtze, C., and Weitz, D.A. (2007) Dripping, jetting, drops, and wetting: the magic of microfluidics. *MRS Bull.*, **32**, 702–708.
8 Shah, R.K., Shum, H.C., Rowat, A.C., Lee, D., Agresti, J.J., Utada, A.S., Chu, L.Y., Kim, J.W., Fernandez-Nieves, A., Martinez, C.J., and Weitz, D.A. (2008) Designer emulsions using microfluidics. *Mater. Today*, **11**, 18–27.
9 Dendukuri, D. and Doyle, P.S. (2009) The synthesis and assembly of polymeric microparticles using microfluidics. *Adv. Mater.*, **21**, 4071–4086.
10 Tumarkin, E. and Kumacheva, E. (2009) Microfluidic generation of microgels from synthetic and natural polymers. *Chem. Soc. Rev.*, **38**, 2161–2168.
11 Theberge, A., Courtois, F., Schaerli, Y., Fischlechner, M., Abell, C., Hollfelder, F., and Huck, W. (2010) Microdroplets in microfluidics: an evolving platform for discoveries in chemistry and biology. *Angew. Chem. Int. Ed.*, **49**, 5846–5868.
12 Marre, S. and Jensen, K.F. (2010) Synthesis of micro and nanostructures in microfluidic systems. *Chem. Soc. Rev.*, **39**, 1183–1202.
13 Chu, L.Y., Utada, A.S., Shah, R.K., Kim, J.W., and Weitz, D.A. (2007) Controllable monodisperse multiple emulsions. *Angew. Chem. Int. Ed.*, **46**, 8970–8974.

14 Chu, L.Y., Kim, J.W., Shah, R.K., and Weitz, D.A. (2007) Monodisperse thermoresponsive microgels with tunable volume-phase transition kinetics. *Adv. Funct. Mater.*, **17**, 3499–3504.

15 Wang, W., Liu, L., Ju, X.J., Zerrouki, D., Xie, R., Yang, L.H., and Chu, L.Y. (2009) A novel thermo-induced self-bursting microcapsule with magnetic-targeting property. *ChemPhysChem*, **10**, 2405–2409.

16 Zhou, M.Y., Xie, R., Ju, X.J., Zhao, Z.L., and Chu, L.Y. (2009) Flow characteristics of thermo-responsive microspheres in microchannel during the phase transition. *AIChE J.*, **55**, 1559–1568.

17 Liu, L., Yang, J.P., Ju, X.J., Xie, R., Yang, L.H., Liang, B., and Chu, L.Y. (2009) Microfluidic preparation of monodisperse ethyl cellulose hollow microcapsules with non-toxic solvent. *J. Colloid Interface Sci.*, **336**, 100–106.

18 Zhang, H., Ju, X.J., Xie, R., Cheng, C.J., Ren, P.W., and Chu, L.Y. (2009) A microfluidic approach to fabricate monodisperse hollow or porous poly(HEMA-MMA) microspheres using single emulsions as templates. *J. Colloid Interface Sci.*, **336**, 235–243.

19 Liu, L., Wang, W., Ju, X.J., Xie, R., and Chu, L.Y. (2010) Smart thermo-triggered squirting capsules for nanoparticle delivery. *Soft Matter*, **6**, 3759–3763.

20 Ren, P.W., Ju, X.J., Xie, R., and Chu, L.Y. (2010) Monodisperse alginate microcapsules with oil core generated from a microfluidic device. *J. Colloid Interface Sci.*, **343**, 392–395.

21 Pi, S.W., Ju, X.J., Wu, H.G., Xie, R., and Chu, L.Y. (2010) Smart responsive microcapsules capable of recognizing heavy metal ions. *J. Colloid Interface Sci.*, **349**, 512–518.

22 Yu, Y.L., Xie, R., Zhang, M.J., Li, P.F., Yang, L.H., Ju, X.J., and Chu, L.Y. (2010) Monodisperse microspheres with poly(N-isopropylacrylamide) core and poly(2-hydroxyethyl methacrylate) shell. *J. Colloid Interface Sci.*, **346**, 361–369.

23 Wei, J., Ju, X.J., Xie, R., Mou, C.L., Lin, X., and Chu, L.Y. (2011) Novel cationic pH-responsive poly(N,N-dimethylaminoethyl methacrylate) microcapsules prepared by a microfluidic technique. *J. Colloid Interface Sci.*, **357**, 101–108.

24 Liu, L., Yang, J.P., Ju, X.J., Xie, R., Liu, Y.M., Wang, W., Zhang, J.J., Niu, C.H., and Chu, L.Y. (2011) Monodisperse core-shell chitosan microcapsules for pH-responsive burst release of hydrophobic drugs. *Soft Matter*, **7**, 4821–4827.

25 Liu, L., Wu, F., Ju, X.J., Xie, R., Wang, W., Niu, C.H., and Chu, L.Y. (2013) Preparation of monodisperse calcium alginate microcapsules via internal gelation in microfluidic-generated double emulsions. *J. Colloid Interface Sci.*, **404**, 85–90.

26 Zhang, M.J., Wang, W., Xie, R., Ju, X.J., Liu, L., Gu, Y.Y., and Chu, L.Y. (2013) Microfluidic fabrication of monodisperse microcapsules for glucose-response at physiological temperature. *Soft Matter*, **9**, 4150–4159.

27 Wang, W., Yao, C., Zhang, M.J., Ju, X.J., Xie, R., and Chu, L.Y. (2013) Thermo-driven microcrawlers fabricated via a microfluidic approach. *J. Phys. D: Appl. Phys.* (Special Issue on Microfluidics), **46**, 114007.

28 Wang, W., Zhang, M.J., Xie, R., Ju, X.J., Yang, C., Mou, C.L., Weitz, D.A., and Chu, L.Y. (2013) Hole-shell microparticles from controllably evolved double emulsions. *Angew. Chem. Int. Ed.*, **52**, 8084–8087.

29 Wang, W., Luo, T., Ju, X.J., Xie, R., Liu, L., and Chu, L.Y. (2012) Microfluidic preparation of multicompartment microcapsules for isolated co-encapsulation and controlled release of diverse components. *Int. J. Nonlinear Sci. Numer. Simul.* (Special Issue on Microfluidics), **13**, 325–332.

30 Liu, Y.M., Wang, W., Zheng, W.C., Ju, X.J., Xie, R., Zerrouki, D., Deng, N.N., and Chu, L.Y. (2013) Hydrogel-based micro-actuators with remote-controlled locomotion and fast Pb^{2+}-response for micromanipulation. *ACS Appl. Mater. Interfaces*, **5**, 7219–7226.

31 Wang, W., Xie, R., Ju, X.J., Luo, T., Liu, L., Weitz, D.A., and Chu, L.Y. (2011) Controllable microfluidic production of multicomponent multiple emulsions. *Lab Chip*, **11**, 1587–1592.

32 Wang, W., Zhang, M.J., and Chu, L.Y. (2014) Functional polymeric microparticles engineered from controllable microfluidic emulsions. *Acc. Chem. Res.*, **47**, 373–384.

33 Sun, Y.M., Wang, W., Wei, Y.Y., Deng, N.N., Liu, Z., Ju, X.J., Xie, R., and Chu, L.Y. (2014) *In situ* fabrication of temperature- and ethanol-responsive smart membrane in microchip. *Lab Chip*, **14**, 2418–2427.

34 He, X.H., Wang, W., Liu, Y.M., Jiang, M., Wu, F., Deng, K., Liu, Z., Ju, X.J., Xie, R., and Chu, L.Y. (2015) Microfluidic fabrication of bio-inspired microfibers with controllable magnetic spindle-knots for 3D assembly and water collection. *ACS Appl. Mater. Interfaces*, **7**, 17471–17481.

35 Lin, S., Wang, W., Ju, X.J., Xie, R., Liu, Z., Yu, H.R., Zhang, C., and Chu, L.Y. (2016) Ultrasensitive microchip based on smart microgel for real-time on-line detection of trace threat analytes. *Proc. Natl. Acad. Sci. U. S. A.*, **113**, 2023–2028.

36 Kenis, P.J.A., Ismagilov, R.F., and Whitesides, G.M. (1999) Microfabrication inside capillaries using multiphase laminar flow patterning. *Science*, **285**, 83–85.

37 Hisamoto, H., Shimizu, Y., Uchiyama, K., Tokeshi, M., Kikutani, Y., Hibara, A., and Kitamori, T. (2003) Chemicofunctional membrane for integrated chemical processes on a microchip. *Anal. Chem.*, **75**, 350–354.

38 Lan, W.J., Li, S.W., Lu, Y.C., Xu, J.H., and Luo, G.S. (2009) Controllable preparation of microscale tubes with multiphase co-laminar flow in a double co-axial microdevice. *Lab Chip*, **9**, 3282–3288.

39 Chu, L.Y. (2011) *Smart Membrane Materials and Systems*, Springer, Berlin and Heidelberg.

40 Uozumi, Y., Yamada, Y.M.A., Beppu, T., Fukuyama, N., Ueno, M., and Kitamori, T. (2006) Instantaneous carbon–carbon bond formation using a microchannel reactor with a catalytic membrane. *J. Am. Chem. Soc.*, **128**, 15994–15995.

41 Luo, X.L., Berlin, D.L., Betz, J., Payne, G.F., Bentley, W.E., and Rubloff, G.W. (2010) In situ generation of pH gradients in microfluidic devices for biofabrication of freestanding, semi-permeable chitosan membranes. *Lab Chip*, **10**, 59–65.

42 Jeong, W., Kim, J., Kim, S., Lee, S., Mensing, G., and Beebe, D.J. (2004) Hydrodynamic microfabrication via "on the fly" photopolymerization of microscale fibers and tubes. *Lab Chip*, **4**, 576–580.

43 Dittrich, P.S., Heule, M., Renaud, P., and Manz, A. (2006) On-chip extrusion of lipid vesicles and tubes through microsized apertures. *Lab Chip*, **6**, 488–493.

44 Shin, S.J., Park, J.Y., Lee, J.Y., Park, H., Park, Y.D., Lee, K.B., Whang, C.M., and Lee, S.H. (2007) "On the fly" continuous generation of alginate fibers using a microfluidic device. *Langmuir*, **23**, 9104–9108.

45 Hwang, C.M., Khademhosseini, A., Park, Y., Sun, K., and Lee, S.H. (2008) Microfluidic chip-based fabrication of PLGA microfiber scaffolds for tissue engineering. *Langmuir*, **24**, 6845–6851.

46 Puigmarti-Luis, J., Schaffhauser, D., Burg, B.R., and Dittrich, P.S. (2010) A microfluidic approach for the formation of conductive nanowires and hollow hybrid structures. *Adv. Mater.*, **22**, 2255–2259.

47 Lan, W.J., Li, S.W., Xu, J.H., and Luo, G.S. (2012) Controllable synthesis of microscale titania fibers and tubes using co-laminar micro-flows. *Chem. Eng. J.*, **181–182**, 828–833.

2

Shear-Induced Generation of Controllable Multiple Emulsions in Microfluidic Devices

2.1 Introduction

Multiple emulsions are complex nested liquid systems, containing liquid droplets of decreasing sizes placed one inside another. They are widely used as encapsulation systems in myriad applications, including drug delivery [1–3], foods [4, 5], cosmetics [6, 7], chemical separations [8, 9], and as templates for syntheses of microspheres and microcapsules [10–16]. Accurate control of the size monodispersity and internal structure of multiple emulsions are critical for their versatility because these features enable precise manipulation of the loading levels and the release and transport kinetics of the encapsulated substances [10–21]. Although there have been several reports on the preparation of monodisperse multiple emulsions [13–23], accurate control of both the size and structure of the emulsions remains difficult to achieve, and current techniques are not scalable for fabricating higher order multiple emulsions.

Typically, multiple emulsions can be generated via sequential bulk emulsification using shear. This method usually produces emulsions with polydisperse size. Controlling the shear with constrained geometry [24–28], porous membranes [17, 21], or microchannels [18] can produce emulsions with nearly monodisperse size; these can then themselves be emulsified to produce multiple emulsions. However, although the volume fractions of both the initial and final emulsions can be manipulated [17, 18, 21, 24–27], accurate control over the number of inner droplets in multiple emulsions remains difficult by using these techniques. However, there are many cases where control of the inner droplet number is more important than control of their volume fraction. For example, with precise control of the number and size of inner droplets, transport kinetics of encapsulated substances in the emulsions can be precisely manipulated. Control of the inner droplet number is also critical for engineering colloidal assemblies to produce nonspherical particles [29]. Moreover, it is still challenging for co-encapsulation of droplets containing distinct contents in multiple emulsions, with accurate control of the number, ratio, and size of different inner droplets within each level. With such control on the structure of multicomponent multiple emulsions, these emulsions can offer advanced platforms for design of more complex multicompartment materials and provide synergistic delivery systems or chemical microreactors for incompatible actives or chemicals with versatile encapsulation and flexible mass-transfer kinetics.

Microfluidics for Advanced Functional Polymeric Materials, First Edition. Liang-Yin Chu and Wei Wang.
© 2017 Wiley-VCH Verlag GmbH & Co. KGaA. Published 2017 by Wiley-VCH Verlag GmbH & Co. KGaA.

Microfluidic techniques provide an alternate route to generate monodisperse multiple emulsions. Cascading two T-junction geometries [19, 23] can generate monodisperse double emulsions with some control over the number and size of inner droplets. However, the microfluidic device with standard two-dimensional (2D) microchannels requires precise, localized modification of the microchannel wettability for producing multiple emulsions. This ultimately restricts their utility. By contrast, coaxial flow-focusing [13, 14, 16] geometries relax the constraints for wettability modification, but are still limited in the range of fluids that can be used and in the accurate control afforded over the number and size of inner droplets. Moreover, co-encapsulation of multicomponent droplets in multiple emulsions with precise control of their number, ratio, and size within each level, cannot be easily achieved with these devices. In addition, current microfluidic approaches cannot be scaled up for producing higher order multiple emulsions, such as triple emulsions.

The authors' group developed a highly scalable and versatile microfluidic strategy for controllable shear-induced generation of monodisperse multiple emulsions. In this chapter, the microfluidic devices for controllable emulsion generation and the controllable microfluidic fabrication of single emulsions, double emulsions, triple emulsions, and multicomponent multiple emulsions with even more complex structures are introduced.

2.2 Microfluidic Strategy for Shear-Induced Generation of Controllable Emulsion Droplets

Usually, microfluidic devices for generating emulsion droplets consist of droplet-making units and connecting units. The droplet-making units usually contain microchannels with co-flow (Figure 2.1a) [31], flow-focusing (Figure 2.1b) [32], and T-junction (Figure 2.1c) [19] geometries for generating emulsion droplets, while the connecting units usually contain microchannels for manipulating the generated droplets.

Typically, the microchannels in microfluidic device can be constructed by coaxially inserting cylinder glass capillaries into square glass tubes. The inner microchannel of the inserted capillary and the interstice space between the inserted capillary and square tube create three-dimensional (3D) microchannels for flowing different fluids. After fixing such assembly structures of glass capillaries and tubes on glass plates, microfluidic device with microchannels for emulsion generation can be fabricated [33, 34]. The glass-capillary microfluidic device can be used for emulsion generation without microchannel surface modification due to their 3D microchannel structure, but the fabrication process of these devices requires troublesome manual assembling. Alternatively, based on soft lithography, 2D microchannels can be etched on polydimethylsiloxane (PDMS) plates for fabrication of microfluidic devices [35]. These PDMS microfluidic devices allow flexible construction of microchannel networks for generation and manipulation of droplets. Meanwhile, their fabrication process based on the well-established soft lithography technique allows massive production of the devices. However, these PDMS devices require troublesome

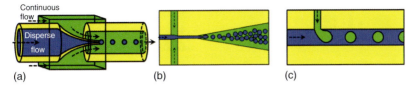

(a) (b) (c)

Figure 2.1 Microfluidic devices for controllable production of monodisperse emulsion droplets. (a–c) Droplet-making units with co-flow (a), flow-focusing (b), and T-junction (c) geometries for generating monodisperse single emulsion droplets. (Zhang et al. 2016 [30]. Reproduced with permission of Elsevier.)

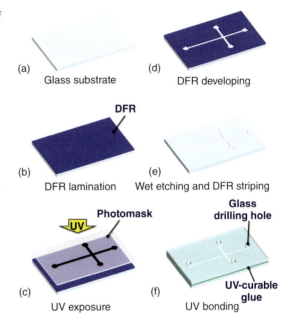

Figure 2.2 Fabricating process of microfluidic device with dry film photoresist (DFR) and soda-lime glass slide: First, laminating the DFR (b) on the glass substrate (a) using an office laminator, followed by UV exposure with a printed photomask on the DFR (c). Then, developing and transferring the microchannel patterns onto the DFR (d). Next, wet etching and DFR striping for transferring microchannel patterns onto the glass substrate (e). Finally, bonding the cover glass with drilled holes onto the glass substrate with etched microchannels using UV-curable glue (f). (Zhang et al. 2015 [36]. Reproduced with permission of The Royal Society of Chemistry.)

process for modifying the microchannel wettability for emulsion generation and exhibit poor chemical resistance to organic solvents.

Besides, microfluidic devices with 2D microchannels can also be fabricated by using dry film photoresist (DFR) and glass slide (Figure 2.2) [36]. Such fabrication technique requires no expensive facilities and materials, and the resultant glass-based devices are featured with flexibly designed microchannels, good chemical compatibility, and mass producibility. Microfluidic device based on glass slide with much lower fabrication cost can be achieved by manually patterning coverslips on glass slides [37], with the gaps between the coverslips as the microchannels for emulsion generation (Figure 2.3). The fabricated microfluidic devices are also featured with flexible design of microchannels, easy spatial patterning of surface wettability, and good chemical compatibility.

Generation of emulsion droplets, in all of the abovementioned microfluidic devices, usually requires two fluids flowing in the device, with one fluid sheared by another one to produce emulsion droplets. Typically, liquid phases, which are respectively used as the disperse fluid and continuous fluid, are injected into the

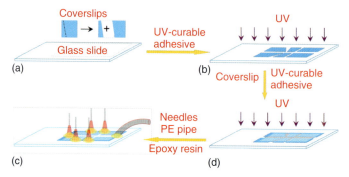

Figure 2.3 Fabrication process of simple microfluidic device based on microscope glass slides and coverslips. First, cutting the coverslips into different shapes according to the design of microchannel structure (a). Then, bonding the tailored coverslips to a glass slide in a designed geometrical arrangement with a UV-curable adhesive (b). Next, covering a new coverslip on them and bonding them with a UV-curable adhesive (c), and followed with connecting needles and a PE pipe to the inlets and outlet, respectively, by epoxy resin (d). Each time after using the UV-curable adhesive, the device is treated with 2-min UV exposure to cure the adhesive. (Deng et al. 2011 [37]. Reproduced with permission of The Royal Society of Chemistry.)

microfluidic devices by using constant-flow pumps or constant-pressure pumps for flow rate control. In the droplet-making units, single emulsion droplets can be produced by using the continuous fluid to shear off the dispersed fluid (Figure 2.1a–c). This can periodically break up the disperse fluid into uniform drops (Figure 2.1a). Since the emulsion droplets can be generated with one droplet at each time, accurate control over their size and size distribution can be achieved. Based on this general strategy for generating emulsion droplets, in the following sections, controllable microfluidic fabrication of single emulsions and scale-up of the microfluidic devices for generating controllable double emulsions, triple emulsions, and multicomponent multiple emulsions with even more complex structures are discussed in detail.

2.3 Shear-Induced Generation of Controllable Monodisperse Single Emulsions

Single emulsions are mixed systems of two immiscible liquids, with droplets of one liquid dispersed in the continuous phase of another one. Recent advances in microfluidics enable controllable production of single emulsion drops, with great control over their size and monodispersity. The generation of controllable single emulsions from 3D and 2D microfluidic devices is introduced by using the glass-capillary microfluidic device and PDMS microfluidic device as typical examples. To form single emulsions with glass-capillary microfluidic device, inner fluid (IF), the dispersed phase, which flows through the injection tube, is broken up into droplets at the tapered orifice by outer fluid (OF), the continuous phase, which co-flows through the square tube (Figure 2.4a,b). Alternatively, in a typical PDMS device with flow-focusing cross-junction geometry (Figure 2.4c,d), liquid (IF) that flows in microchannel A is broken up into droplets at the orifice D by two flows of a second liquid (OF) from two wing microchannels B and C.

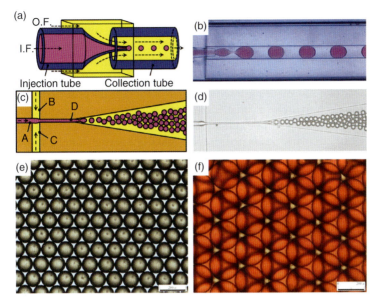

Figure 2.4 Microfluidic generation of monodisperse single emulsions. (a–d) 3D glass-capillary microfluidic device with co-flow geometry (a) and 2D PDMS microfluidic device with cross-junction microchannel (c) for generating single emulsions (b, d). (e, f) Monodisperse oil-in-water (O/W) [38] (e) and water-in-oil (W/O) [37] (f) single emulsions generated from microfluidics. Scale bars are 200 μm. (Reproduced with permission from Refs [37–39]. Copyright (2011), Royal Society of Chemistry. Copyright (2013), Wiley. Copyright (2014), American Chemical Society.)

Both microfluidic devices can produce droplets with highly monodisperse size (Figure 2.4e,f).

Typically, microfluidic devices can produce monodisperse emulsion drops with sizes in the range from millimeter scale [40] to micrometer scale [31], or even to submicrometer scale [41]. The coefficient of variation (CV, defined as the ratio of the standard deviation of the size distribution to its arithmetic mean) for the droplet size is usually less than 5%. The physical properties of different fluids, geometries of microchannel, and processing conditions are important factors for the formation of emulsion droplets. As a typical example, for droplets generated in a co-flow geometry, the factors that influence the droplet sizes and CV values can be the viscosity (η_c) and flow rate (u_c) of the continuous fluid, the density (ρ_d) and flow rate (u_d) of the dispersed fluid, the inner diameter (d_{tip}) of the tapered cone tip of glass capillary (Figure 2.4a), and the interfacial tension (γ) of the two fluids. These factors can be consolidated into two numbers: the capillary number of the continuous fluid ($C_c = \eta_c u_c/\gamma$) and the Weber number of the dispersed fluid ($W_d = \rho_d d_{tip} u_d^2/\gamma$) [42]. These two numbers determine the transition between the dripping and jetting models for generating emulsion droplets. Usually, generation of emulsion droplets with a jetting model can lead to smaller droplet size but wider size distributions, associated with larger CV values, as compared with the droplets generated with dripping model [31, 42]. Alternatively, a decrease in microchannel dimension and an increase in flow rate of continuous fluid can also result in reduced droplet size. Besides, viscosity and interfacial tension are also two important parameters that can influence the droplet generation, because the droplet formation in microfluidic devices involves a balance between the surface tension and the viscous shear stress on the droplet [42, 43]. Moreover, the microfluidic emulsion generation can be scaled up by utilizing multiple droplet-making microchannels to parallelly produce emulsion droplets [44, 45] or using microchannels with regular post-arrays to split primary droplets into uniform smaller droplets [46] for massive production. For example, the disk-shaped microfluidic device with size of 4 cm × 4 cm, which contains 128 circularly patterned droplet-making microchannels, allows production of monodisperse emulsion droplets at a throughput of 320 ml h^{-1} [45]. Alternatively, the post-array-containing microfluidic device, each with dimensions of 24 mm × 6 mm × 5 mm, can produce emulsion droplets at a rate of ~250 l h^{-1} m^{-2}. Thus, with ~1500 of such microfluidic devices, which occupy only 1 l, massive production of more than 2000 t of emulsion droplets per year can be achieved [46]. These results show the good controllability and application of microfluidic techniques for emulsion generation.

2.4 Shear-Induced Generation of Controllable Multiple Emulsions

One of the significant advantages of microfluidic techniques over other emulsification techniques is their excellent extendibility based on flexible combination of droplet-making geometries for controllable production of multiple emulsions. In this section, the scale-up of microfluidic devices for shear-induced generation of controllable double emulsions and triple emulsions is introduced. Meanwhile, to demonstrate the extendibility of microfluidic device for controllable generation of multiple emulsions, glass-capillary microfluidic device is used as a typical example.

2.4.1 Shear-Induced Generation of Controllable Double Emulsions

Double emulsions are usually dispersed droplets encapsulating smaller droplets inside. For generation of double emulsions, as shown in Figure 2.5a, glass-capillary microfluidic device that combines two co-flow geometries for two-step sequential emulsification is used. The two-stage microfluidic device comprises the injection tube and transition tube, both of which are cylindrical capillaries. The tapered end of the injection tube is inserted into the transition tube, with an inner diameter of D_2. Both cylindrical capillaries are coaxially aligned within a larger square tube. The outer diameters of the cylindrical capillaries match the inner dimension of the square one. The other tapered end of the transition tube is inserted into another coaxially aligned cylindrical capillary (collection tube), with inner diameter of D_3. With such a device, monodisperse double emulsions can be generated in two emulsification steps: the IF is emulsified into droplets at the first stage of the device by coaxially flowing the middle fluid (MF) (Figure 2.5a,b). The MF flow that carries single emulsion droplets is then emulsified into the double emulsions at the second

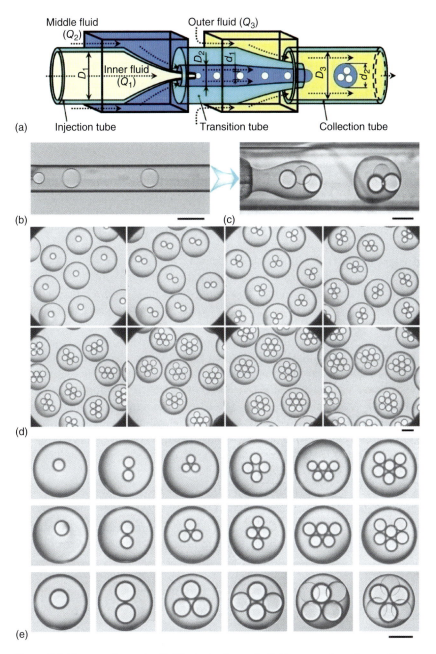

Figure 2.5 Glass-capillary microfluidic device for controllable generation of monodisperse double emulsions. (a) Schematic illustration of the microfluidic device. (b, c) High-speed optical micrographs showing the (b) first and (c) second emulsification steps. (d, e) Optical micrographs of monodisperse double emulsions containing inner droplets with precisely controlled number (d) and size (e). The flow rates in (b) and (c) are as follows: $Q_1 = 350\,\mu l\,h^{-1}$, $Q_2 = 2000\,\mu l\,h^{-1}$, and $Q_3 = 5000\,\mu l\,h^{-1}$. The flow rates in (d) are as follows: $Q_2 = 2000\,\mu l\,h^{-1}$ and $Q_3 = 5000\,\mu l\,h^{-1}$, and $Q_1 = 20, 55, 70, 85, 150, 200, 225$, and $240\,\mu l\,h^{-1}$, from the left top to the right bottom. The flow rates in (e) are as follows: $Q_1 = 20$–$600\,\mu l\,h^{-1}$, $Q_2 = 1600$–$5000\,\mu l\,h^{-1}$, and $Q_3 = 2000$–$8000\,\mu l\,h^{-1}$; in each case, Q_3 is always larger than Q_2. Scale bars are 200 μm. (Chu et al. 2007 [34]. Reproduced with permission of John Wiley & Sons.)

stage through coaxial flow of the OF. The OF is injected in the outer stream through the square tube (Figure 2.5a,c). The separated two-step emulsification process affords independent control of the droplet formation in each step. Such a control is achieved by tuning the microchannel dimensions as well as the flow rates of IF (Q_1), MF (Q_2), and OF (Q_3), respectively. The number of inner droplets in double emulsions can be precisely controlled, as demonstrated by the double emulsions containing 1–8 inner droplets shown in Figure 2.5d. Similarly, precise control of the size of the inner droplets is demonstrated by producing three sets of double emulsions with different inner droplet sizes, as shown in Figure 2.5e. In each case, all droplets formed are produced with highly monodisperse size. The CV values for diameters of inner droplets and double emulsions are always less than 2.3% and 1.6%, respectively. In addition, no wettability modification of microchannel is necessary for such a device, due to the coaxial 3D structure of their microchannels. Thus, with same glass-capillary microfluidic device, it allows the preparation of either water-in-oil-in-water (W/O/W) or the inverse, oil-in-water-in-oil (O/W/O), double emulsions.

The high degree of control allows quantitative evaluation of the flow-rate-dependent diameters of the inner and outer droplets, d_1 and d_2, respectively, and prediction of the number of inner droplets, N_1. For droplet generation in microfluidic device with fixed microchannel dimensions and solution conditions, in the dripping regime, the diameter of droplets in coaxial flow is inversely proportional to the velocity of the surrounding flow [43]. The dependence of d_1/D_2 on Q_1/Q_2 is calibrated to find the linear dependence as depicted in Figure 2.6a. A similar linear dependence of d_2/D_3 on $(Q_1 + Q_2)/Q_3$ is also found, as shown in Figure 2.6b. Based on these empirical relations, the number of inner droplets can be quantitatively predicted by $N_1 = f_1/f_2$, where f_1 and f_2 are the formation rates of the inner and outer droplets. From mass conservation for each flow,

$$N_1 = \frac{f_1}{f_2} = \frac{\frac{Q_1}{(\pi d_1^3/6)}}{\frac{(Q_1+Q_2)}{(\pi d_2^3/6)}} = \frac{Q_1}{Q_1 + Q_2} \cdot \frac{d_2^3}{d_1^3} \tag{2.1}$$

Using the linear fits to the experimental data in Figure 2.6a,b, the number of encapsulated droplets can be predicted as a function of the flow rates Q_1, Q_2, and Q_3,

$$N_1 = \frac{Q_1}{Q_1 + Q_2} \cdot \frac{D_3^3}{D_2^3} \cdot \left(\frac{a_2(Q_1 + Q_2)/Q_3 + b_2}{a_1(Q_1/Q_2) + b_1} \right)^3 \tag{2.2}$$

where a_1 and a_2 are the slopes, and b_1 and b_2 are the intercepts obtained from the fits in Figure 2.6a,b, respectively. The N_1 that is experimentally measured can be well described by Equation 2.2, as shown by the solid line in Figure 2.6c, allowing easy calibration of the device. When the predicted number is an integer, the number (N_1) of inner droplets in double emulsions can be precisely controlled. When the predicted number is between two integers, the emulsion generation is in a transition zone, with N_1 at either integral value. It is worth noting that the precise control of N_1 can always be achieved via correct adjustment of the flow rates.

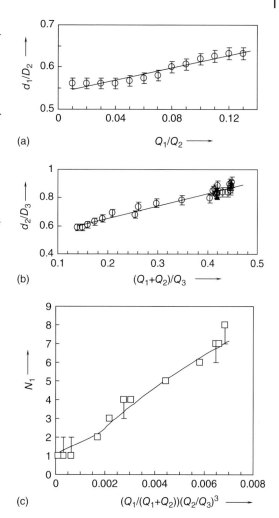

Figure 2.6 Effect of the flow rates on the diameters of inner and outer droplets and the number of inner droplets in double emulsions. (a) Scaled droplet diameter, d_1/D_2, as a function of Q_1/Q_2. The solid line is a linear fit with slope, $a_1 = 0.72$, and intercept, $b_1 = 0.54$. (b) Scaled outer droplet diameter, d_2/D_3, as a function of $(Q_1 + Q_2)/Q_3$. The solid line is a linear fit with slope, $a_2 = 0.88$, and intercept, $b_2 = 0.48$. (c) Experimentally measured number (squares) of inner droplets in each double emulsion droplet as a function of the flow rates. The line depicts the number of inner droplets predicted using Equation 2.2. The slopes and intercepts from (a) and (b) are used in Equation 2.2 for calculation. The variation ranges for Q_1, Q_2, and Q_3 are 20–260, 2000, and 5000–15 000 μl h^{-1}. (Chu et al. 2007 [34]. Reproduced with permission of John Wiley & Sons.).

2.4.2 Shear-Induced Generation of Controllable Triple Emulsions

Triple emulsions are multiple emulsions with droplets containing smaller double emulsion droplets inside. The microfluidic technique for generating controllable double emulsions possesses a significant advantage that the device can be very easily extended for controllable generation of higher order multiple emulsions with additional hierarchical levels. This is demonstrated by controllable production of monodisperse triple emulsions, consisting of water-in-oil-in-water-in-oil (W/O/W/O) droplets. This is achieved by adding a second transition tube at the tapered outlet of the first one, and then injecting the outermost fluid to coaxially flow around this tube to form a third emulsification process at its outlet (Figure 2.7a). The individual steps for generating droplet allow the production of controllable monodisperse triple emulsions, as shown in Figure 2.7b–d. Although large deformations of the outermost droplets occur as they flow

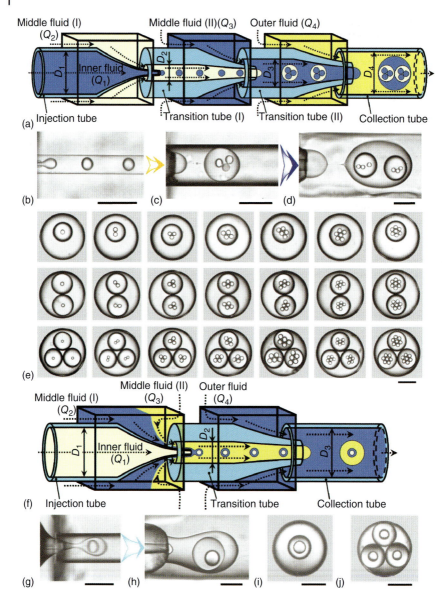

Figure 2.7 Microfluidic generation of controllable monodisperse triple emulsions. (a) Schematic illustration of the extended glass-capillary microfluidic device. (b–d) High-speed snapshots showing the first (b), second (c), and third (d) emulsification stages for generating triple emulsions. (e) Optical micrographs of triple emulsions containing inner and middle droplets with controlled number. (f) Schematic illustration showing an alternate method for generating triple emulsions. (g, h) High-speed snapshots showing the formation of double emulsions (g) in a one-step process in the transition tube and the subsequent formation of triple emulsions (h) in the collection tube. (i, j) Optical micrographs of triple emulsions containing double emulsions with different numbers. Scale bars are 200 μm. (Chu et al. 2007 [34]. Reproduced with permission of John Wiley & Sons.)

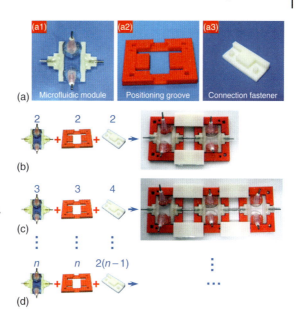

Figure 2.8 Plug-n-play microfluidic devices from flexible assembly of droplet-making modules. (a) Photographs of three basic functional units: the (a1) flow-control module, (a2) positioning groove, and (a3) connection fastener. (b–d) Flexible assembly of (b) two-stage ($n = 2$), (c) three-stage ($n = 3$), and (d) n-stage microfluidic devices from n droplet-making modules, n positioning grooves, and $2(n − 1)$ connection fasteners ($n = 2, 3, …$). (Meng et al. 2015 [47]. Reproduced with permission of The Royal Society of Chemistry.)

through the tapered regions during their formation, the emulsification process remains stable enough for generating triple emulsions. Again, both the diameter and number of the inner droplets, at every level of the triple emulsions, can be accurately controlled. This is demonstrated by the series of triple emulsions containing innermost droplets with numbers varying from 1 to 7 and middle droplets with numbers varying from 1 to 3, as shown in Figure 2.7e. Meanwhile, the *CV* values of the diameters of every batch of triple emulsions are always less than 1.5%.

The simple scale-up strategy also allows the incorporation of alternate emulsification schemes in the microfluidic device for generating triple emulsions. For example, a second fluid can be injected at the inlet of the transition tube to generate double emulsions, by using flow-focusing of a coaxial flow. Followed by an additional emulsification step at the outlet of the transition tube, triple emulsions can then be generated, as shown in Figure 2.7f–h. Again, both the size and number of inner and middle droplets in the generated triple emulsions can be individually and precisely controlled (Figure 2.7i,j).

By adding more additional stages of droplet-making geometry, microfluidic devices can be scaled up for controllable generation of multiple emulsions with even higher levels. For example, by combining the glass-plate-based droplet-making units with 3D printing techniques, plug-n-play microfluidic devices with five stages of droplet-making units can be constructed for controllable generation of quadruple emulsions (Figure 2.8) [47]. Even with such complex high-order multiple structures, the emulsions still remain highly controllable (Figure 2.9).

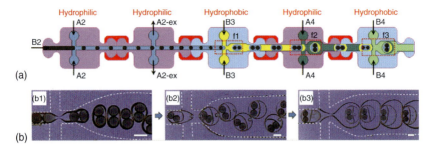

Figure 2.9 Plug-n-play microfluidic device (a) for controllable generation of oil-in-water-in-oil-in-water-in-oil (O/W/O/W/O) quadruple emulsions (b). Scale bars are 200 μm. (Meng et al. 2015 [47]. Reproduced with permission of The Royal Society of Chemistry.)

2.5 Shear-Induced Generation of Controllable Multicomponent Multiple Emulsions

2.5.1 Shear-Induced Generation of Controllable Quadruple-Component Double Emulsions

Based on the concepts of microfluidics for emulsion generation and scale up, a more hierarchical and scalable glass-capillary microfluidic device is developed for controllable generation of multicomponent multiple emulsions, where the number, ratio, and size of the co-encapsulated distinct droplets are accurately controlled at each level. Such a microfluidic device comprises three basic building blocks: a droplet maker, a connector, and a liquid extractor. The droplet maker is designed for generating emulsion droplets (Figure 2.10a1), the connector is designed for merging droplets that are generated from different droplet makers (Figure 2.10a2), and the liquid extractor is designed for removing unwanted fluid from the continuous phase (Figure 2.10a3). Using different combinations of these building blocks enables a flexible and controllable method to create microfluidic devices for the formation of multicomponent multiple emulsions with highly controlled yet exceptionally diverse structures. The scalability and controllability of the microfluidic device are demonstrated by starting with controllable production of quadruple-component double emulsions, and scaling up by simply adding building blocks for controllable production of quintuple-component double emulsions, quintuple-component triple emulsions, and even sextuple-component triple emulsions. In each case of the multicomponent multiple emulsions, the number, ratio, and size of co-encapsulated different droplets are individually controlled with unprecedented accuracy.

First, for controllable generation of quadruple-component double emulsions, three droplet makers, two connectors, and one liquid extractor are used to construct the microfluidic device (Figure 2.10b). At the beginning, droplets of two different inner fluids (F_{1-1} and F_{1-2}) are separately generated in droplet makers from different branch microchannels. Then, converged by the connectors, these two droplet-containing streams flow into the main microchannel and form a linear array of droplets. The two types of droplets can be regularly and alternately

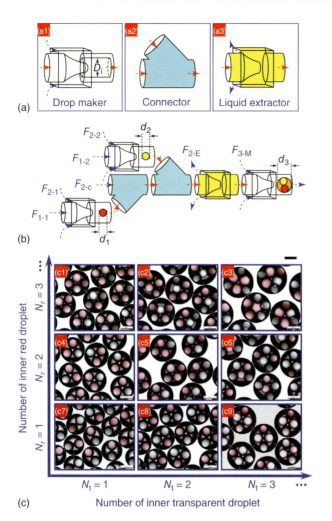

Figure 2.10 Controllable production of quadruple-component double emulsions. (a) Schematic illustration of functional building blocks. (b) Schematic illustration of microfluidic device constructed from the building blocks for controllable generation of quadruple-component double emulsions. d_1, d_2, and d_3 are the diameters of inner red, inner transparent, and outer droplets, respectively. $F_{i\text{-}j}$ is the fluid phase for each level of multiple emulsions, where the subscript i refers to the level number where $F_{i\text{-}j}$ exists in multiple emulsions by counting from the inside outward, and the subscript j refers to the injection position of $F_{i\text{-}j}$ in the microfluidic device. The subscript j can be number or letter, where number (1, 2, …) means the number of branch microchannel, and letters C, E, and M stand for connector, liquid extractor, and main microchannel, respectively. (c) Optical micrographs of quadruple-component double emulsions with precise control over the number and ratio of inner red and transparent droplets. Scale bars are 200 μm. (Wang et al. 2011 [33]. Reproduced with permission of The Royal Society of Chemistry.)

aligned in the main microchannel, due to their periodic droplet formation process in each droplet maker. The distance between every two droplets can be controlled by using a connector ($F_{2\text{-}C}$) to inject fluid and a separate extractor ($F_{2\text{-}E}$) to remove the MF as required. The MF that carries these two types of droplets is further emulsified by outermost fluid in the droplet maker ($F_{3\text{-}M}$) downstream. This can produce monodisperse double emulsions with two different types of inner droplets. The resultant controllable quadruple-component O/W/O double emulsions are shown in Figure 2.10c. The number and ratio of inner red and transparent droplets in these double emulsions can be accurately controlled. Precise manipulation of their number and ratio is achieved by fine adjustment of the flow rates. The excellent controllability of the microfluidic device is demonstrated by independently and precisely changing the number of the encapsulated inner red (N_r) and transparent (N_t) droplets; this also results in accurate control over the ratio (R_{12}) of red droplets to transparent droplets (Figure 2.10c). This control is illustrated by showing the resultant double emulsions with controlled N_r and N_t, each of which varies from 1 to 3, as shown in Figure 2.10c. Independent control over N_r and N_t also enables the precise manipulation of R_{12} with desired values. Because the dripping mechanism is used to provide high control over the droplet formation process, the generated droplets are with good reproducibility and monodispersity.

Excellent control over inner droplet size (d_i) can also be achieved by changing the flow rates or the inner diameter (D_i) of the collection tube of droplet makers. Such a size manipulation is demonstrated by controllable generation of quadruple-component double emulsions that contain smaller red droplets and larger blue droplets with precisely controlled numbers (Figure 2.11). These results demonstrate successful co-encapsulation of droplets with different contents in double emulsions, with precise and independent control over the number, ratio, and size of different inner droplets. These co-encapsulated

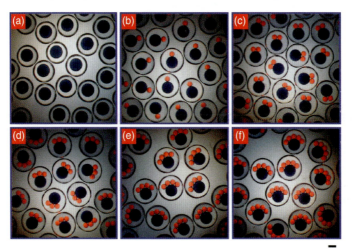

Figure 2.11 Optical micrographs showing controllable production of quadruple-component double emulsions containing smaller red droplets and larger blue droplets with precisely controlled numbers. Scale bar is 200 μm. (Wang et al. 2011 [33]. Reproduced with permission of The Royal Society of Chemistry.)

different droplets can be employed as containers for separate loading of incompatible actives or chemicals and as compartment templates for creating multicompartment materials. The excellent control over number, ratio, and size of different inner droplets enables precise and independent manipulation of the encapsulation of each active or chemical and the internal structure of multicompartment materials. Moreover, because the coaxial 3D structure of the device does not require wettability modification of microchannel for forming droplets from different fluids, generation of quadruple-component W/O/W double emulsions can also be produced with the same device.

Precise and independent control over the size of outer droplets in the quadruple-component double emulsions can also be achieved based on the functionality of the connectors and liquid extractors. Besides merging the main microchannel with branch microchannels, the connector can also be used to adjust the distance between droplets in the main microchannel by injecting additional fluid. When increasing the flow rate ($Q_{2\text{-}C}$) of injecting fluid ($F_{2\text{-}C}$) into the connector while keeping the other flow rates fixed, the distance between droplets as well as the size of the outer droplets (d_3) increases, due to the encapsulation of more fluid in the middle aqueous layer. Such a size control of outer droplets is shown in Figure 2.12a. Increasing the injecting flow rate ratio $Q_{2\text{-}C}/(Q_{1\text{-}1}+Q_{2\text{-}1}+Q_{1\text{-}2}+Q_{2\text{-}2}-Q_{2\text{-}E})$ from ~0.250 to ~1.333 can lead to increase in the volume ratio V_i/V_0 of outer droplets from 1 to ~1.683. Importantly, because these inner droplets are individually formed in branch channels, the size of inner red droplets (d_1) and transparent droplets (d_2) remains unchanged, in spite of the increase in d_3. By contrast, when increasing the flow rate through the liquid extractor ($Q_{2\text{-}E}$) while keeping the other flow rates fixed, the distance between droplets decreases and the outer droplets become smaller due to the removal of continuous fluid. With the increase in the extracting flow rate ratio $Q_{2\text{-}E}/(Q_{1\text{-}1}+Q_{2\text{-}1}+Q_{1\text{-}2}+Q_{2\text{-}2}+Q_{2\text{-}C})$ from ~0.034 to ~0.241, the volume ratio V_i/V_0 of outer droplets decreases from 1 to ~0.736, with d_1 and d_2 unchanged (Figure 2.12b). These results confirm the precise and independent size control over the outer droplets of the quadruple-component double emulsions.

The excellent control over the number, ratio, and size of co-encapsulated inner droplets enables the quantitative prediction of the inner structure of quadruple-component double emulsions for these microfluidic devices. The total number of both inner red droplets and transparent droplets (N) encapsulated in quadruple-component double emulsions is controlled by matching the formation rates of inner red ($f_{1\text{-}1}$) and transparent ($f_{1\text{-}2}$) droplets with that of the outer droplets (f_3). This formation rate that depends on flow rates can be manipulated by adjusting the flow rates of the dispersed (Q_d) and continuous (Q_c) fluids in the droplet maker [24]. For microfluidic device with fixed microchannel dimensions (D_i) and solution conditions, there is a linear dependence of d_i/D_i on Q_d/Q_c [24]. Based on these relations, the value of N can be quantitatively predicted using Equation 2.3:

$$N = \frac{f_{1\text{-}1}+f_{1\text{-}2}}{f_3} = \frac{\frac{Q_{1\text{-}1}}{D_1^3(a_1(Q_{1\text{-}1}/Q_{2\text{-}1})+b_1)^3} + \frac{Q_{1\text{-}2}}{D_2^3(a_2(Q_{1\text{-}2}/Q_{2\text{-}2})+b_2)^3}}{\frac{Q_3}{D_3^3(a_3(Q_3/Q_{3\text{-}M})+b_3)^3}} \tag{2.3}$$

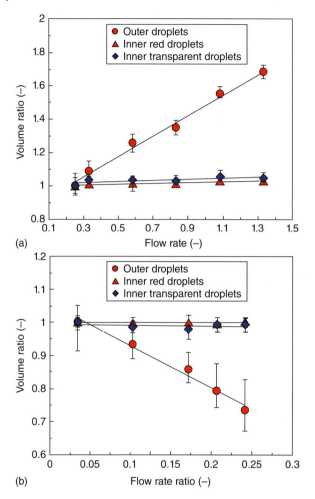

Figure 2.12 Effects of flow rates of connector (a) and liquid extractor (b) on the sizes of inner and outer droplets in quadruple-component double emulsions. (a) Droplet volume ratio V_i/V_0 as a function of flow rate ratio $Q_{2\text{-}C}/(Q_{1\text{-}1} + Q_{2\text{-}1} + Q_{1\text{-}2} + Q_{2\text{-}2} - Q_{2\text{-}E})$. The initial values of d_1, d_2, and d_3 are 163.2, 163.7, and 353.1 μm, respectively. (b) Droplet volume ratio V_i/V_0 as a function of flow rate ratio $Q_{2\text{-}E}/(Q_{1\text{-}1} + Q_{2\text{-}1} + Q_{1\text{-}2} + Q_{2\text{-}2} + Q_{2\text{-}C})$. V_0 is the initial droplet volume, and V_i is the droplet volume after varying $Q_{2\text{-}C}$ (a) and $Q_{2\text{-}E}$ (b). The initial values of d_1, d_2, and d_3 are 169.6, 168.5, and 386 μm, respectively. (Wang et al. 2011 [33]. Reproduced with permission of The Royal Society of Chemistry.)

where $Q_{i\text{-}j}$ is the flow rate of fluid $F_{i\text{-}j}$ and Q_3 is the sum of $Q_{1\text{-}1}$, $Q_{2\text{-}1}$, $Q_{2\text{-}1}$, $Q_{2\text{-}2}$, $Q_{2\text{-}C}$, and $Q_{2\text{-}E}$ (Figure 2.12b). The coefficients a_i and b_i in Equation 2.3 are the slopes and intercepts that are obtained experimentally from the empirical relations for d_i/D_i and Q_d/Q_c. The ratio of inner red droplets to transparent droplets (R_{12}) depends on the number of each kind of droplet encapsulated in the double emulsions; this can be manipulated by adjusting $f_{1\text{-}1}$ and $f_{1\text{-}2}$. Thus, the value of

R_{12} can be quantitatively determined using Equation 2.4:

$$R_{12} = \frac{f_{1\text{-}1}}{f_{1\text{-}2}} = \frac{\frac{Q_{1\text{-}1}}{\pi d_1^3/6}}{\frac{Q_{1\text{-}2}}{\pi d_2^3/6}} = \frac{Q_{1\text{-}1}}{Q_{1\text{-}2}} \cdot \frac{D_2^3\left(a_2 \frac{Q_{1\text{-}2}}{Q_{2\text{-}2}} + b_2\right)^3}{D_1^3\left(a_1 \frac{Q_{1\text{-}1}}{Q_{2\text{-}1}} + b_1\right)^3} \quad (2.4)$$

In Figure 2.13, the calculated values of N and R_{12} obtained from Equations 2.3 and 2.4 are compared with the experimentally measured values. The good agreement between the calculated and experimental values of both N and R_{12} indicates the successful prediction of the inner structure of quadruple-component double emulsions. Moreover, further equations can be developed to predict the inner structure of higher order multicomponent multiple emulsions, based on the mass conservation of each fluid.

2.5.2 Extended Microfluidic Device for Controllable Generation of More Complex Multicomponent Multiple Emulsions

Based on the versatile combinations of the building blocks, simple scale-up of the microfluidic device can be achieved for controlled generation of higher order multicomponent multiple emulsions with more complex structures. This is demonstrated by controllable fabrication of quintuple-component double emulsions, quintuple-component triple emulsions, and even sextuple-component triple emulsions.

For controllable generation of quintuple-component double emulsions, the microfluidic device in Figure 2.10b is scaled up by adding an additional branch microchannel upstream of the main microchannel to introduce a third kind of oil droplets into the quadruple-component double emulsions, as shown in Figure 2.14a. The resultant quintuple-component O/W/O double emulsions contain controllable oil droplets of three colors, where the number of each kind of droplets increases from 1 to 5, as shown in Figure 2.14b.

For controllable generation of quintuple-component triple emulsions, an additional droplet maker is added downstream in the main microchannel of the microfluidic device in Figure 2.10b to further emulsify the quadruple-component double emulsions, as shown in Figure 2.14c. The resultant quintuple-component triple emulsions with oil-in-water-in-oil-in-water (O/W/O/W) type are generated with red and transparent oil droplets in the innermost level surrounded by a blue aqueous layer, which is further enveloped in an oil droplet. The number and ratio of innermost red and transparent oil droplets as well as the number of blue aqueous drops can be precisely controlled (Figure 2.14d).

For controllable generation of more complex sextuple-component triple emulsions, two droplet makers are placed in each branch microchannel of the device shown in Figure 2.10b to generate double emulsions (Figure 2.15a). Due to the different components of $F_{1\text{-}1}$, $F_{2\text{-}1}$, $F_{1\text{-}2}$, and $F_{2\text{-}2}$, two kinds of O/W/O double emulsions can be formed in each branch microchannel. Each of the double emulsions differs in compositions of both inner and middle phases. After further encapsulation of the two types of double emulsions in the droplet maker located

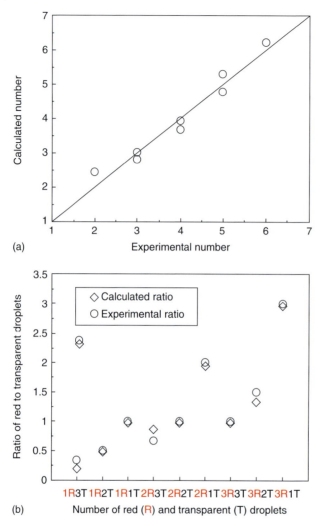

Figure 2.13 Prediction for the number (N) (a) and ratio (R_{12}) (b) of inner red and transparent droplets in quadruple-component double emulsions. (a) Prediction of N. The calculated N value is obtained from Equation 2.3. (b) Prediction of R_{12}. The calculated R_{12} value is obtained from Equation 2.4. For the device and quadruple-component double emulsions shown in Figure 2.8, the values of coefficients in Equations 2.3 and 2.4 are as follows: $a_1 = 0.3957$, $b_1 = 0.9742$, $a_2 = 0.2482$, $b_2 = 1.0137$, $a_3 = 0.8451$, and $b_3 = 0.4894$. (Wang et al. 2011 [33]. Reproduced with permission of The Royal Society of Chemistry.)

downstream in the main microchannel, monodisperse sextuple-component triple emulsions with highly controllable structures can be generated, as shown in Figure 2.15b. These controllable O/W/O/W triple emulsions contain two different types of double emulsions inside, one of which contains red oil droplets in each transparent aqueous droplet while the other contains transparent oil droplets in each blue aqueous droplet (Figure 2.15b1,b2). Interestingly, when injection of one of the innermost fluid (F_{1-2}) is stopped, a novel type of triple

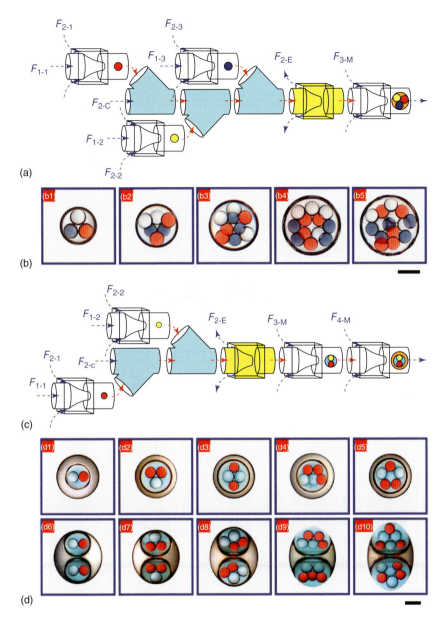

Figure 2.14 Controllable production of quintuple-component multiple emulsions. (a) Schematic illustration of microfluidic device for controllable generation of quintuple-component double emulsions. (b) Optical micrographs of quintuple-component double emulsions containing three different kinds of inner droplets with precisely controlled number. (c) Schematic illustration of microfluidic device for controllable generation of quintuple-component triple emulsions. (d) Optical micrographs of quintuple-component triple emulsions containing two different kinds of droplets in the innermost level, with the number and ratio of different innermost droplets and the number of the middle droplets precisely controlled. Scale bars are 200 μm. (Wang et al. 2011 [33]. Reproduced with permission of The Royal Society of Chemistry.)

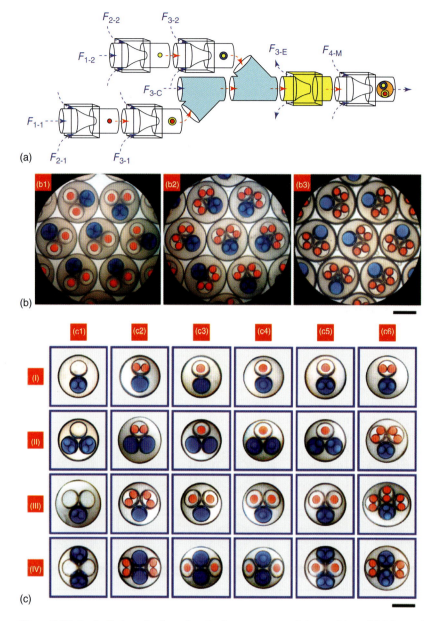

Figure 2.15 Controlled production of sextuple-component triple emulsions. (a) Schematic illustration of microfluidic device for controllable generation of sextuple-component triple emulsions. (b) Optical micrographs of monodisperse sextuple-component triple emulsions. (c) Optical micrographs of sextuple-component triple emulsions containing different droplets in each internal level, with precisely controlled number and ratio. Scale bars are 400 μm. (Wang et al. 2011 [33]. Reproduced with permission of The Royal Society of Chemistry.)

emulsion, which contains both double and single emulsions in the same internal level, is generated (Figure 2.15b3). Meanwhile, accurate control over the internal structure of the sextuple-component triple emulsions with such complex structures is still achievable (Figure 2.15c). By matching the droplet formation rates of the relevant fluids, the number and ratio of the transparent and blue aqueous droplets in the second internal level can be accurately manipulated (Figure 2.15c, from row I to row IV). Meanwhile, by selectively stopping the injection of either of the innermost fluids, sextuple-component triple emulsions with no red oil droplet in the encapsulated transparent aqueous droplets (column c1 in Figure 2.15c), or with no transparent oil droplet in the encapsulated blue aqueous droplets (columns c2 and c3 in Figure 2.15c), can be generated. We can also achieve precise control over the number of different innermost droplets as well as the ratio of innermost transparent to red droplets (columns c4–c6 in Figure 2.15c). These controllably produced higher order multicomponent multiple emulsions highlight the remarkable scalability of the microfluidic technique, which enables flexible arrangement of the building blocks to engineer multicomponent multiple emulsions with versatile complex structures.

2.6 Summary

In this chapter, a highly scalable microfluidic technique is introduced for shear-induced generation of controllable multiple emulsions. The high degree of controllability and scalability provided by the microfluidic technique makes it a versatile and powerful platform for engineering multiple emulsions with highly controlled structures. The microfluidic approach circumvents the difficulties in controllable formation of multiphase structures. It enables the optimization of the structure and composition of multiple emulsion systems for myriad applications in fields such as pharmaceutics, foods, cosmetics, and separations. Furthermore, its generality allows the fabrication of novel materials containing complex internal structures.

References

1 Vasiljevic, D., Parojcic, J., Primorac, M., and Vuleta, G. (2006) An investigation into the characteristics and drug release properties of multiple W/O/W emulsion systems containing low concentration of lipophilic polymeric emulsifier. *Int. J. Pharm.*, **309**, 171–177.
2 Nakano, M. (2000) Places of emulsions in drug delivery. *Adv. Drug Delivery Rev.*, **45**, 1–4.
3 Davis, S.S. and Walker, I.M. (1987) Multiple emulsions as targetable delivery systems. *Methods Enzymol.*, **149**, 51–64.
4 Lobato-Calleros, C., Rodriguez, E., Sandoval-Castilla, O., Vernon-Carter, E.J., and Alvarez-Ramirez, J. (2006) Reduced-fat white fresh cheese-like products obtained from W1/O/W2 shear-induced generation of controllable multiple emulsions: viscoelastic and high-resolution image analyses. *Food Res. Int.*, **39**, 678–685.

5 Weiss, J., Scherze, I., and Muschiolik, G. (2005) Polysaccharide gel with multiple emulsion. *Food Hydrocolloids*, **19**, 605–615.
6 Lee, M.H., Oh, S.G., Moon, S.K., and Bae, S.Y. (2001) Preparation of silica particles encapsulating retinol using O/W/O multiple emulsions. *J. Colloid Interface Sci.*, **240**, 83–89.
7 Yoshida, K., Sekine, T., Matsuzaki, F., Yanaki, T., and Yamaguchi, M. (1999) Stability of vitamin A in oil-in-water-in-oil-type multiple emulsions. *J. Am. Oil Chem. Soc.*, **76**, 195–200.
8 Chakraborty, M., Ivanova-Mitseva, P., and Bart, H.-J. (2006) Selective separation of toluene from n-heptane via emulsion liquid membranes containing substituted cyclodextrins as carrier. *Sep. Sci. Technol.*, **41**, 3539–3552.
9 Chakravarti, A.K., Chowdhury, S.B., and Mukherjee, D.C. (2000) Liquid membrane multiple emulsion process of separation of copper(II) from waste waters. *Colloids Surf., A*, **166**, 7–25.
10 Rizkalla, N., Range, C., Lacasse, F.X., and Hildgen, P. (2006) Effect of various formulation parameters on the properties of polymeric nanoparticles prepared by multiple emulsion method. *J. Microencapsulation*, **23**, 39–57.
11 Koo, H.Y., Chang, S.T., Choi, W.S., Park, J.H., Kim, D.Y., and Velev, O.D. (2006) Emulsion-based synthesis of reversibly swellable, magnetic nanoparticle-embedded polymer microcapsules. *Chem. Mater.*, **18**, 3308–3313.
12 Zoldesi, C.I. and Imhof, A. (2005) Synthesis of monodisperse colloidal spheres, capsules, and microballoons by emulsion templating. *Adv. Mater.*, **17**, 924–928.
13 Utada, A.S., Lorenceau, E., Link, D.R., Kaplan, P.D., Stone, H.A., and Weitz, D.A. (2005) Monodisperse double emulsions generated from a microcapillary device. *Science*, **308**, 537–541.
14 Nie, Z.H., Xu, S.Q., Seo, M., Lewis, P.C., and Kumacheva, E. (2005) Polymer particles with various shapes and morphologies produced in continuous microfluidic reactors. *J. Am. Chem. Soc.*, **127**, 8058–8063.
15 Lorenceau, E., Utada, A.S., Link, D.R., Cristobal, G., Joanicot, M., and Weitz, D.A. (2005) Generation of polymerosomes from double-emulsions. *Langmuir*, **21**, 9183–9186.
16 Bocanegra, R., Sampedro, J.L., Ganan-Calvo, A.M., and Marquez, M. (2005) Monodisperse structured multi-vesicle microencapsulation using flow-focusing and controlled disturbance. *J. Microencapsulation*, **22**, 745–759.
17 Vladisavljevic, G.T., Shimizu, M., and Nakashima, T. (2006) Production of multiple emulsions for drug delivery systems by repeated SPG membrane homogenization: influence of mean pore size, interfacial tension and continuous phase viscosity. *J. Membr. Sci.*, **284**, 373–383.
18 Sugiura, S., Nakajima, M., Yamamoto, K., Iwamoto, S., Oda, T., Satake, M., and Seki, M. (2004) Preparation characteristics of water-in-oil-in-water multiple emulsions using microchannel emulsification. *J. Colloid Interface Sci.*, **270**, 221–228.
19 Okushima, S., Nisisako, T., Torii, T., and Higuchi, T. (2004) Controlled production of monodisperse double emulsions by two-step droplet breakup in microfluidic devices. *Langmuir*, **20**, 9905–9908.

20 Loscertales, I.G., Barrero, A., Guerrero, I., Cortijo, R., Marquez, M., and Ganan-Calvo, A.M. (2002) Micro/nano encapsulation via electrified coaxial liquid jets. *Science*, **295**, 1695–1698.

21 Nakashima, T., Shimizu, M., and Kukizaki, M. (2000) Particle control of emulsion by membrane emulsification and its applications. *Adv. Drug Delivery Rev.*, **45**, 47–56.

22 Huang, S.-H., Tan, W.-H., Tseng, F.-G., and Takeuchi, S. (2006) A monolithically three-dimensional flow-focusing device for formation of single/double emulsions in closed/open microfluidic systems. *J. Micromech. Microeng.*, **16**, 2336–2344.

23 Nisisako, T., Okushima, S., and Torii, T. (2005) Controlled formulation of monodisperse double emulsions in a multiple-phase microfluidic system. *Soft Matter*, **1**, 23–27.

24 Schmitt, V., Leal-Calderon, F., and Bibette, J. (2003) Preparation of monodisperse particles and emulsions by controlled shear. *Top. Curr. Chem.*, **227**, 195–215.

25 Pays, K., Giermanska-Kahn, J., Pouligny, B., Bibette, J., and Leal-Calderon, F. (2002) Double emulsions: how does release occur? *J. Controlled Release*, **79**, 193–205.

26 Pays, K., Giermanska-Kahn, J., Pouligny, B., Bibette, J., and Leal-Calderon, F. (2001) Double emulsions: a tool for probing thin-film metastability. *Phys. Rev. Lett.*, **87**, 178304.

27 Goubault, C., Pays, K., Olea, D., Gorria, P., Bibette, J., Schmitt, V., and Leal-Calderon, F. (2001) Shear rupturing of complex fluids: application to the preparation of quasi-monodisperse water-in-oil-in-water double emulsions. *Langmuir*, **17**, 5184–5188.

28 Mason, T.G. and Bibette, J. (1997) Shear rupturing of droplets in complex fluids. *Langmuir*, **13**, 4600–4613.

29 Manoharan, V.N., Elsesser, M.T., and Pine, D.J. (2003) Dense packing and symmetry in small clusters of microspheres. *Science*, **301**, 483–487.

30 Zhang, M., Wang, W., Xie, R., Ju, X., Liu, Z., Jiang, L., Chen, Q., and Chu, L. (2016) Controllable microfluidic strategies for fabricating microparticles using emulsions as templates. *Particuology*, **24**, 18–31.

31 Shah, R.K., Shum, H.C., Rowat, A.C., Lee, D., Agresti, J.J., Utada, A.S., Chu, L.-Y., Kim, J.-W., Fernandez-Nieves, A., Martinez, C.J., and Weitz, D.A. (2008) Designer emulsions using microfluidics. *Mater. Today*, **11**, 18–27.

32 Seo, M., Paquet, C., Nie, Z., Xu, S., and Kumacheva, E. (2007) Microfluidic consecutive flow-focusing droplet generators. *Soft Matter*, **3**, 986–992.

33 Wang, W., Xie, R., Ju, X.-J., Luo, T., Liu, L., Weitz, D.A., and Chu, L.-Y. (2011) Controllable microfluidic production of multicomponent multiple emulsions. *Lab Chip*, **11**, 1587–1592.

34 Chu, L.-Y., Utada, A.S., Shah, R.K., Kim, J.-W., and Weitz, D.A. (2007) Controllable monodisperse multiple emulsions. *Angew. Chem. Int. Ed.*, **46**, 8970–8974.

35 McDonald, J.C. and Whitesides, G.M. (2002) Poly(dimethylsiloxane) as a material for fabricating microfluidic devices. *Acc. Chem. Res.*, **35**, 491–499.

36 Zhang, L., Wang, W., Ju, X.-J., Xie, R., Liu, Z., and Chu, L.-Y. (2015) Fabrication of glass-based microfluidic devices with dry film photoresists as pattern transfer masks for wet etching. *RSC Adv.*, **5**, 5638–5646.

37 Deng, N.-N., Meng, Z.-J., Xie, R., Ju, X.-J., Mou, C.-L., Wang, W., and Chu, L.-Y. (2011) Simple and cheap microfluidic devices for the preparation of monodisperse emulsions. *Lab Chip*, **11**, 3963–3969.

38 Wang, W., Zhang, M.-J., Xie, R., Ju, X.-J., Yang, C., Mou, C.-L., Weitz, D.A., and Chu, L.-Y. (2013) Hole-shell microparticles from controllably evolved double emulsions. *Angew. Chem. Int. Ed.*, **52**, 8084–8087.

39 Wang, W., Zhang, M.-J., and Chu, L.-Y. (2014) Functional polymeric microparticles engineered from controllable microfluidic emulsions. *Acc. Chem. Res.*, **47**, 373–384.

40 Steinbacher, J.L., Lui, Y., Mason, B.P., Olbricht, W.L., and McQuade, D.T. (2012) Simplified mesofluidic systems for the formation of micron to millimeter droplets and the synthesis of materials. *J. Flow Chem.*, **2**, 56–62.

41 Jeong, W.-C., Lim, J.-M., Choi, J.-H., Kim, J.-H., Lee, Y.-J., Kim, S.-H., Lee, G., Kim, J.-D., Yi, G.-R., and Yang, S.-M. (2012) Controlled generation of submicron emulsion droplets via highly stable tip-streaming mode in microfluidic devices. *Lab Chip*, **12**, 1446–1453.

42 Utada, A.S., Fernandez-Nieves, A., Stone, H.A., and Weitz, D.A. (2007) Dripping to jetting transitions in coflowing liquid streams. *Phys. Rev. Lett.*, **99**, 094502.

43 Umbanhowar, P.B., Prasad, V., and Weitz, D.A. (2000) Monodisperse emulsion generation via drop break off in a coflowing stream. *Langmuir*, **16**, 347–351.

44 Romanowsky, M.B., Abate, A.R., Rotem, A., Holtze, C., and Weitz, D.A. (2012) High throughput production of single core double emulsions in a parallelized microfluidic device. *Lab Chip*, **12**, 802–807.

45 Nisisako, T. and Torii, T. (2008) Microfluidic large-scale integration on a chip for mass production of monodisperse droplets and particles. *Lab Chip*, **8**, 287–293.

46 Amstad, E., Datta, S.S., and Weitz, D.A. (2014) The microfluidic post-array device: high throughput production of single emulsion drops. *Lab Chip*, **14**, 705–709.

47 Meng, Z.-J., Wang, W., Liang, X., Zheng, W.-C., Deng, N.-N., Xie, R., Ju, X.-J., Liu, Z., and Chu, L.-Y. (2015) Plug-n-play microfluidic systems from flexible assembly of glass-based flow-control modules. *Lab Chip*, **15**, 1869–1878.

3

Wetting-Induced Generation of Controllable Multiple Emulsions in Microfluidic Devices

3.1 Introduction

Multiple emulsions, or "emulsions of emulsions," are complex liquid systems in which dispersed droplets themselves contain smaller droplets inside. Typically, multiple emulsions can be generated by sequential shear-induced emulsifications. For example, for the generation of double emulsions, the inner droplets are produced at the first emulsification step and then encapsulated by the outer droplets at the second emulsification step. This two-step emulsification process leads to double emulsions with highly polydisperse structures because of the randomness intrinsic to each step. Double emulsions with more uniform structures can be produced by using Couette devices [1] or porous membranes [2] to provide relatively well-controlled shear for both emulsification steps. However, precise control over the number and size of the inner droplets still remains a challenge. Alternatively, microfluidic techniques provide a powerful platform for controllable generation of multiple emulsions with perfect monodisperse and diverse structures, with the number, size, and composition of the inner droplets precisely and independently controlled [3–6], as introduced in Chapter 2. Such a high degree of controllability enables the creation of various important structures, including polymersomes [7], colloidosomes [8], Janus particles [9], nonspherical particles [10], microcapsules [11], and microgels [5], for myriad applications. However, current microfluidic strategies for generating multiple emulsions are typically achieved in tailored devices; these devices are usually complicated to construct or difficult to modify, such as manually assembled glass-capillary devices [3, 12] or lithographically fabricated devices [6, 13]. Meanwhile, accurate control of the flows at each emulsification stage is crucial to these microfluidic methods, because such controls can ensure a stable emulsification process, either sequentially [5, 6] or simultaneously [4, 14], for emulsion generation. Moreover, current microfluidic methods still possess certain restrictions, for example, in generation of double emulsions with ultrathin shell containing different inner droplets that are less resistant to mass transfer of multicomponent ingredients. Thus, development of simpler methods for controllable production of monodisperse multiple emulsions is of great value and significance.

The authors' group developed new microfluidic strategies based on wetting phenomenon for controllable generation of monodisperse multiple emulsions.

36 | *3 Multiple Emulsions via Wetting-Induced Droplet Coalescing*

The emulsions are created via wetting-induced spreading of one droplet on another droplet or wetting-induced coalescing of two droplets. This allows the production of complex multiple emulsions from simpler emulsion droplets, with accurate control over their structures and compositions. In this chapter, the controllability and flexibility of these two strategies for generating multiple emulsions are introduced.

3.2 Microfluidic Strategy for Wetting-Induced Production of Controllable Emulsions

3.2.1 Strategy for Wetting-Induced Production of Controllable Emulsions via Wetting-Induced Spreading

When droplets of two immiscible liquids that are dispersed in a third immiscible liquid come into contact, three equilibrated configurations can be formed as follows [15, 16]: first is the nonengulfing configuration, where the two droplets remain separated by the third liquid phase; second is the partial engulfing configuration, where the two droplets form a dumbbell-like structure; and third is the complete engulfing configuration, where one droplet completely engulfs the other one. The three different configurations involve the synergistic effect of interfacial energies between every two liquid phases; this is the basis for the method of wetting-induced generation of multiple emulsions directly from two droplets.

As shown in Figure 3.1, the criterion for formation of multiple emulsions can be characterized by the spreading coefficient of the engulfing droplet A over the engulfed droplet B. It is assumed that the engulfing process is determined by the interfacial tensions between every two liquid phase and the droplet sizes, thus gravity, fluid motion, and forces between particles are ignored [16]. When two immiscible droplets dispersed in a third immiscible liquid are brought into contact, one droplet A can completely engulf the other one B only if it can lead to a reduced total interfacial energy. As shown in Figure 3.1a–c, when the relationship

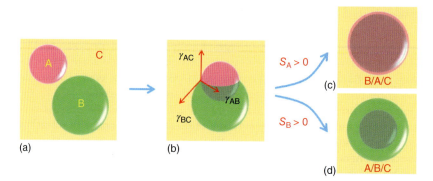

Figure 3.1 Schematic illustration of the wetting-induced engulfing process of two immiscible droplets (A and B) in a third immiscible fluid (C) via interfacial energy adjustment. (Deng *et al.* 2013 [17]. Reproduced with permission of The Royal Society of Chemistry.)

of the three interfacial energies fits in with Equation 3.1, droplet A with radius R_A can completely engulf droplet B with radius R_B:

$$4\pi R_A^2 \gamma_{AC} + 4\pi R_B^2 \gamma_{BC} > 4\pi R_B^2 \gamma_{BC} + 4\pi \left(R_A^3 + R_B^3\right)^{2/3} \gamma_{AC} \tag{3.1}$$

where γ_{ij} is the interfacial tension between fluids i and j and R_i is the radius of droplet i.

Then, Equation 3.1 can be derived as follows:

$$\gamma_{BC} - \left(\alpha \cdot \gamma_{AC} + \gamma_{AB}\right) > 0 \tag{3.2}$$

where

$$\alpha = \left[1 + \left(\frac{R_A}{R_B}\right)^3\right]^{2/3} - \left(\frac{R_A}{R_B}\right)^2, \quad \alpha \in (0, 1) \tag{3.3}$$

The spreading coefficient can be defined as

$$S_A = \gamma_{BC} - \left(\alpha \cdot \gamma_{AC} + \gamma_{AB}\right) \tag{3.4}$$

Thus, forming a complete engulfing configuration from the two droplets requires a positive spreading coefficient ($S_A > 0$). This indicates that, as compared to the continuous fluid C, the fluid of the engulfing droplet A preferentially wets the interface of the engulfed droplet B. Similarly, if S_B becomes positive with changing interfacial energies, droplet B will completely engulf droplet A to form an A/B/C double emulsion (Figure 3.1b,d). This wetting-induced generation method of multiple emulsions requires three different liquids, each of which is immiscible with the other two. The oil phases that are immiscible with many of other oil phases and water phase can be employed as the continuous phase. Typical examples are families of silicone oil and fluorinated oil, especially the commonly used fluorocarbon oils. The interfacial energies of the three phase systems can be easily adjusted to achieve different spreading coefficients via addition of surfactants or other reagents or by controlling the droplet sizes. This allows the creation of multiple emulsions with versatile structures. The partial engulfing configuration via control of spreading coefficients has already been used to create a variety of structures [15, 18, 19]. In this section, by ensuring the complete engulfing configuration, one droplet can fully engulf the other droplet to form a multiple emulsion droplet. The excellent manipulation of microdroplets in microfluidics enables the utilization of such a strategy to produce multiple emulsions from simpler emulsion droplets.

3.2.2 Strategy for Wetting-Induced Production of Controllable Emulsions via Wetting-Induced Coalescing

In this section, a simple and novel surgery-like strategy for controllable generation of complex multiple emulsions with controlled compositions and structures from simpler emulsion droplets via wetting-induced droplet coalescence is introduced. A microlancet with suitable surface wettability to wet the droplets is designed in the converging microchannel of a microfluidic device for droplet coalescence (Figure 3.2a). When pairs of the generated droplets simultaneously flow across the microlancet, the microlancets scratch the surfactant-stabilized surfaces of the droplets, and temporarily relocate the surfactant arrangement;

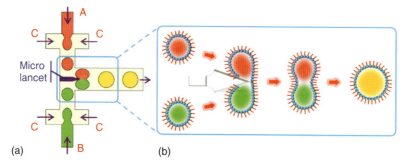

Figure 3.2 Schematic illustration of microlancet in microchannels (a) for wetting-induced coalescence of surfactant-stabilized emulsion droplets (b). (Deng et al. 2013 [20]. Reproduced with permission of The Royal Society of Chemistry.)

this can trigger the coalescence of the paired droplets by joining up their scratched wounds (Figure 3.2b). Thus, this microsurgery-like process draws cracks on the surfaces of the droplets to promote their coalescence. When multiple emulsion droplets are used for such coalescence, more complex multiple emulsions with controllable compositions and structures can be generated. Such a strategy provides new route to controllably form emulsions from simpler droplets in microfluidics.

3.3 Generation of Controllable Multiple Emulsions via Wetting-Induced Spreading

3.3.1 Wetting-Induced Generation of Monodisperse Controllable Double Emulsions

For generation of multiple emulsions via wetting-induced spreading, three immiscible phases are chosen: soybean oil (SO) or a mixture of SO and octanol, with volume ratio of 3:1, containing 1–2 wt% polyglycerol polyricinoleate (PGPR 90); an aqueous solution with 0.5–1 wt% sodium dodecyl sulfate (SDS); and silicone oil (SiO) (10 cSt) with 0.5–1 wt% Dow Corning 749 (DC749). A glass microfluidic device [22] with two flow-focusing geometries is utilized to create two different drops for wetting-induced generation of double emulsions. One type of the droplets comprises soybean oil (A1) and the other one comprises water (B), in a continuous phase of silicone oil (C). An expanded microchamber (Figure 3.3a,b) is created in the collection microchannel of the device to trigger the contact between the neighboring two droplets via slow down of their local flow rates. With the proper combination of surfactant or stabilizer in the three phases as mentioned earlier, their interfacial tensions become $\gamma_{A1B} = 0.18$ mN m^{-1}, $\gamma_{BC} = 3.07$ mN m^{-1}, and $\gamma_{A1C} = 1.37$ mN m^{-1}. Such a condition ensures a positive spreading coefficient for the soybean oil over the water, in spite of droplet sizes ($S_{A1} > 1.52$). Therefore, when a droplet of soybean oil (A1) meets a droplet of water (B) in the expanded microchamber, the droplet A1 can completely engulf the droplet B, resulting in a B/A1/C water-in-oil-in-oil

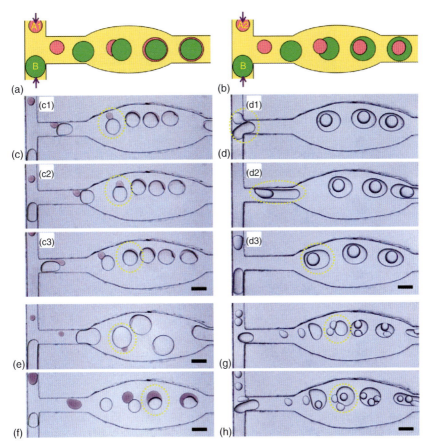

Figure 3.3 Wetting-induced generation of controllable double emulsions. (a–d) Schematic illustrations and high-speed snapshots of wetting-induced formation of double emulsions. In one case, water droplets (B) are completely engulfed by oil droplets (A1) dyed with LR300 to form B/A1/C double emulsions (a, c). In the other case, oil droplets (A2) are completely engulfed by water droplets (B) to form A2/B/C double emulsions (b, d). (e, f) High-speed snapshots of the engulfing formation of B/A1/C double emulsions with an ultrathin shell (e) and a relatively thick shell (f). (g, h) High-speed snapshots of the engulfing formation of A2/B/C double emulsions with two (g) and three (h) inner droplets. Scale bars are 200 μm. (Deng et al. 2013 [17]. Reproduced with permission of The Royal Society of Chemistry.)

(W/O/O) double emulsion (Figure 3.3c). This demonstrates the principle for wetting-induced formation of multiple emulsions from two independent single emulsion droplets via wettability control. This strategy shows great flexibility and controllability for generation of versatile multiple emulsions. For example, by simply adding 25 vol% octanol to the SO (A2), the interfacial tensions can be adjusted as $\gamma_{A2B} = 0.58\,\mathrm{mN\,m^{-1}}$, $\gamma_{BC} = 3.07\,\mathrm{mN\,m^{-1}}$, and $\gamma_{A2C} = 3.16\,\mathrm{mN\,m^{-1}}$. This can lead to a positive spreading coefficient (S_B) for the water phase (B) over the SO phase (A2), if the droplet size ratio of $R_B/R_{A2} > 0.49$. Thus, the complete engulfing process can be reversed to form A2/B/C oil-in-water-in-oil (O/W/O) double emulsions (Figure 3.3b,d). By adjusting the relative droplet

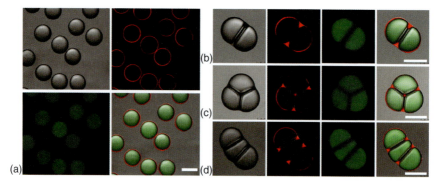

Figure 3.4 Ultrathin-shelled double emulsions. (a–d) CLSM images of ultrathin-shelled W/O/O double emulsions with one (a), two (b), and three (c, d) inner droplets. Scale bars are 200 μm. (Deng et al. 2013 [17]. Reproduced with permission of The Royal Society of Chemistry.)

sizes, the thickness of the middle phase layer in the double emulsions can be controlled to achieve a very thin shell (Figure 3.3e) and relatively thick shell (Figure 3.3f). Alternatively, size control of the multiple emulsions can also be realized by simply tuning the precursor droplet sizes. Moreover, the number of inner droplets can be easily manipulated by tuning the number ratio of the droplets that contact in the microchamber (Figure 3.3g,h).

Next, the ability of this strategy to create double emulsions with ultrathin shells is discussed in detail, since such emulsions show great potential for template synthesis of ultrathin-shelled materials. This is demonstrated by generation of W/O/O double emulsions with ultrathin shells via wetting-induced spreading of SO droplets over water droplets in SiO phase. The obtained double emulsions exhibit good size monodispersity (Figure 3.4a) and diverse structures (Figure 3.4b–d). For example, the value of coefficient of variation (CV) for the W/O/O double emulsions with ultrathin shells (Figure 3.4a) is only 1.92%. The ultrathin-shelled double emulsions can show diverse structures depending on the assembly patterns of multiple inner droplets (Figure 3.4b–d). Besides the ultrathin shell, shells with adjustable thickness can also be achieved for double emulsions (Figure 3.5a–d). For double emulsions with core–shell structures, their shell thickness (δ) can be precisely manipulated by controlling the diameter (D_1) of the engulfing droplet and the diameter (D_2) of the engulfed droplet. The relationship among δ, D_1, and D_2 can be described by Equation 3.5 as follows:

$$\delta = \frac{\left[(D_1^3 + D_2^3)^{1/3} - D_2\right]}{2} \tag{3.5}$$

Based on Equation 3.5, with fixed D_1, decrease in D_2 can lead to increase in δ; on the contrary, with fixed D_2, increase in D_1 can result in increase in δ. For example, when D_2 is fixed at 200 μm, the value of δ increases from 33 nm to 26 μm, with increasing D_1 from 20 to 200 μm (Figure 3.5b). Figure 3.5c,d shows the CLSM (confocal laser scanning microscopy) images of the resultant double emulsions with an ultrathin shell and a relatively thick shell. These emulsions provide excellent templates for fabricating microcapsules with a desired shell thickness.

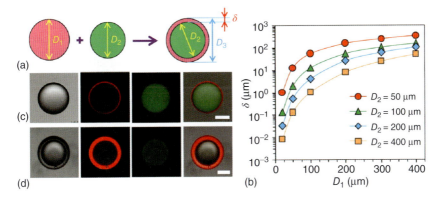

Figure 3.5 Double emulsions with shell of controllable thickness. (a, b) The relationship between the shell thickness (δ) and the diameters of the inner (D_1) and outer (D_2) droplets. The shell volume of double emulsions equals the volume of the outer droplet. (c, d) CLSM images of double emulsions with an ultrathin shell (c) and a relatively thick shell (d). Scale bars are 100 µm. (Deng et al. 2013 [17]. Reproduced with permission of The Royal Society of Chemistry.)

3.3.2 Wetting-Induced Generation of Monodisperse Higher Order Multiple Emulsions

Similar to the shear-induced emulsion generation, the wetting-induced strategy for generating multiple emulsions can also be easily scaled up to produce higher order multiple emulsions. This concept is illustrated by controllable generation of monodisperse O/W/O double emulsions containing distinct inner oil droplets (Figure 3.6) and monodisperse triple emulsions consisting of oil-in-water-in-oil-in-oil (O/W/O/O) droplets (Figure 3.7). Such emulsion generations are achieved by spreading O/W/O double emulsion droplets over O/O droplets (Figure 3.6a) or by the reverse manner (Figure 3.7a). A microfluidic device containing two single-stage and double-stage flow-focusing geometries is employed to simultaneously generate the precursor droplets of single and double emulsions (Figure 3.6b). To form (A + D)/B/C ((O1 + O2)/W/O) double emulsions containing two different inner oil droplets (Figure 3.6a3,b), SiO containing 1 wt% DC749 is used as the continuous phase (C) in both flow-focusing geometries; an aqueous solution containing 1 wt% SDS and SO containing 1 wt% PGPR 90 are, respectively, employed as the middle phase (B) and inner phase (D) in the double-stage flow-focusing geometry to generate D/B/C (O/W/O) emulsions (Figure 3.6c), and a mixture of SO and octanol, with volume ration of 3:1, which contains 1 wt% PGPR 90 and 1 mg ml^{-1} LR300, is employed as the disperse phase (A2) in the single-stage flow-focusing geometry to generate A2/C (O/O) emulsions (Figure 3.6d). This combination of fluids ensures a positive spreading coefficient for phase B over phase A. Thus, when paired droplets of the single and double emulsions come into contact in the expanded microchamber, the water shell of the D/B/C double emulsion droplet can completely spread over the A2/C single emulsion droplet to form (A2 + D)/B/C (O1 + O2)/W/O double emulsions (Figure 3.6e). By changing the disperse oil phase of the A/C single emulsion, a positive spreading coefficient

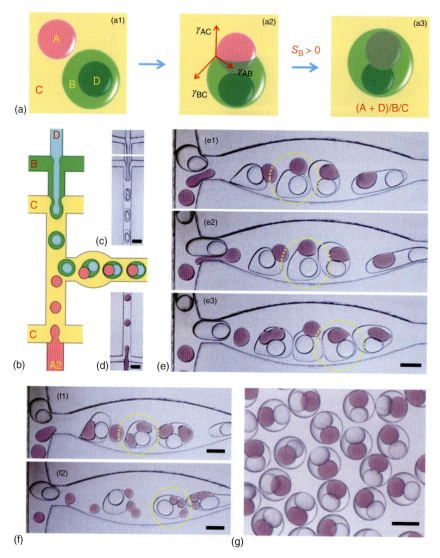

Figure 3.6 Wetting-induced generation of double emulsions containing different inner droplets. (a) Schematic illustration showing the complete engulfing of one O/O single emulsion droplet by one O/W/O double emulsion droplet to form a (A2 + D)/B/C double emulsion containing different inner droplets under a positive spreading coefficient S_B. (b–e) Schematic illustration and high-speed snapshots showing the formation of precursor droplets (c, d) and double emulsions with two different inner droplets (e). (f) One O/W/O double emulsion droplet completely engulfs O/O single emulsion droplets with controlled numbers to form a double emulsion droplet controllably containing different inner droplets. (g) Optical micrographs of the resultant monodisperse double emulsions with two different inner droplets. Scale bars are 200 μm. (Deng *et al.* 2013 [17]. Reproduced with permission of The Royal Society of Chemistry.)

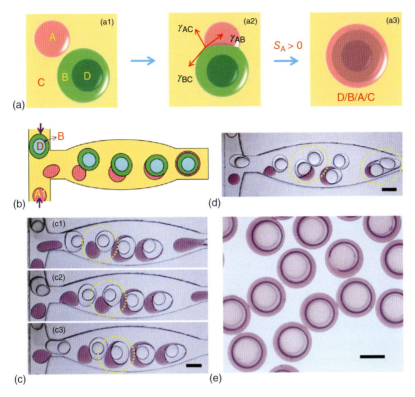

Figure 3.7 Wetting-induced generation of triple emulsions. (a) Schematic illustration of wetting-induced completely engulfing of one O/W/O double emulsion droplet by one O/O single emulsion droplet to form a D/B/A1/C triple emulsion under a positive spreading coefficient S_A. (b, c) Schematic illustration and high-speed snapshots of the generation of triple emulsions. (d) One O/O single emulsion droplet completely engulfs two O/W/O double emulsion droplets to form an O/W/O/O triple emulsion containing two double emulsion droplets. (e) Optical micrographs of the resultant monodisperse triple emulsions. Scale bars are 200 μm. (Deng et al. 2013 [17]. Reproduced with permission of The Royal Society of Chemistry.)

for phase A over phase B can be obtained to achieve reverse engulfing process (Figure 3.7a3). This is demonstrated by using SO containing 2 wt% PGPR 90 and 1 mg ml^{-1} LR300 as the disperse phase (A1) of the A/C single emulsion to completely engulf the D/B/C double emulsion droplet for generation of D/B/A1/C oil-in-water-in-oil-in-oil (O/W/O/O) triple emulsions (Figure 3.7b,c). Both of the resultant (O1 + O2)/W/O double emulsions and O/W/O/O triple emulsions are stable and exhibit highly uniform structures (Figures 3.6g and 3.7e). Meanwhile, (W1 + W2)/O/O double emulsions and W/O/W/O triple emulsions can also be produced via the engulfing of W/O/O double emulsions and W/O single emulsions, based on the different combinations of fluids and interfacial energies. The simplicity of the strategy also enables the controllable generation of multiple emulsions with more complex structures. For example, multiple emulsions that contain inner single emulsion droplets (Figure 3.6f) or double emulsion droplets (Figure 3.7d) with controlled number can be generated by simply adjusting the number ratio of the precursor droplets that converge in the microchamber. All the produced multiple emulsions highlight the power of this strategy for controllable generation of multiple emulsions.

3.3.3 Wetting-Induced Generation of Monodisperse Multiple Emulsions via Droplet-Triggered Droplet Pairing

Besides synchronizing the formation of different droplets for pairing for wetting-induced emulsion generation, alternatively, such a generation process can also be realized by employing droplet-triggered droplet formation for pairing. First, droplet pairs are accurately generated by droplet-triggering formation of droplets for precise pairing. Then, such droplet pairs can be used for wetting-induced generation of multiple emulsions via manipulation of their interfacial energies.

Typically, to generate multiple emulsion droplets, three immiscible phases are used: SiO (10 cSt) with 1% (w/v) DC749 is used as outer continuous oil phase; water with 1% (w/v) SDS is used as inner aqueous phase; SO with 2% (w/v) PGPR 90 is used as inner oil phase for preparing W/O/O double emulsions, while a mixture of SO and octanol, with volume ratio of 3 : 1, containing 2% (w/v) PGPR 90, is used as inner oil phase for preparing O/W/O double emulsions. To realize droplet-triggered droplet forming and pairing for wetting-induced emulsion generation, a glass-plate microfluidic device with two hydrophobic flow-focusing geometries (Figure 3.8a) is used to synchronously generate oil jet of SO (A) (Figure 3.8b) and water droplets (B) (Figure 3.8c) in continuous phase of SiO (C). As the oil jet and water droplets meet in the collection microchannel, the tip of the oil jet is broken into uniform oil droplets due to the shear of the water droplets, forming pairs of one oil droplet and one water droplet. The size of oil droplets can be controlled not only by manipulating the flow rate conditions

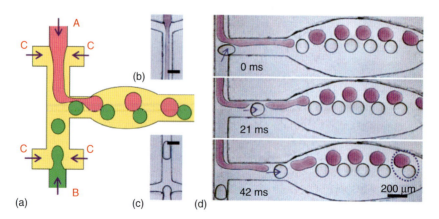

Figure 3.8 Droplet-triggered formation of droplets for droplet pairing in microchannel. (a) Schematic illustration showing the generation of droplet pairs via droplet-triggered droplet formation. (b–d) High-speed snapshots showing the generation of an O/O jet (b) and W/O emulsion droplets (c), and the droplet-triggered formation of droplets for droplet pairing (d). Scale bars are 200 µm. (Deng et al. 2014 [21]. Reproduced with permission of Springer.)

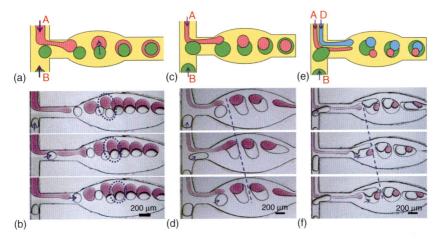

Figure 3.9 Schematic illustrations and high-speed snapshots showing the droplet-triggered droplet pairing for wetting-induced generation of double emulsions. (a, b) Generation of W/O/O double emulsions by using the oil droplets (A) formed from the jet tip to completely engulf water droplets (B). (c–f) Generation of O/W/O double emulsions with one inner droplets (c, d) and two different inner droplets (e, f) by engulfing of water droplets (B) over oil droplets (A, D) formed from the jet tips. Scale bars are 200 μm. (Deng et al. 2014 [21]. Reproduced with permission of Springer.)

for generating the oil jets but also by adjusting distance between every two water droplets. An expanded microchamber (Figure 3.8a) is created downstream the collection microchannel to decrease the flow rates of droplets for bringing them into contact (Figure 3.8d). This droplet-pairing process can be further employed for wetting-induced generation of multiple emulsions. Typically, by adding 2% (w/v) PGPR 90, 1% (w/v) SDS, and 1% (w/v) DC749, respectively, to the SO (A), water (B), and SiO (C) phases, the interfacial tensions are, respectively, $\gamma_{AB} = 0.28$ mN m^{-1}, $\gamma_{BC} = 8.82$ mN m^{-1}, and $\gamma_{AC} = 1.21$ mN m^{-1}. This can make the spreading coefficient for the oil droplet over the water droplet always positive ($S_A = 7.33$). As shown in Figure 3.9a,b, when the paired water droplet and SO droplet come into contact in the expanded microchamber, the oil droplet entirely engulfs water droplet to form a B/A/C double emulsion. This method also possesses great flexibility and controllability for emulsion generation. For example, the engulfing process can be simply inverted by adding 25% (v/v) octanol to the SO to achieve a positive spreading coefficient of water phase over SO phase. This can make the water droplets to fully engulf the oil droplets to form A/B/C double emulsions (Figure 3.9c,d). Moreover, by easily adjusting the number of the disperse jet phases, double emulsions with distinct inner droplets can be generated. As shown in Figure 3.9e,f, by adding SO and octanol, with volume ratio of 3 : 1, containing 2% (w/v) PGPR 90, as another disperse oil phase for generating jet, O/W/O emulsions with two different inner droplets can be prepared. The two oil jets can be broken by one water droplet to generate two different oil droplets, which are then completely engulfed by the water droplet to generate double emulsions with distinct inner droplets.

This strategy can also be easily scalable for generation of higher order multiple emulsions. This is demonstrated by pairing single and double emulsion droplets for generating monodisperse O/W/O/O triple emulsions (Figure 3.10a). A glass-plate microfluidic device with both one-stage and two-stage flow-focusing geometries is used to generate oil jets and double emulsions. To generate O/W/O/O triple emulsions, SiO with 1% (w/v) DC749 is used as continuous phase (C) for both single and double emulsions; aqueous solution with 1% (w/v) SDS and SO with 2% (w/v) PGPR 90 are, respectively, used as middle phase (B) and inner phase (D) in the two-stage flow-focusing geometry to generate D/B/C double emulsions (Figure 3.10b); SO with 2% (w/v) PGPR 90 and 1 mg ml^{-1} LR300 is used as disperse phase (A) in the one-stage flow-focusing geometry to generate A/C jets (Figure 3.10c). This combination of fluid phases ensures a positive spreading coefficient for phase A over phase B. In the microchannel, the tip of jet of phase A is broken into droplets by the double emulsion droplets, forming one-to-one droplet pairs in the expanded microchamber. Similarly, upon contact between the paired droplets, the single emulsion droplets can completely engulf the double emulsion droplets, forming O/W/O/O triple emulsions (Figure 3.10d). The generated O/W/O/O triple emulsions exhibit monodisperse sizes (Figure 3.10e), with *CV* values for inner, middle, and outer droplets, respectively, at 4.2%, 2.8%, and 1.7%. All the results

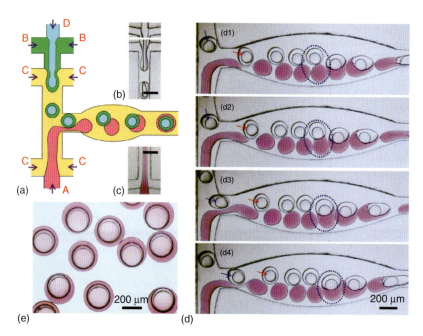

Figure 3.10 Wetting-induced generation of O/W/O/O triple emulsions from pairs of O/O single emulsion droplets and O/W/O double emulsion droplets. (a–d) Schematic illustration (a–c) and high-speed snapshots (d) showing the generation of O/W/O emulsions (b) and O/O droplets from oil jet (c) for preparing O/W/O/O triple emulsions (d). (e) Optical micrograph of the generated O/W/O/O triple emulsions. Scale bars are 200 μm. (Deng *et al.* 2014 [21]. Reproduced with permission of Springer.)

show the power of this strategy for wetting-induced generation of controllable multiple emulsions. Moreover, this strategy that utilizes droplet-triggered formation and pairing of droplets for emulsion generation is very efficient and robust, and more resistant to environmental interference.

3.4 Generation of Controllable Multiple Emulsions via Wetting-Induced Droplet Coalescing

In this section, the simple and novel surgery-like strategy for controllable generation of complex multiple emulsions with controlled compositions and structures from simpler emulsion droplets via wetting-induced droplet coalescence is introduced. To demonstrate this new concept for emulsion construction, the microchannels are designed with two flow-focusing geometries to generate two different droplets, and with a converging microchannel to synchronously pair the two droplets (Figure 3.2a). A properly treated filament, such as plastic wire (Ø 120 μm), copper wire (Ø 60–90 μm), and aluminum wire (Ø 100 μm), is inserted into the converging microchannel as the microlancet. Before insertion, the microlancet tips are tailored into different shapes, including sharp tip and flat tip, and then modified with different surface wettabilities. To achieve a hydrophilic surface, the tailored filaments are dipped into an aqueous solution with 1% (w/v) SDS or 1% (w/v) hydroxyethylcellulose ($M_W = 500\,000$) for ~1 min, then dried using a hair drier, and the process is repeated for more than three times. To achieve a hydrophobic surface, the microlancets are immersed in a hexane solution with 10 vol% chlorotrimethylsilane for 10 min. All the microchannels are fabricated by bonding patterned coverslips to a glass slide by an UV-curable adhesive [22].

To demonstrate the surgery-like approach for droplet coalescence, a copper microlancet with sharp tip and hydrophilic surface is first utilized to trigger the coalescence of W/O emulsion droplets. To generate the W/O emulsions, aqueous solution containing 1% (w/v) SDS and SiO containing 1% (w/v) DC749 are, respectively, used as the disperse phase (A and B) and continuous phase (C) (Figure 3.2a). Under proper flow rates, the emulsion droplets generated from the two flow-focusing geometries can simultaneously flow into the converging microchannel for pairing. When the paired W/O emulsion droplets flow across the microlancet, the water droplets wet the microlancet due to their hydrophilic property, leading to scratched droplet surfaces with temporarily relocated surfactants. Thus, coalescence occurs when the scratched wounds of the paired droplets contact with each other at the microlancet tip for minimization of the total interfacial energies (Figures 3.2b and 3.11a). Meanwhile, microlancets made with different materials, such as plastic and aluminum wires, can also achieve droplet coalescence when they are modified with hydrophilic surfaces. However, when no microlancet is incorporated in the microchannel, two droplets can collide with each other but without coalescence due to the stabilization of surfactants (Figure 3.11b). Meanwhile, droplet coalescence cannot be achieved by using hydrophobic microlancets (Figure 3.11c,d). Because oil phase preferentially wets the hydrophobic microlancets as compared to water phase, the

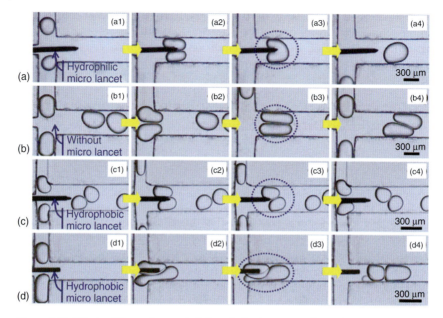

Figure 3.11 Effects of the surface wettability of microlancets on the coalescence behavior of W/O droplets. The microlancets include copper microlancet with sharp tip and hydrophilic surface (a), sharp tip and hydrophobic surface (c), and flat tip and hydrophobic surface (d). No microlancet is used in (b). The flow rates are $Q_A, Q_B = 100–500\,\mu l\,h^{-1}$, $Q_{C(upper)} = 150–1200\,\mu l\,h^{-1}$, and $Q_{C(lower)} = 300–1000\,\mu l\,h^{-1}$. (Deng et al. 2013 [20]. Reproduced with permission of The Royal Society of Chemistry.)

water droplets cannot be scratched by hydrophobic microlancets due to the presence of thin oil film in between. Thus, the proper surface wettability of the microlancet plays a key role in this novel surgery-like strategy.

Moreover, the effects of the tip shape and length of the hydrophilic microlancets, the oil phases, and surfactants on the coalescences of W/O emulsion droplets are also investigated. As shown in Figures 3.12 and 3.13a, hydrophilic microlancets with sharp tip and flat tip can successfully trigger the droplet coalescence. Besides, the length of the hydrophilic microlancets can be flexible for such droplet coalescences, as long as the microlancet is long enough to make contact with the droplets. To further demonstrate the feasibility of this strategy, droplet coalescences are conducted by using different continuous oil phases with different types and concentrations of surfactants. With all the SiO, SO, and benzyl benzoate (BB) as the continuous oil phases and all the DC749, calcium carbonate nanoparticles, and PGPR90 as surfactants, the coalescence of W/O emulsion droplets can always be realized with hydrophilic microlancets but can never be achieved without microlancet or with hydrophobic microlancets.

The droplet coalescence induced by the microlancet is very reliable and requires no exactly synchronized droplet generation. Previous works on passive droplet fusion largely depend on the synchronized generation of droplets, thus limiting their long-term operations. The microlancet-based wetting-induced approach provides more flexibility, because the microlancet can hold on the

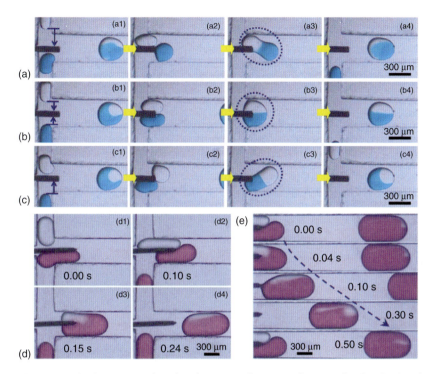

Figure 3.12 (a–c) Wetting-induced coalescences of two asynchronous droplets (a, c) and two synchronous droplets (b) via microlancet. (d, e) The mixing processes of W/O emulsion droplets. (Deng et al. 2013 [20]. Reproduced with permission of The Royal Society of Chemistry.)

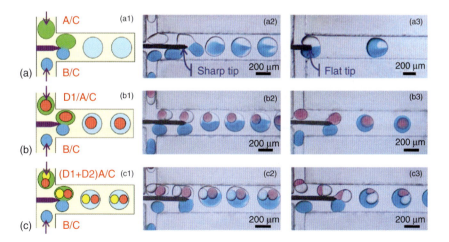

Figure 3.13 Schematic illustration (a1, b1, c1) and high-speed snapshots of coalescence of different W/O emulsion droplets induced by hydrophilic copper microlancets, and coalescence between W/O emulsion droplets and O/W/O emulsion droplets containing single (b) or dual (c) inner droplets, with either relatively thick shells (b2, c2) or ultrathin shells (b3, c3). (Deng et al. 2013 [20]. Reproduced with permission of The Royal Society of Chemistry.)

first droplet for a certain time to achieve its coalescence with the later one (Figure 3.12a–c). Such a flexible strategy enables control of the composition of the generated emulsion drops via droplet coalescence. Figure 3.12d,e shows the mixing process during the droplet coalescence by dyeing one of the droplets with 0.1% (w/v) rhodamine B. Under flow rate condition of $Q_A = 200\,\mu l\,h^{-1}$, $Q_B = 100\,\mu l\,h^{-1}$, $Q_{C(upper)} = 600\,\mu l\,h^{-1}$, and $Q_{C(lower)} = 300\,\mu l\,h^{-1}$, the coalesced droplet can achieve an expeditious mixing within 0.5 s (Figure 3.12e). Such a coalescence-based composition control is promising for use in microreactions.

Besides coalescence of single emulsion droplets, wetting-induced coalescence between more complex emulsion droplets can also be achieved. Microfluidic device with one single-stage and one double-stage flow-focusing geometries is used for wetting-induced coalescence of single and double emulsions. SiO containing 1% (w/v) DC749, aqueous solution containing 1% (w/v) SDS, and SO containing 1% (w/v) PGPR 90 and 1 mg ml^{-1} LR300 are, respectively, used as continuous phase (C), middle phase (A), and inner phase (D1) in the double-stage flow-focusing geometries to generate O/W/O emulsions. W/O emulsions dyed with methylene blue trihydrate are generated as mentioned earlier. The surfaces of the single aqueous droplet and the aqueous shell of the double emulsion droplet are scratched upon contacting with the hydrophilic microlancet. Then, the cracks of the two paired droplets contact at the microlancet tip and then induce the coalescence (Figure 3.13b1,b2). Similarly, coalescence between single W/O emulsions and double emulsions with two different oil droplets can also be realized (Figure 3.13c). Moreover, by controlling the flow rates in the double-stage flow-focusing geometries, double emulsions with ultrathin shells can be generated for their wetting-induced coalescence with the single emulsion droplets (Figure 3.13b3,c3). These results show the power of the wetting-induced coalescence strategy for controlling the composition of multiple emulsions.

Furthermore, this wetting-induced strategy can also be applied for controllable generation of emulsions with more complex structures from the coalescence of simpler droplets. For example, O/W/O double emulsions with two inner droplets can be generated by coalescence of two neighboring O/W/O emulsion droplets that contain single inner droplet (Figure 3.14a–c). Moreover, coalescence among three droplets can also be achieved by coalescing one O/W/O emulsion droplet with two neighboring single emulsion droplets (Figure 3.14d). Similarly, coalescences between two O/W/O emulsion droplets and one single emulsion droplet (Figure 3.14e) and between two (O1 + O2)/W/O emulsion droplets and one single emulsion droplet (Figure 3.14f) can also be realized to generate more complex multiple emulsions.

3.5 Summary

In this chapter, flexible microfluidic strategies for wetting-induced generation of controllable multiple emulsions via wetting-induced spreading and wetting-induced coalescing are introduced. For the strategy based on wetting-induced spreading, the multiple emulsions are formed in microchannels from wetting-induced spreading of one emulsion droplet over another emulsion droplet by

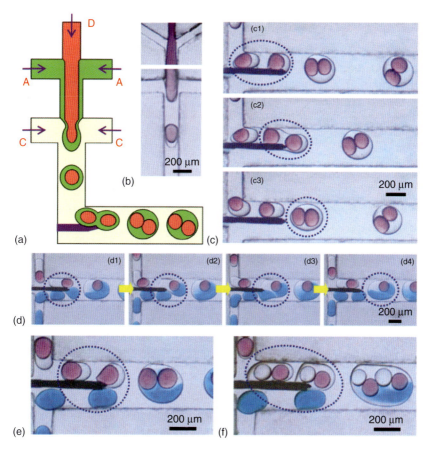

Figure 3.14 Wetting-induced coalescence of emulsion droplets for controllable generation of more complex multiple emulsions. (a–c) Schematic illustration (a) and high-speed snapshots showing the generation of O/W/O emulsions (b) and coalescence of two single-core O/W/O emulsion droplets to form dual-core O/W/O emulsions (c). (d) Coalescence of two W/O single emulsion droplets and one O/W/O double emulsion droplet. (e, f) Coalescence of one W/O single emulsion droplet and two O/W/O double emulsion droplets containing one (e) and two (f) different inner droplets. (Deng et al. 2013 [20]. Reproduced with permission of The Royal Society of Chemistry.)

controlling the interfacial energies. For the strategy based on wetting-induced coalescing, the multiple emulsions are formed in microchannels by triggering the coalescence of two emulsion droplets via wetting on microlancet with suitable surface wettability. Both strategies allow controllable generation of multiple emulsions from simpler emulsion droplets, with structures and compositions precisely controlled. These strategies exhibit exceptional controllability and flexibility and provide valuable concepts to prepare multiple emulsions. Thus, these microfluidic strategies show high potential in myriad applications such as microreaction, template synthesis, high-throughput injection, and multiple emulsion formation.

References

1 Goubault, C., Pays, K., Olea, D., Gorria, P., Bibette, J., Schmitt, V., and Leal-Calderon, F. (2001) Shear rupturing of complex fluids: application to the preparation of quasi-monodisperse water-in-oil-in-water double emulsions. *Langmuir*, **17**, 5184–5188.

2 van der Graaf, S., Schroen, C., and Boom, R.M. (2005) Preparation of double emulsions by membrane emulsification – a review. *J. Membr. Sci.*, **251**, 7–15.

3 Chu, L.-Y., Utada, A.S., Shah, R.K., Kim, J.-W., and Weitz, D.A. (2007) Controllable monodisperse multiple emulsions. *Angew. Chem. Int. Ed.*, **46**, 8970–8974.

4 Utada, A.S., Lorenceau, E., Link, D.R., Kaplan, P.D., Stone, H.A., and Weitz, D.A. (2005) Monodisperse double emulsions generated from a microcapillary device. *Science*, **308**, 537–541.

5 Nie, Z.H., Xu, S.Q., Seo, M., Lewis, P.C., and Kumacheva, E. (2005) Polymer particles with various shapes and morphologies produced in continuous microfluidic reactors. *J. Am. Chem. Soc.*, **127**, 8058–8063.

6 Okushima, S., Nisisako, T., Torii, T., and Higuchi, T. (2004) Controlled production of monodisperse double emulsions by two-step droplet breakup in microfluidic devices. *Langmuir*, **20**, 9905–9908.

7 Shum, H.C., Y-j, Z., Kim, S.-H., and Weitz, D.A. (2011) Multicompartment polymersomes from double emulsions. *Angew. Chem. Int. Ed.*, **50**, 1648–1651.

8 Dinsmore, A.D., Hsu, M.F., Nikolaides, M.G., Marquez, M., Bausch, A.R., and Weitz, D.A. (2002) Colloidosomes: selectively permeable capsules composed of colloidal particles. *Science*, **298**, 1006–1009.

9 Nie, Z., Li, W., Seo, M., Xu, S., and Kumacheva, E. (2006) Janus and ternary particles generated by microfluidic synthesis: design, synthesis, and self-assembly. *J. Am. Chem. Soc.*, **128**, 9408–9412.

10 Dendukuri, D. and Doyle, P.S. (2009) The synthesis and assembly of polymeric microparticles using microfluidics. *Adv. Mater.*, **21**, 4071–4086.

11 Liu, L., Wang, W., Ju, X.-J., Xie, R., and Chu, L.-Y. (2010) Smart thermo-triggered squirting capsules for nanoparticle delivery. *Soft Matter*, **6**, 3759–3763.

12 Wang, W., Xie, R., Ju, X.-J., Luo, T., Liu, L., Weitz, D.A., and Chu, L.-Y. (2011) Controllable microfluidic production of multicomponent multiple emulsions. *Lab Chip*, **11**, 1587–1592.

13 Abate, A.R. and Weitz, D.A. (2009) High-order multiple emulsions formed in poly(dimethylsiloxane) microfluidics. *Small*, **5**, 2030–2032.

14 Kim, S.-H. and Weitz, D.A. (2011) One-step emulsification of multiple concentric shells with capillary microfluidic devices. *Angew. Chem. Int. Ed.*, **50**, 8731–8734.

15 Guzowski, J., Korczyk, P.M., Jakiela, S., and Garstecki, P. (2012) The structure and stability of multiple micro-droplets. *Soft Matter*, **8**, 7269–7278.

16 Torza, S. and Mason, S.G. (1969) Coalescence of 2 immiscible liquid drops. *Science*, **163**, 813–814.

References

17 Deng, N.N., Wang, W., Ju, X.J., Xie, R., Weitz, D.A., and Chu, L.Y. (2013) Wetting-induced formation of controllable monodisperse multiple emulsions in microfluidics. *Lab Chip*, **13**, 4047–4052.

18 Barikbin, Z., Rahman, M.T., and Khan, S.A. (2012) Fireflies-on-a-chip: (ionic liquid)-aqueous microdroplets for biphasic chemical analysis. *Small*, **8**, 2152–2157.

19 Pannacci, N., Bruus, H., Bartolo, D., Etchart, I., Lockhart, T., Hennequin, Y., Willaime, H., and Tabeling, P. (2008) Equilibrium and nonequilibrium states in microfluidic double emulsions. *Phys. Rev. Lett.*, **101**, 164502.

20 Deng, N.N., Sun, S.X., Wang, W., Ju, X.J., Xie, R., and Chu, L.Y. (2013) A novel surgery-like strategy for droplet coalescence in microchannels. *Lab Chip*, **13**, 3653–3657.

21 Deng, N.-N., Mou, C.-L., Wang, W., Ju, X.-J., Xie, R., and Chu, L.-Y. (2014) Multiple-emulsion formation from controllable drop pairs in microfluidics. *Microfluid. Nanofluid.*, **17**, 967–972.

22 Deng, N.N., Meng, Z.-J., Xie, R., Ju, X.-J., Mou, C.-L., Wang, W., and Chu, L.-Y. (2011) Simple and cheap microfluidic devices for the preparation of monodisperse emulsions. *Lab Chip*, **11**, 3963–3969.

4

Microfluidic Fabrication of Monodisperse Hydrogel Microparticles

4.1 Introduction

Stimuli-sensitive hydrogel microparticles or microgels are polymeric particles consisting of cross-linked three-dimensional hydrophilic polymeric networks. They shrink or swell significantly by expelling or absorbing large amounts of water in response to external stimuli such as changes in temperature, pH, and certain chemicals [1–3]. The chemical composition of the microgel determines the stimulus that can trigger this volume-phase transition. The dramatic response and stimuli-specific behavior make these materials extremely valuable for numerous applications, including drug delivery, sensors and actuators, and chemical separation [2, 3].

Monodispersity is very important for the stimuli-responsive microparticles to improve their performances in various applications [3]. In this chapter, microfluidic strategies are introduced to controllably prepare monodisperse stimuli-responsive hydrogel microparticles, including monodisperse poly(N-isopropylacrylamide) (PNIPAM) microgels for sensing tannic acid (TA), core–shell PNIPAM microcapsules for sensing ethyl gallate (EG), and core–shell microspheres with magnetic core and ion-recognizable shell for selective adsorption and separation of Pb^{2+}.

4.2 Microfluidic Fabrication of Monodisperse PNIPAM Hydrogel Microparticles for Sensing Tannic Acid (TA)

PNIPAM is a popular thermo-responsive polymer, which shows a distinct and reversible phase transition at the lower critical solution temperature (LCST) around 32 °C [4]. When the environmental temperature is lower than the LCST, the PNIPAM can bind to plenty of water molecules on its amide groups through the hydrogen-bonding interaction, and thus it is in the swollen and hydrophilic state; however, when the temperature is higher than the LCST, the PNIPAM is dehydrated because of the cleavage of the hydrogen bonding, and thus it is in the shrunken and hydrophobic state. That is, the conformation change of the PNIPAM can result in volume-phase transition of PNIPAM hydrogels from swollen and hydrophilic state at temperatures below the LCST to shrunken

Microfluidics for Advanced Functional Polymeric Materials, First Edition. Liang-Yin Chu and Wei Wang.
© 2017 Wiley-VCH Verlag GmbH & Co. KGaA. Published 2017 by Wiley-VCH Verlag GmbH & Co. KGaA.

and hydrophobic state at temperatures above the LCST, which makes PNIPAM hydrogels extremely valuable for numerous applications [2].

TA is a plant secondary metabolite and a popular polyphenol that can be found in approximately 80% of woody and 15% of herbaceous dicotyledonous species. It has been known that TA is an important protein precipitator, antioxidant, and curative for body injuries. It has been reported that TA molecules could bind to PNIPAM chains through the hydroxyls on TA aromatic groups [5]. Because cross-linked PNIPAM hydrogels are more visible than linear PNIPAM chains during their phase transition process, cross-linked PNIPAM hydrogels are easier to be used to study the effect of TA on the phase transition behavior of PNIPAM compared with linear PNIPAM chains. Furthermore, microgels can respond much faster than macroscale hydrogels to environmental stimuli because of their small dimensions [3, 4]; therefore, it should be more prompt and accurate to study the phase transition behavior of PNIPAM by using PNIPAM hydrogel microparticles rather than macroscale hydrogels. Here, we show the microfluidic fabrication of monodisperse PNIPAM microgels for studying the TA-induced phase transition behaviors of PNIPAM microgels [6].

4.2.1 Microfluidic Fabrication of Monodisperse PNIPAM Hydrogel Microparticles

A capillary microfluidic device is used to prepare monodisperse emulsions, which are then used as templates for fabricating monodisperse PNIPAM microgels via free radical polymerization (Figure 4.1). The disperse phase for droplets is aqueous solution containing monomer N-isopropylacrylamide (NIPAM), cross-linker N,N-methylene-bis-acrylamide (MBA), and initiator ammonium persulfate (APS). The continuous phase is soybean oil containing polyglycerol polyricinoleate (PGPR 90) as surfactant and N,N,N,N-tetramethylethylenediamine (TEMED) as reaction accelerator. The aqueous phase is broken into droplets at

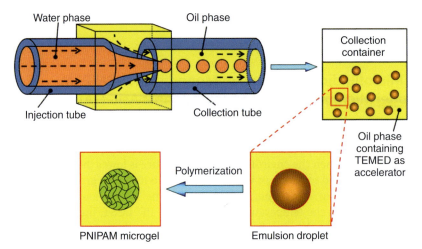

Figure 4.1 Schematic illustration of preparation of monodisperse PNIPAM microgels by the microfluidic approach. (Chen et al. 2010 [6]. Reproduced with permission of Elsevier.)

the tip of the injection tube, and the resultant droplets are then carried away by continuous phase flow as monodisperse water-in-oil (W/O) emulsion drops. The W/O emulsion drops are then collected in a container with soybean oil phase containing PGPR 90 and TEMED. TEMED is both oil and water soluble. When TEMED in the continuous phase diffuses into the aqueous droplets and meets the initiator APS, a redox reaction is started to polymerize the monomers.

The monodispersity of emulsion droplets or PNIPAM microspheres is evaluated based on an index called coefficient of variation (CV), which is defined as the ratio of the standard deviation of size distribution to its arithmetic mean ($\bar{D}$).

$$\bar{D} = \sum_{i=1}^{N} D_i/N \tag{4.1}$$

$$CV = \frac{100\% \times \left\{ \sum_{i=1}^{N} \frac{(D_i - \bar{D})}{N-1} \right\}}{\bar{D}} \tag{4.2}$$

where D_i is the diameter of the droplet or microsphere and N is the total number of the microparticles counted. Usually, more than 100 microspheres are counted for obtaining each CV value. The smaller the CV value, the better is the monodispersity of particles.

Figure 4.2a,b shows typical optical micrographs of microfluidic-prepared W/O emulsion droplets and the resultant PNIPAM microspheres in water at 25 °C [7]. Both the W/O emulsion droplets and PNIPAM microspheres look quite uniform and the size distributions are rather narrow. Figure 4.2c presents the corresponding size distributions of W/O emulsion droplets and PNIPAM microspheres shown in Figure 4.2a,b. The mean diameter of the W/O emulsion droplets and PNIPAM microspheres are 315 and 377 μm, respectively. Both the prepared W/O emulsion droplets and PNIPAM microspheres are reasonably monodisperse [8], with CV values of 1.4% and 1.9%, respectively, which indicate that the prepared emulsion droplets and PNIPAM microspheres are highly monodisperse. The PNIPAM microspheres are somewhat larger than the emulsion templates, because the PNIPAM microspheres are hydrophilic and swelled to some extent in water at 25 °C. The PNIPAM microspheres are transparent when they are in a swollen state in water at 25 °C, which indicates that the polymerization inside PNIPAM microspheres is homogeneous.

4.2.2 Volume-Phase Transition Behaviors of PNIPAM Microgels Induced by TA

The dynamic volume-phase transition behaviors of PNIPAM microgels are characterized by measuring the sizes of PNIPAM microgels at 25 °C and certain time intervals. A batch of PNIPAM microgels (more than 10) are immersed in 2 ml aqueous TA solution for at least 4 h at 25 °C (below the LCST) that is controlled by a thermostatic stage system to ensure equilibration of isothermal phase transition induced by TA. Before investigating the thermo-sensitive phase transition behaviors of PNIPAM microgels in aqueous solutions with different TA concentrations (10^{-8}, 10^{-7}, 10^{-6}, 10^{-5}, and 10^{-4} mol l^{-1}, respectively), the PNIPAM microgels are

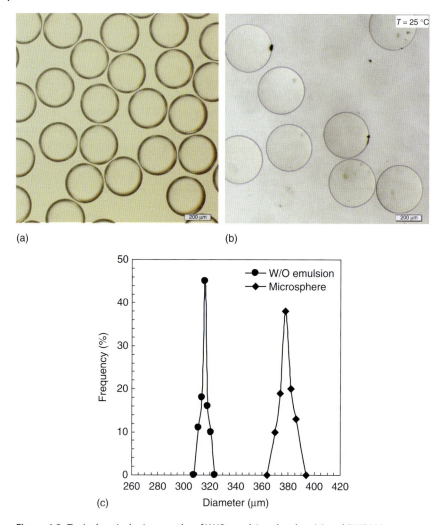

Figure 4.2 Typical optical micrographs of W/O emulsion droplets (a) and PNIPAM microspheres (b), and their size distributions (c) in water at 25 °C. Scale bars are 200 μm. (Mou et al. 2012 [7]. Reproduced with permission of Elsevier.)

immersed in the aqueous TA solution in a transparent sample holder at 25 °C for more than 4 h to ensure equilibration of isothermal phase transition. The temperature of TA solution containing PNIPAM microgels is increased step by step from 25 to 44.6 °C to beyond the LCST of PNIPAM. At each predetermined temperature, the temperature is kept constant for more than 15 min to ensure the full equilibrium state of thermo-responsive phase transition.

Figure 4.3 shows the dynamic isothermal volume-phase transition behaviors of PNIPAM microgels in solutions with different TA concentrations at 25 °C. The results demonstrate that the volumes of PNIPAM microgels do not change much at lower TA concentrations such as 10^{-8}, 10^{-7}, and 10^{-6} mol l^{-1}. As the TA

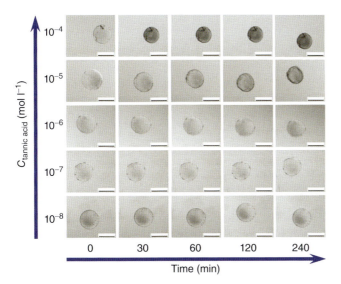

Figure 4.3 Optical micrograms of PNIPAM microgels at different time intervals during the deswelling process in TA solutions with different concentrations at 25 °C. Scale bar is 200 μm. (Chen et al. 2010 [6]. Reproduced with permission of Elsevier.)

concentration increases to 10^{-5} and 10^{-4} mol l^{-1}, the PNIPAM microgels exhibit more significant isothermal volume change. The higher the TA concentration, the faster is the isothermal volume-phase transition of PNIPAM microgels induced by TA. During the isothermal phase transition process of PNIPAM microgels induced by TA, the response rate of PNIPAM microgels to TA presence is mainly governed by diffusion-limited transport of TA molecules into PNIPAM polymeric networks. At the initial stage of the isothermal phase transition, the rate is mainly controlled by external diffusion in which TA molecules are transferred from the bulk solution to the interface of microgels/liquid. Then, the TA molecules further go through a slower internal diffusion process within PNIPAM polymeric networks in the microgels to approach the binding sites and finally reach adsorption equilibrium. Therefore, the more the TA molecules in the aqueous solution, the more are the TA molecules diffusing into the PNIPAM microgels within the same time interval, and the faster is the deswelling rate of PNIPAM microgels. At TA concentrations of 10^{-4} and 10^{-5} mol l^{-1}, although the PNIAPM microgels reach equilibrium state finally with almost the same phase transition degrees, it takes less than 50 min for the microgels to reach the equilibrium state in the case of 10^{-4} mol l^{-1} while it takes more than 200 min in the case of 10^{-5} mol l^{-1}.

Figure 4.4 shows the optical micrograms of PNIPAM microgels before and after the equilibrium volume-phase transition in aqueous solutions with different TA concentrations at 25 °C, from which it is obviously found that the size of PNIPAM microgels decreases with the increasing TA concentration. The equilibrium isothermal volume-phase transition degree of PNIPAM microgels is obviously concentration dependent. When the TA concentration is lower than 10^{-5} mol l^{-1},

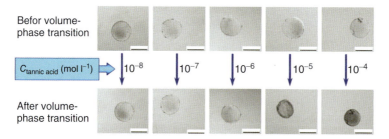

Figure 4.4 Optical micrograms of PNIPAM microgels before and after volume-phase transition in TA solutions with different concentrations at 25 °C. Scale bar is 200 μm. (Chen et al. 2010 [6]. Reproduced with permission of Elsevier.)

the phase transition degree of PNIPAM microgels increases remarkably with the increasing TA concentration; however, when the TA concentration is higher than 10^{-5} mol l^{-1}, the phase transition degrees are almost the same. The results indicate that 10^{-5} mol l^{-1} is a critical TA concentration for PNIPAM microgels to reach the state of full isothermal phase transition.

TA is a kind of polyphenol with plenty of hydroxyl groups. It is well known that the hydrogen bonding could be formed between the hydroxyl groups of phenols and amide groups of PNIPAM, and it has been reported that TA molecules could interact with PNIPAM polymeric chains through hydrogen-bonding and hydrophobic interaction in aqueous solutions [5]. Due to the presence of a lot of hydroxyl groups in a TA molecule, it is possible for one TA molecule to occupy a number of PNIPAM binding sites (i.e., amide groups) and to act as physical cross-linker for the polymeric networks to some extent. During the adsorption process, the TA molecules bind to PNIPAM polymeric networks and expel the water molecules originally located on the adsorption sites. As a result, when enough TA molecules are adsorbed onto the polymeric networks, the PNIPAM microgels shrink. In addition, the aromatic groups in TA molecules strengthen the hydrophobic interaction in polymeric networks and break the balance between the hydrophilic and hydrophobic interaction, which also leads to the shrinking of PNIPAM microgels. As the TA concentration increases, more TA molecules diffuse into the PNIPAM microgels and bind to polymeric networks and enhance the equilibrium isothermal volume-phase transition of the PNIPAM microgels.

Figure 4.5 shows the optical micrograms of PNIPAM microgels at different temperatures after being equilibrated in TA solutions with different concentrations. Obvious differences between the thermo-sensitive phase transition behaviors of PNIPAM microgels could be seen at different TA concentrations. When the TA concentration is higher than 10^{-5} mol l^{-1}, most of the hydrogen-bonding sites inside the PNIPAM microgels are firmly occupied by TA molecules and most of the water molecules inside the PNIPAM microgels are repelled out from the polymeric networks at 25 °C; as a result, the PNIPAM microgels do not show significant thermo-responsive phase transition behavior anymore.

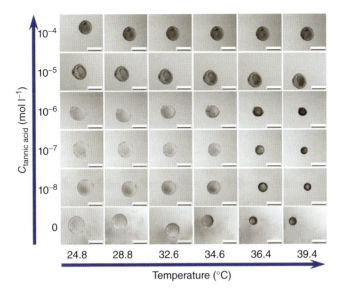

Figure 4.5 Optical micrograms of PNIPAM microgels at different temperatures after being equilibrated in TA solutions with different concentrations. Scale bar is 200 μm. (Chen *et al.* 2010 [6]. Reproduced with permission of Elsevier.)

When the TA concentration is lower than 10^{-6} mol l^{-1}, the isothermal phase transition of PNIPAM microgels induced by TA at 25 °C is not so significant; that is, there are still plenty of water molecules inside the PNIPAM microgels at 25 °C; therefore, the PNIPAM microgels still show remarkable thermo-responsive phase transition behaviors. The lower the TA concentration, the larger is the thermo-induced deswelling ratio of the PNIPAM microgels. Because the TA molecules act as physical cross-linkers to some extent as mentioned earlier, more TA molecules result in a more serious physical cross-linkage between the polymeric networks inside the PNIPAM microgels; as a result, the thermo-induced shrinking degree decreases at a higher TA concentration. Another interesting phenomenon is that, when the TA concentration is lower than 10^{-6} mol l^{-1}, the LCST of PNIPAM microgels slightly shifts to higher temperatures with the increasing TA concentration. For example, the LCST of PNIPAM microgel in pure water is about 33 °C, while it increases to about 35.5 °C when the TA concentration is 10^{-6} mol l^{-1}. Although part of hydroxyl groups in the TA molecule can bind to the amide groups on the PNIPAM polymeric networks through hydrogen bonding, there are still parts of free hydroxyl groups that cannot bind to the amide groups on the PNIPAM polymeric networks because of steric hindrance. These free hydroxyl groups of the TA molecule entrapped within the polymeric networks of PNIPAM microgels enhance the hydrophilicity of the PNIPAM microgels. As a result, the LCST of PNIPAM microgel shifts to higher temperatures due to the binding of TA onto PNIPAM polymeric networks.

4.3 Microfluidic Fabrication of Monodisperse Core–Shell PNIPAM Hydrogel Microparticles for Sensing Ethyl Gallate (EG)

The alkane esters of gallic acid are phenols that are abundant in medicinal plants and fermented beverages, such as *Lagerstroemia speciosa* and red wine, and are known to display a wide variety of biological functions in addition to their antioxidant activity [9]. EG, 3,4,5-trihydroxybenzoic acid ethyl ester, is one of the typical gallic acid alkane esters and has a wide range of biological activities including antioxidant, antimicrobial, and anti-inflammatory functions [7, 10].

PNIPAM hydrogel is a well-known thermo-responsive material for its dramatic phase transition property when environmental temperature changes across its LCST or volume-phase transition temperature (VPTT). The phase transition of PNIPAM could be affected by hydroxybenzenes due to the existence of hydroxyl and phenyl structures of phenols and its strong ability to form hydrogen-bonding and hydrophobic interactions [6]. The abovementioned results show that the volume-phase transition of PNIPAM could be induced by addition of phenols to a certain critical concentration at a constant temperature below the LCST, and the PNIPAM has remarkable concentration–response property to phenol molecules. That is, PNIPAM could be used as a candidate material for sensors that isothermally respond to the phenol molecules.

The authors' group prepared monodisperse core–shell microcapsules with a PNIPAM shell and a colored oil core via a microfluidic approach and studied the change in size and structure of monodisperse PNIPAM microcapsules in response to varying temperatures and EG concentrations [7]. To systematically investigate the VPTT values of PNIPAM in aqueous solutions with varying EG concentrations, monodisperse PNIPAM microspheres are prepared via the microfluidic approach in order to eliminate the effect of oil core on the VPTT values.

4.3.1 Microfluidic Fabrication of Monodisperse Core–Shell PNIPAM Hydrogel Microparticles

A capillary microfluidic device is assembled for the fabrication of double emulsion droplets according to a well-established method [11], as shown in Figure 4.6 [7]. The outer diameters of all the cylindrical capillary tubes are 0.99 mm. The square capillary tubes had an inner dimension of 1.0 mm. The inner diameters of the injection tube, the transition tube, and the collection tube are 500, 200, and 450 μm, respectively. A micropuller and a microforge are used to fabricate the capillary microfluidic device. The inner diameters of the tapered end of the injection and transition tubes are 60 and 140 μm, respectively. To generate oil-in-water-in-oil (O/W/O) emulsion droplets as templates for the preparation of core–shell PNIPAM microcapsules, soybean oil containing PGPR 90 and fluorescent dye Lumogen Red 300 (LR300) is used as the inner oil phase. The middle water phase is an aqueous solution containing monomer NIPAM, emulsifier Pluronic F127, cross-linker MBA, initiator APS, and viscosity adjustor glycerol. Soybean oil containing emulsifier PGPR 90 is used as the outer oil

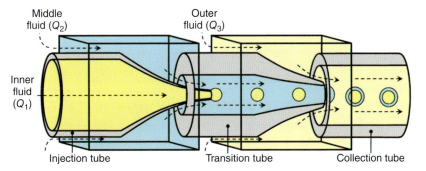

Figure 4.6 Schematic illustration of the microfluidic device for fabricating monodisperse double emulsion templates. (Mou et al. 2012 [7]. Reproduced with permission of Elsevier.)

phase. The flow rates of inner oil phase, water phase, and oil phase are 500, 800, and 2000 μl h^{-1}, respectively. The oil phase in a container for collecting the generated O/W/O emulsion droplets is soybean oil containing PGPR 90 and photoinitiator dimethoxy-2-phenylacetophenone (BDK). The formation process of O/W/O emulsion droplets in the capillary microfluidic device is recorded with a high-speed compact camera. The polymerization of the water phase of O/W/O emulsion droplets is initialized via UV irradiation in an ice bath for 12 min. A 250-W UV lamp with an illuminance spectrum of 250–450 nm is employed to produce UV light. The ice-water bath is used to ensure that the polymerization is carried out at a low temperature. After polymerization for 12 h, the PNIPAM microcapsules are washed with DI water for several times to remove the outer oil phase.

The optical micrographs of the formation processes of O/W/O emulsion droplets in the capillary microfluidic device, taken by a high-speed camera, are shown in Figure 4.7a,b. At the first stage of the capillary microfluidic device, droplets of the colored oil are emulsified by a coaxial flow of water phase to generate O/W single emulsion droplets (Figure 4.7a). The formed droplets are subsequently emulsified at the second stage by a coaxial flow of the outer oil phase to form O/W/O emulsion droplets (Figure 4.7b). Optical micrographs of the monodisperse O/W/O double emulsion droplets and resultant core–shell PNIPAM microcapsules in water at 25 °C are shown in Figure 4.7c,d. When we observe the double emulsions from the top view, the oil core seems to be in the center, even if there is a density difference between the inner oil phase and middle water phase. It is likely that the oil core is at the top of each emulsion droplet. Consequently, the core in the resultant microcapsule is eccentric and is squeezed out easily at the point with the thinnest thickness when the PNIPAM microcapsules shrink. Figure 4.7e shows the corresponding size distributions of O/W/O emulsion droplets (Figure 4.7c) and microcapsules (Figure 4.7d). The mean diameter of the inner oil cores is 239 μm ($CV = 1.1\%$), and the mean outer diameter of O/W/O emulsion droplets is 366 μm ($CV = 1.3\%$). The mean outer diameter of core–shell PNIPAM microcapsules is 422 μm ($CV = 3.6\%$), which is significantly larger than that of the O/W/O emulsion droplets, due to the hydrophilicity of their shell material.

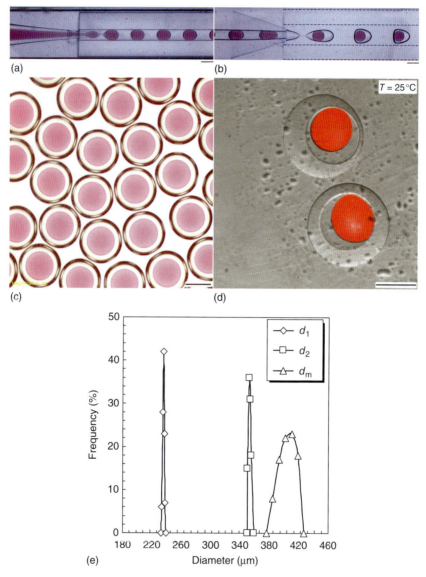

Figure 4.7 Preparation of O/W/O double emulsion droplets and core–shell microcapsules. (a, b) High-speed optical micrographs of the generation processes of O/W emulsion droplets at the first stage (a) and O/W/O double emulsion droplets at the second stage (b). Scale bars are 250 μm. (c, d) Optical micrographs of monodisperse O/W/O double emulsion droplets (c) and resultant core–shell PNIPAM microcapsules in water at 25 °C (d). Scale bars are 200 μm. (e) Size distributions of the emulsion droplets corresponding to (c) and microcapsules (d), in which d_1 and d_2 are, respectively, the diameter of inner oil cores and the outer diameter of O/W/O double emulsion droplets, and d_m is the outer diameter of the prepared PNIPAM microcapsules in water at 25 °C. (Mou et al. 2012 [7]. Reproduced with permission of Elsevier.)

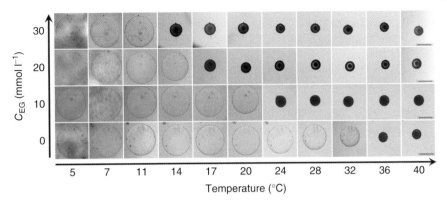

Figure 4.8 Optical micrographs of the PNIPAM microspheres at different temperatures after being equilibrated in aqueous solution with varying EG concentrations. Scale bars are 200 μm. (Mou et al. 2012 [7]. Reproduced with permission of Elsevier.)

4.3.2 Thermo-Responsive Phase Transition Behaviors of PNIPAM Microspheres in EG Solution

Figure 4.8 shows the optical micrographs of PNIPAM microspheres at different temperatures after being equilibrated in aqueous solution with varying EG concentrations. The PNIPAM microspheres are transparent and swollen when the temperature of EG aqueous solution is lower than the corresponding VPTT value (e.g., $\leq 11\,°C$ for an EG concentration of 30 mmol l^{-1}), but opaque and shrunken when the temperature of EG solution is higher than the corresponding VPTT value ($\geq 14\,°C$ in that case). According to the abovementioned studies on the interaction of phenols with PNIPAM hydrogels, the shift of the VPTT value of PNIPAM to lower temperature in EG aqueous solution can be explained by the hydrophilic and hydrophobic interactions of EG with PNIPAM. Hydrogen bonding could be formed between the hydroxyl groups of EG molecules and amide groups of PNIPAM, and benzene rings of EG molecules and isopropyl groups of PNIPAM form hydrophobic interaction with each other. As a result, EG disrupts the mechanism of water ordering around the PNIPAM polymer chains, which leads to the shift of VPTT of PNIPAM in EG aqueous solution to lower temperature.

4.3.3 The Intact-to-Broken Transformation Behaviors of Core–Shell PNIPAM Microcapsules in Aqueous Solution with Varying EG Concentrations

Figure 4.9 shows the confocal laser scanning microscope (CLSM) micrographs of dynamic phase transition behaviors of core–shell PNIPAM microcapsules in aqueous solution with varying EG concentrations and at different temperatures. When the surrounding water is replaced by the EG solution, the PNIPAM microcapsules shrink slightly in the first few seconds, and then remain unchanged afterward. Figure 4.9b shows that the PNIPAM microcapsules shrink slowly and squirt out the colored oil core when the temperature is increased to 25 °C. At time $t = 0$, the diameters of the PNIPAM microcapsules in water at 25 °C are smaller

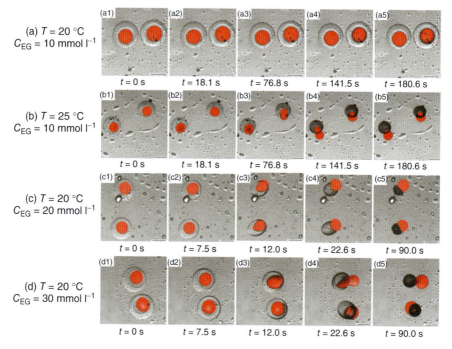

Figure 4.9 CLSM micrographs of dynamic phase transition behaviors of PNIPAM microcapsules with colored oil core in aqueous solution with varying EG concentrations and at different temperatures. The volume ratio between the inner core and the whole microcapsule is about 0.27. (a) $T = 20\,°C$, $C_{EG} = 10\,mmol\,l^{-1}$; (b) $T = 25\,°C$, $C_{EG} = 10\,mmol\,l^{-1}$; (c) $T = 20\,°C$, $C_{EG} = 20\,mmol\,l^{-1}$; (d) $T = 20\,°C$, $C_{EG} = 30\,mmol\,l^{-1}$. Scale bars are 200 μm. (Mou et al. 2012 [7]. Reproduced with permission of Elsevier.)

than that at 20 °C, because the PNIPAM shells shrink slightly when the temperature rises from 20 to 25 °C. When the environmental water is replaced by aqueous solution with an EG concentration of 20 or 30 mmol l^{-1} at 20 °C, the shell of PNIPAM microcapsules shrinks rapidly within 12 s, and the core–shell microcapsules are transformed from the intact state to the broken state because the PNIPAM shell cannot contain the colored oil core anymore, as shown in Figure 4.9c,d.

When the concentration of EG aqueous solution is fixed at 10 mmol l^{-1}, the microcapsules can transform from the intact state to the broken state on increasing the temperature from 20 °C (below the VPTT, which is approximately 23 °C) to 25 °C (above the VPTT). When the EG concentration is 20 or 30 mmol l^{-1}, the environmental temperature 20 °C is higher than the corresponding VPTT values (~16 or 12 °C, respectively). Therefore, under these conditions the microcapsules can transform from the intact state to the broken state. The larger the difference between the operation temperature and the corresponding VPTT, the faster is the shrinkage of the core–shell microcapsules. For example, at 20 °C the PNIPAM microcapsules in aqueous solution with EG concentration of 30 mmol l^{-1} shrink faster than that in aqueous solution with EG concentration of 20 mmol l^{-1} because the difference between the operation temperature and

Figure 4.10 State diagram of the intact-to-broken transformation of core–shell PNIPAM microcapsules in aqueous solution as a function of temperature and EG concentration. (Mou et al. 2012 [7]. Reproduced with permission of Elsevier.)

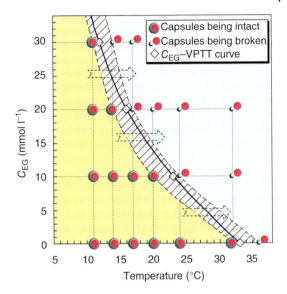

the corresponding VPTT under the former condition is 8 °C, while that under the later condition is just 4 °C.

Figure 4.10 illustrates the state diagram of the intact-to-broken transformation of core–shell PNIPAM microcapsules in aqueous solutions as a function of temperature and EG concentration. The intact-state region is located in the lower left region and the broken-state region is located in the upper right region, which are separated by a narrow region near the C_{EG}–VPTT curve. From the state diagram, the state of PNIPAM microcapsules is related to the temperature and the EG concentration. The results indicate that the core–shell PNIPAM microcapsules may be highly potential to be used as sensors and/or indicators for some simple detection of EG concentration if the EG concentration falls in the range from 0 to 35 mmol l^{-1}. When the core–shell PNIPAM microcapsules are put in EG aqueous solution and are heated up gradually, their size and the structure will change and transition from an intact state to a burst state. Therefore, the EG concentration can be determined simply by measuring the corresponding critical temperature for the microcapsules to squirt out their colored oil cores.

4.4 Microfluidic Fabrication of Monodisperse Core–Shell Hydrogel Microparticles for the Adsorption and Separation of Pb^{2+}

Lead (Pb^{2+}) is one of the most abundant and toxic heavy metal pollutants. Due to its non-biodegradability, Pb^{2+} can accumulate in the body from contaminated food and water. Even very low levels of lead intake will cause serious effects on the nervous system, reproductive system, immune system, liver, and kidneys of human beings, especially in infants and children [12]. However, Pb^{2+} is commonly found in industrial wastewater because of their wide applications in smelting, battery industries, painting, mining, and so on. Therefore, effective removal of Pb^{2+} from industrial wastewater will contribute to environmental sustainability.

18-Crown-6 and its derivatives are one of the most promising substances for designing Pb^{2+} detection and adsorption materials because of their specific Pb^{2+}-recognition ability based on the formation of supramolecular host–guest complexes [13]. The authors' group developed a novel kind of core–shell microspheres, each with an Fe_3O_4 magnetic nanoparticle (MNP) core and an ion-recognizable poly(N-isopropylacrylamide-co-benzo-18-crown-6-acrylamide) (PNB) shell, which combine the advantages of excellent Pb^{2+} selective adsorption and convenience of magnetic separation [14]. B18C6Am units act as active receptors for Pb^{2+} recognition and adsorption, which can selectively capture Pb^{2+} to form stable 1 : 1 (ligand:ion) B18C6Am/Pb^{2+} host–guest complexes [13]. PNIPAM networks act as actuators with thermo-responsive swelling/shrinking configuration change for convenient purification and regeneration of the adsorbents. The inner magnetic core enables the adsorbent with a magnetically guided aggregation to be separated from the decontaminated water, and even with a magnetically guided movement in some Pb^{2+}-contaminated microenvironments in a remotely controlled manner. Core–shell structure guarantees that the core and shell materials act independently and that the PNB shell can also protect the magnetic core from leaking.

4.4.1 Microfluidic Fabrication of Monodisperse Core–Shell Microparticles with Magnetic Core and Hydrogel Shell

The magnetic PNB core–shell microspheres are prepared using O/W/O double emulsions as templates, which are fabricated from a two-stage microfluidic device (Figure 4.11a). First, the solid magnetic cores are generated inside the emulsions by solvent evaporation of the inner oil phase (Figure 4.11b,c). Then, the middle aqueous phase layer (W) is converted to PNB shell through UV-initiated polymerization, from which the magnetic PNB core–shell microspheres, each with an independent magnetic core and an ion-recognizable shell, are formed (Figure 4.11d). After washing with deionized water and freeze-drying, the magnetic PNB core–shell microspheres are used as adsorbents for the separation of Pb^{2+} (Figure 4.11e) from the environment at room temperature (Figure 4.11f). After the adsorption reaches equilibrium, the microspheres are separated from the decontaminated water by an external magnet (Figure 4.11g,h). Pb^{2+} ions are then desorbed from 18-crown-6 units by increasing the operation temperature and washing repeatedly with deionized water (Figure 4.11i), which is caused by the thermo-induced decrease of the inclusion constant of B18C6Am/Pb^{2+} complexes. That is, the magnetic PNB core–shell microspheres can be regenerated simply by changing the operation temperature and washing with deionized water (Figure 4.11i,j). Therefore, the proposed magnetic PNB core–shell microspheres can be used as a selectively separable and efficiently reusable Pb^{2+} adsorbent [14].

Monodisperse O/W/O double emulsions as templates for the synthesis of magnetic PNB core–shell microspheres are generated from the microfluidic device (Figure 4.12a). The magnetic cores are formed inside the emulsions

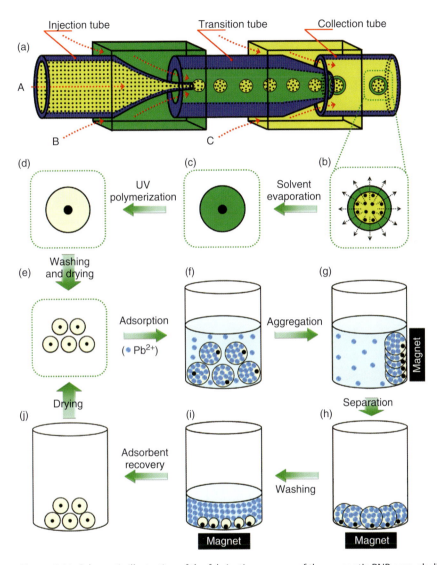

Figure 4.11 Schematic illustration of the fabrication process of the magnetic PNB core–shell microspheres and the Pb^{2+} adsorption and regeneration processes. (a) Microfluidic device for generating O/W/O emulsion templates, in which A, B, and C stand for the inner, middle, and outer fluid phases, respectively. (b, c) Formation of the magnetic core through solvent evaporation. (d) Fabrication of the magnetic PNB core–shell microspheres through UV-initiated polymerization of the emulsion templates. (e) Purification and dehydration of the magnetic PNB core–shell microspheres by washing and freeze-drying. (f) Adsorption of Pb^{2+} onto the magnetic PNB core–shell microspheres. (g, h) Separation of the adsorbents from the decontaminated solution via magnetic-induced aggregation. (i, j) Removal of the adsorbed Pb^{2+} from the microspheres and regeneration of the adsorbents by increasing the environmental temperature and washing with deionized water. (Liu et al. 2014 [14]. Reproduced with permission of American Chemical Society.)

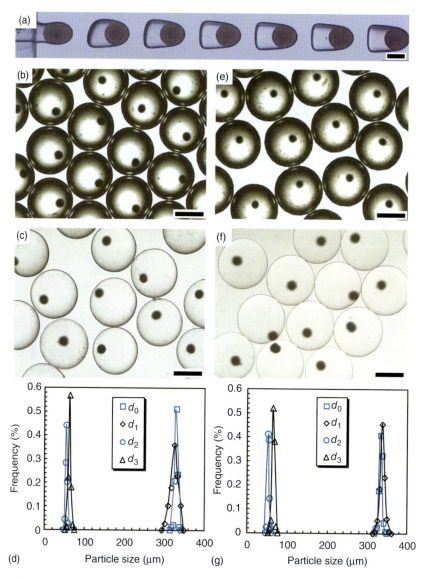

Figure 4.12 Template synthesis of core–shell microspheres from O/W/O double emulsions. (a) High-speed optical micrograph showing the generation of O/W/O double emulsions in microfluidic device. (b, e) Optical micrographs of the emulsion templates used for synthesis of magnetic PNB (b) and PNIPAM (e) core–shell microspheres at about 10 min after formation. (c, f) The resultant magnetic PNB (c) and PNIPAM (f) core–shell microspheres in deionized water at 25 °C. (d, g) Size distributions of the magnetic PNB (d) and PNIPAM (g) core–shell microspheres in deionized water at 25 °C and their corresponding emulsion templates, where d_0, d_1, d_2, and d_3 are, respectively, the diameters of the O/W/O double emulsions, the core–shell microspheres, the magnetic oil droplets, and the magnetic cores. Scale bars are 200 μm. (Liu et al. 2014 [14]. Reproduced with permission of American Chemical Society.)

through the evaporation of inner ethyl acetate. The generated emulsions, which are used as templates for the synthesis of magnetic PNB and PNIPAM core–shell microspheres, are shown in Figure 4.12b,e. The size of the inner oil droplets greatly decreases after being collected for 10 min. Figure 4.12c,f shows the optical micrographs of the prepared magnetic PNB and PNIPAM core–shell microspheres in deionized water at 25 °C. The magnetic core is completely embedded in the hydrogel shell after polymerization of the middle aqueous layer. The size distributions of the emulsion templates and the resultant magnetic PNB and PNIPAM core–shell microspheres dispersed in deionized water at 25 °C are shown in Figure 4.12d,g, where d_0, d_1, d_2, and d_3 represent the diameters of the O/W/O double emulsions, the core–shell microspheres, the magnetic oil droplets, and the magnetic cores, respectively. As shown in Figure 4.12d,g, the sizes of the two kinds of microspheres and their corresponding inner magnetic cores are almost the same, which results from the similar size of the used emulsion templates. It also indicates that the introduction of B18C6Am units has almost no effect on the size of the resulted core–shell microspheres. The average diameter of the magnetic oil droplets is a little smaller than the solid magnetic cores, because the magnetic oil droplets are difficult to be focused in the photograph. The CV, which is defined as the ratio of the standard deviation of the size distribution to its arithmetic mean, is used to characterize the size monodispersity. The emulsion templates, core–shell microspheres, and inner magnetic cores, whose CV values are all smaller than 5%, all have good monodispersity due to the microfluidic method.

4.4.2 Pb^{2+} Adsorption Behaviors of Magnetic PNB Core–Shell Microspheres

The ion-adsorption behaviors of magnetic PNB core–shell microspheres in a mixed heavy metal ion solution containing Pb^{2+}, Cd^{2+}, Co^{2+}, Cr^{3+}, Cu^{2+}, Ni^{2+}, and Zn^{2+} with the same initial ion concentration and equilibration time at different pH conditions are shown in Figure 4.13. The variation of pH has nearly no obvious effect on removal efficiencies of all heavy metal ions. For Pb^{2+}, there exists only 8% difference in its removal efficiency over a pH range from 2 to 7. It is worth noting that the removal efficiencies of magnetic PNB core–shell microspheres toward Pb^{2+} are always much higher than that toward the other metal ions at different pH conditions in the range of 2–7. The results indicate that the magnetic PNB core–shell microspheres can effectively and selectively remove Pb^{2+} even though the external pH value is changed. The slight decrease in the Pb^{2+} removal efficiency with a decrease of pH value from 5 to 2 may be caused by the slight protonation of B18C6Am at lower pH, which results in a reduction of complex ability of Pb^{2+} [15]. The slight decrease in the Pb^{2+} removal efficiency with an increase of pH value from 5 to 7 may be caused by the formation of lead hydroxide, because Pb^{2+} will precipitate slightly when pH is higher than 5.5 [16]. Therefore, the pH value of the aqueous solution in the subsequent experiments is adjusted to 5 to obtain optimal adsorption.

To examine the specific and selective Pb^{2+} adsorption of magnetic PNB core–shell microspheres toward Pb^{2+}, the batch adsorption performances of magnetic PNB core–shell microspheres in a mixed metal ion solution containing

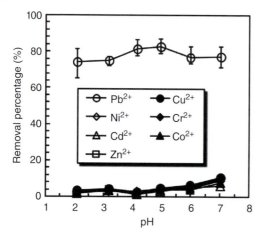

Figure 4.13 Effects of pH on ion removal efficiencies of magnetic PNB core–shell microspheres in a mixed heavy metal ion solution containing Pb^{2+}, Cu^{2+}, Ni^{2+}, Cr^{3+}, Cd^{2+}, Co^{2+}, and Zn^{2+}. The initial concentrations of the heavy metal ions are all about 40 mg l^{-1}. The amount of PNB microspheres is 9 g l^{-1}, and operation temperature is 25 °C. (Liu et al. 2014 [14]. Reproduced with permission of American Chemical Society.)

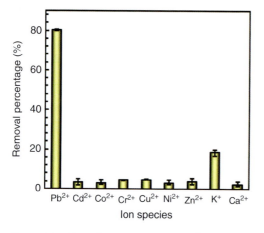

Figure 4.14 The selective and specific Pb^{2+} adsorption of magnetic PNB core–shell microspheres in a mixed metal ion solution containing Pb^{2+}, Cd^{2+}, Co^{2+}, Cr^{3+}, Cu^{2+}, Ni^{2+}, Zn^{2+}, K^+, and Ca^{2+}. The initial ion concentrations of Pb^{2+}, Cd^{2+}, Co^{2+}, Cr^{3+}, Cu^{2+}, Ni^{2+}, and Zn^{2+} are all about 40 mg l^{-1}. And the initial ion concentrations of K^+ and Ca^{2+} are both 4 mmol l^{-1}, which are 20 times higher than that of Pb^{2+} (0.2 mmol l^{-1}). The amount of microspheres is 9 g l^{-1}, operation temperature is 25 °C, and the pH value of the solution is 5. (Liu et al. 2014 [14]. Reproduced with permission of American Chemical Society.)

Pb^{2+}, Cd^{2+}, Co^{2+}, Cr^{3+}, Cu^{2+}, Ni^{2+}, Zn^{2+}, K^+, and Ca^{2+} as the contaminated water are investigated and their removal efficiencies at pH 5 are also studied by the batch adsorption technique. The removal efficiencies are calculated from the batch adsorption experiment carried out at 25 °C for 12 h. The magnetic PNB core–shell microspheres exhibit a much higher Pb^{2+} removal efficiency than that toward the other metal ions as shown in Figure 4.14. Pb^{2+} removal efficiency is nearly not affected by the addition of Cd^{2+}, Co^{2+}, Cr^{3+}, Cu^{2+}, Ni^{2+}, Zn^{2+}, and Ca^{2+}. Even though the initial concentration of K^+ (4 mmol l^{-1}) is 20 times higher than that of Pb^{2+} (0.2 mmol l^{-1}), the removal efficiency of Pb^{2+} is still about 4 times higher than that of K^+. These results also indicate that the magnetic PNB core–shell microspheres can specifically and selectively capture Pb^{2+} from the mixed ion solution of Cd^{2+}, Co^{2+}, Cr^{3+}, Cu^{2+}, Ni^{2+}, Zn^{2+}, K^+, and Ca^{2+} ions. The removal efficiency of K^+ is higher than that of Ca^{2+}. Such specific adsorption is caused by the formation of stable crown-ether/metal-ion complexes, which is dominated by the size/shape fitting or matching between the host and guest molecules. The B18C6Am receptors in the magnetic PNB core–shell microspheres could selectively capture specific metal ions, whose size fits the cavity size of crown ether well and binds with the cavity tightly and effectively. The complexes' stability constant ($\log K$) of benzo-18-crown-6 with metal ions in water is $Pb^{2+} > K^+ >$ the other coexisting metal ions [13]. Therefore, the magnetic PNB core–shell microspheres exhibit selective Pb^{2+} adsorption based on the formation of stable B18C6Am/Pb^{2+} complexes.

Because the skeleton structure is PNIPAM polymeric network, the magnetic PNB core–shell microspheres exhibit obvious temperature-responsive volume change both in deionized water and 20 mmol l^{-1} Pb^{2+} solution, as shown in Figure 4.15a. With increasing temperature, the magnetic PNB core–shell microspheres show a thermo-induced shrinking behavior in both deionized water and Pb^{2+} solution. Obviously, the VPTT of the magnetic PNB core–shell microspheres shifts to a higher temperature in Pb^{2+}-contained solution ($VPTT_2$) than that in deionized water ($VPTT_1$). Such positive VPTT shift is also caused by the formation of stable B18C6Am/Pb^{2+} complexes. The repulsion among charged B18C6Am/Pb^{2+} complexes and the osmotic pressure within the magnetic PNB core–shell microspheres counteract the shrinkage of the hydrogel networks with the increase of temperature, thereby resulting in the VPTT shifting to a higher temperature.

To investigate the effect of temperature on the adsorption capacity of magnetic PNB core–shell microspheres toward Pb^{2+}, three representative temperatures – 25 °C (below $VPTT_1$), 36 °C (between $VPTT_1$ and $VPTT_2$), and 52 °C (higher than $VPTT_2$) – are chosen from Figure 4.15a. The pH values of the Pb^{2+} aqueous solutions are all set at 5. As shown in Figure 4.15b,c, regardless of the variation in the initial concentration of Pb^{2+} or the amount of magnetic PNB core–shell microspheres, the highest equilibrium Pb^{2+}-adsorption capacity is always obtained at 25 °C at each condition. Such thermo-induced decrease of Pb^{2+}-adsorption capacities of magnetic PNB core–shell microspheres is mainly caused by the decrease in the formation constant of B18C6Am/Pb^{2+} complexes as temperature increases. At 25 °C, the formation constant of B18C6Am/Pb^{2+} complex is high, so the cavities of B18C6Am units can capture Pb^{2+} efficiently and tightly. As a result, Pb^{2+} can be effectively removed from the environment. As temperature increases, the formation constant of B18C6Am/Pb^{2+} complex decreases, and thereby part of captured Pb^{2+} desorbs from the magnetic PNB core–shell microspheres at higher temperature. The desorbed Pb^{2+} can be squirted out from the microspheres taking advantage of the thermo-induced dramatic shrinkage of the PNB microspheres. Therefore, the magnetic PNB core–shell microspheres can be regenerated easily by simply increasing the operation temperature and washing repeatedly with deionized water.

Taking cost savings into account, regeneration of adsorbents is very important. The regeneration of Pb^{2+}-adsorbed magnetic PNB core–shell microspheres can

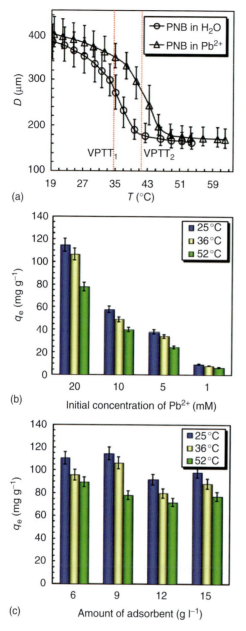

Figure 4.15 Effect of temperature on the Pb^{2+}-adsorption capacity of magnetic PNB core–shell microspheres. (a) Temperature-dependent diameter changes of magnetic PNB core–shell microspheres in Pb^{2+} aqueous solution (20 mmol l^{-1}) and deionized water. (b) Effect of temperature on the Pb^{2+}-adsorption capacity with different initial Pb^{2+} concentrations (the amount of adsorbent is 9 g l^{-1}). (c) Effect of temperature on the Pb^{2+}-adsorption capacity with different amounts of magnetic PNB core–shell microspheres ($C_0 = 20$ mmol l^{-1}). The pH values of the solutions are all 5. (Liu et al. 2014 [14]. Reproduced with permission of American Chemical Society.)

be implemented simply by increasing the operation temperature and washing with deionized water. The Pb^{2+} adsorption and regeneration of magnetic PNB core–shell microspheres are performed for five cycles, and the results are shown in Figure 4.16. The magnetic PNB core–shell microspheres show good thermo-induced reduction of adsorption capacity in each recycle. At 25 °C, only 15% of Pb^{2+}-adsorption capacity is lost after five adsorption–removal cycles. So it can be inferred that more than 85% of the adsorbed Pb^{2+} ions

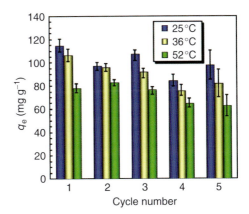

Figure 4.16 Pb^{2+}-adsorption capacities of magnetic PNB core–shell microspheres at different regeneration cycles ($C_0 = 20$ mmol l^{-1}, the amount of microspheres is 9 g l^{-1}, and the pH values of the solutions are all 5). (Liu *et al.* 2014 [14]. Reproduced with permission of American Chemical Society.)

can be removed from the microspheres by increasing the temperature and washing with deionized water. Compared with the regeneration process of many previously reported Pb^{2+} adsorbents under strong acid solution [15, 16], the regeneration of the magnetic PNB core–shell microspheres is much easier and more environmentally friendly. The results indicate that the magnetic PNB core–shell microspheres can be easily and well regenerated and be used repeatedly.

4.5 Summary

In this chapter, microfluidic strategies for controllable fabrication of monodisperse hydrogel microparticles are introduced. Through the microfluidic approach, stimuli-responsive hydrogel microparticles can be engineered with different structures for sensing and actuating, and adsorbing and separating. The strategies are demonstrated by fabricating monodisperse PNIPAM microgels for sensing TA, core–shell PNIPAM microcapsules for sensing EG, and core–shell microspheres with magnetic core and PNB shell for selective adsorption and separation of Pb^{2+}. With these monodisperse PNIPAM microgels or core–shell microcapsules as sensors and/or indicators, some simple and cheap techniques for detecting the TA or EG concentrations could be developed. The magnetic PNB core–shell microspheres exhibit selective and rapid Pb^{2+} adsorption through the formation of stable B18C6Am/Pb^{2+} complexes, and the Pb^{2+}-adsorbed microspheres can be separated from the treated solution through an external magnetic field. The microfluidic approach provides new opportunities for the design and fabrication of novel functional hydrogel microparticles for different applications. Although the product output of microfluidic techniques is still limited at present stage in usual cases, many integrated microfluidic devices have already been developed for mass productions [17], so the microfluidic techniques have great potential in practical mass production of monodisperse emulsions and microparticles for applications at industrial scales.

References

1 Xia, L.W., Xie, R., Ju, X.J., Wang, W., Chen, Q., and Chu, L.Y. (2013) Nano-structured smart hydrogels with rapid response and high elasticity. *Nat. Commun.*, **4**, 2226.

2 Chu, L.Y., Xie, R., Ju, X.J., and Wang, W. (2013) *Smart Hydrogel Functional Materials*, Springer, Berlin, Heidelberg.

3 Chu, L.Y., Kim, J.W., Shah, R.K., and Weitz, D.A. (2007) Monodisperse thermo-responsive microgels with tunable volume-phase transition kinetics. *Adv. Funct. Mater.*, **17**, 3499–3504.

4 Pelton, R. (2000) Temperature-sensitive aqueous microgels. *Adv. Colloid Interface Sci.*, **85**, 1–33.

5 Erel-Unal, I. and Sukhishvili, S.A. (2008) Hydrogen-bonded multilayers of a neutral polymer and a polyphenol. *Macromolecules*, **41**, 3962–3970.

6 Chen, G., Niu, C.H., Zhou, M.Y., Ju, X.J., Xie, R., and Chu, L.Y. (2010) Phase transition behaviors of poly(N-isopropylacrylamide) microgels induced by tannic acid. *J. Colloid Interface Sci.*, **343**, 168–175.

7 Mou, C.L., He, X.H., Ju, X.J., Xie, R., Liu, Z., Liu, L., Zhang, Z., and Chu, L.Y. (2012) Change in size and structure of monodisperse poly(N-isopropylacrylamide) microcapsules in response to varying temperature and ethyl gallate concentration. *Chem. Eng. J.*, **210**, 212–219.

8 Xu, S., Nie, Z., Seo, M., Lewis, P., Kumacheva, E., Stone, H.A., Garstecki, P., Weibel, D.B., Gitlin, I., and Whitesides, G.M. (2005) Generation of monodisperse particles by using microfluidics: control over size, shape, and composition. *Angew. Chem. Int. Ed.*, **44**, 724–728.

9 Zhang, Z., Liao, L., Moore, J., Wu, T., and Wang, Z. (2009) Antioxidant phenolic compounds from walnut kernels (*Juglans regia* L.). *Food Chem.*, **113**, 160–165.

10 Mink, S.N., Jacobs, H., Gotes, J., Kasian, K., and Cheng, Z.Q. (2011) Ethyl gallate, a scavenger of hydrogen peroxide that inhibits lysozyme-induced hydrogen peroxide signaling in vitro, reverses hypotension in canine septic shock. *J. Appl. Physiol.*, **110**, 359–374.

11 Chu, L.Y., Utada, A.S., Shah, R.K., Kim, J.W., and Weitz, D.A. (2007) Controllable monodisperse multiple emulsions. *Angew. Chem. Int. Ed.*, **46**, 8970–8974.

12 Kim, H.N., Ren, W.X., Kim, J.S., and Yoon, J. (2012) Fluorescent and colorimetric sensors for detection of lead, cadmium, and mercury ions. *Chem. Soc. Rev.*, **41**, 3210–3244.

13 Zhang, B., Ju, X.J., Xie, R., Liu, Z., Pi, S.W., and Chu, L.Y. (2012) Comprehensive effects of metal ions on responsive characteristics of P(NIPAM-co-B18C6Am). *J. Phys. Chem. B*, **116**, 5527–5536.

14 Liu, Y.M., Ju, X.J., Xin, Y., Zheng, W.C., Wang, W., Wei, J., Xie, R., Liu, Z., and Chu, L.Y. (2014) A novel smart microsphere with magnetic core and ion-recognizable shell for Pb^{2+} adsorption and separation. *ACS Appl. Mater. Interfaces*, **6**, 9530–9542.

15 Luo, X., Liu, L., Deng, F., and Luo, S. (2013) Novel ion-imprinted polymer using crown ether as a functional monomer for selective removal of Pb(II) ions in real environmental water samples. *J. Mater. Chem. A*, **1**, 8280–8286.

16 Ge, F., Li, M.M., Ye, H., and Zhao, B.X. (2012) Effective removal of heavy metal ions Cd^{2+}, Zn^{2+}, Pb^{2+}, Cu^{2+} from aqueous solution by polymer-modified magnetic nanoparticles. *J. Hazard. Mater.*, **211–212**, 366–372.

17 Romanowsky, M.B., Abate, A.R., Rotem, A., Holtze, C., and Weitz, D.A. (2012) High throughput production of single core double emulsions in a parallelized microfluidic device. *Lab Chip*, **12**, 802–807.

5

Microfluidic Fabrication of Monodisperse Porous Microparticles

5.1 Introduction

Porous structures of microparticles, which provide large specific area for the microparticles and channels for mass transfer inside the microparticles, could improve the performance of the microparticles or even bring new functions for the microparticles. For example, the porous structure can significantly increase the specific surface area of microspheres for several potential applications such as the column packing materials in size exclusion chromatography [1]. The porous structures make the microparticles extremely valuable for numerous applications, including drug delivery systems, cell immobilization carriers, catalysts, sensors and actuators, adsorptions and separations, and so on.

Monodispersity is very important for the porous microparticles to improve their performances in various applications, especially in the field of drug delivery systems [2–4]. Both the distribution of microspheres within the body and the interaction with biological cells are greatly affected by the particle size [3]. In addition, if monodisperse microspheres are available, the drug release kinetics can be manipulated, therefore making it easier to formulate more sophisticated drug release systems [4]. The authors' group developed several microfluidic approaches for fabrication of monodisperse porous microparticles, including monodisperse porous poly(hydroxyethyl methacrylate-methyl methacrylate) (poly(HEMA-MMA)) microspheres using single emulsions as templates and poly(vinyl pyrrolidone) (PVP) molecules as porogens, and monodisperse porous poly(N-isopropylacrylamide) (PNIPAM) microgels using solid nanoparticles and fine oil droplets as porogens [2, 5, 6]. In this chapter, microfluidic strategies are introduced to controllably prepare monodisperse porous microparticles with emulsions as templates and with different types of porogens.

5.2 Microfluidic Fabrication of Monodisperse Porous Poly(HEMA-MMA) Microparticles

Poly(methyl methacrylate) (PMMA) particles have been reported to show slow biodegradability. The biodegradability can be improved through the copolymerization of MMA with 2-hydroxyethyl methacrylate (HEMA), which forms a more biodegradable polymer. On the other hand, polymers based on HEMA attract great interests because of their biocompatibility. Therefore, poly(HEMA-MMA) copolymers have stimulated increasing interest and research attention in recent years due to their biocompatibility and water insolubility. Besides, the poly(HEMA-MMA) copolymers also have excellent chemical stability because of their three-dimensional polymeric networks. All these aforementioned favorable properties make poly(HEMA-MMA) microspheres extremely valuable for various applications in therapeutical and biotechnological fields, such as cell immobilization, drug delivery systems, packing materials in chromatography, and ophthalmology [1, 7]. Porous structure and monodispersity are important and valuable for poly(HEMA-MMA) microspheres to improve their performances in their applications. Here, we show a microfluidic method to controllably prepare monodisperse poly(HEMA-MMA) microspheres with porous structures.

5.2.1 Microfluidic Fabrication Strategy

A microfluidic device is designed to fabricate monodisperse oil-in-water (O/W) emulsions as templates, and UV-initiated free-radical polymerization with different porogens is adopted to prepare the poly(HEMA-MMA) porous microspheres. The microfluidic flow-focusing device is fabricated by assembling borosilicate glass capillary tubes on glass slides [2, 8]. The inner dimension of the square capillary tube is 1.0 mm. The cylindrical capillary had an inner diameter of 580 μm and an outer diameter of 1.0 mm. A micropuller is used to taper the end of cylindrical capillary, and the orifice dimension of the tapered end is adjusted by a microforge. The orifice diameter of the tapered end of the cylindrical capillary is 100 μm. The cylindrical capillary with a tapered end is inserted into the square capillary tube coaxially ensured by matching the outer diameter of the cylindrical tube to the inner dimension of the square one. A transparent epoxy resin is used to seal the tubes where required. The formation of O/W emulsions using flow-focusing within microcapillary device is illustrated in Figure 5.1. Two immiscible liquids are separately supplied to the microfluidic device through polyethylene tubing attached to syringes driven by syringe pumps. The flow rates of the inner oil solution and outer aqueous solution are selected as $Q_i = 300\,\mu l\,h^{-1}$ and $Q_o = 2000\,\mu l\,h^{-1}$, respectively.

Poly(HEMA-MMA) porous microspheres are prepared via free-radical polymerization, which is initiated inside the oil phase. The polymerizations are carried out according to the recipe given in Table 5.1, and the preparation process is illustrated in Figure 5.1b–d. Oil-soluble 2,2-dimethoxy-2-phenylacetophenone (BDK) is used as photoinitiator and added into the oil solution, and PVP is used as porogen. Under the UV irradiation, BDK molecules decompose and generate a great deal of reactive free radicals within the oil emulsion, which initiates the copolymerization of HEMA and MMA molecules inside the oil phase. Finally, the poly(HEMA-MMA) microsphere is formed with PVP porogen dispersed inside. The removal of PVP porogen with water results in the poly(HEMA-MMA) microsphere with porous structure.

It is well known that HEMA is a water-soluble monomer, so it is difficult or even impossible to prepare poly(HEMA)-based microspheres in microfluidic devices

Figure 5.1 Schematic illustration of the capillary microfluidic device and polymerization process for preparing monodisperse poly(HEMA-MMA) porous microspheres initiated inside the single emulsion droplet. "Water phase" is deionized water containing PVA and glycerol; and "Oil phase" is the mixed solution of HEMA and MMA containing EGDMA, PGPR, PVP, and BDK. (Zhang et al. 2009 [5]. Reproduced with permission of Elsevier.)

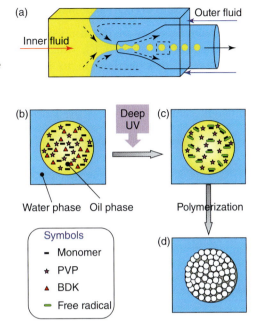

Table 5.1 Recipe for preparing poly(HEMA-MMA) porous microspheres.

Microsphere No.	M-P-I	M-P-II	M-P-III	M-P-IV
[PVP]/([HEMA] + [MMA]) (w/w)	0	3 : 100	5 : 100	7 : 100

Note. (1) Polymerizations are carried out at 20 °C for 30 min initiated by UV irradiation under nitrogen atmosphere. (2) HEMA and MMA comonomers are mixed and directly used as the oil phase (HEMA/MMA = 1/6 (v/v)). The contents of surfactant polyglycerol polyricinoleate (PGPR), cross-linker ethylene glycol dimethacrylate (EGDMA), and initiator BDK in oil phase are 5% (w/w), 5% (v/v), and 0.25% (w/w), respectively, based on the total content of HEMA and MMA. (3) Deionized water is used as water phase and the contents of poly(vinyl alcohol) (PVA) and glycerol are 2% (w/w) and 20% (v/v), respectively, based on the content of deionized water.

by using only HEMA as monomer disperse phase and using aqueous solution as continuous phase. However, in this reaction system, MMA is introduced as a comonomer in the disperse phase. Due to the water insolubility of MMA and the high solubility of HEMA in MMA, the mixed solution of HEMA and MMA could maintain stability against the water phase containing proper stabilizers. In the water phase, glycerol is used as the thickening agent and poly(vinyl alcohol) (PVA) is used as the stabilizer, which could prevent the diffusion of HEMA from oil into aqueous phase to a certain extent. In the oil phase, the content of MMA is much more than that of the HEMA (HEMA/MMA = 1/6 (v/v)), and polyglycerol polyricinoleate (PGPR) is used as surfactant. To prove the stability of this reaction system, the miscibility of oil phase against water phase is tested, and the experimental results show that the HEMA could hardly diffuse into the water phase [5].

5.2.2 Structures of Poly(HEMA-MMA) Porous Microspheres

The monodispersity of microspheres is evaluated based on an index called coefficient of variation (CV), which is defined as the ratio of the standard deviation of size distribution to its arithmetic mean ($\overline{D}$) [5]. Figures 5.2 and 5.3 show the optical microscope images and the size distributions of monodisperse poly(HEMA-MMA) porous microspheres prepared using different contents of PVP as the porogen. With the increasing mass ratio of

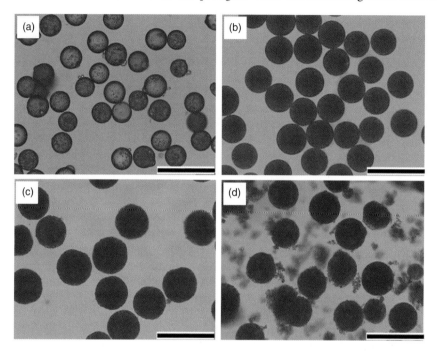

Figure 5.2 Optical microscope images of poly(HEMA-MMA) microspheres prepared with different contents of PVP as the porogen. (a) M-P-I: no PVP, (b) M-P-II: [PVP]/([HEMA] + [MMA]) = 3 : 100, (c) M-P-III: [PVP]/([HEMA] + [MMA]) = 5 : 100, and (d) M-P-IV: [PVP]/([HEMA] + [MMA]) = 7 : 100. Scale bar is 100 μm. (Zhang et al. 2009 [5]. Reproduced with permission of Elsevier.)

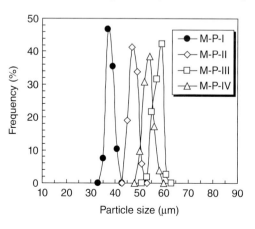

Figure 5.3 Size distribution of poly(HEMA-MMA) porous microspheres (prepared with recipe listed in Table 5.1) dispersed in water. (Zhang et al. 2009 [5]. Reproduced with permission of Elsevier.)

[PVP]/([HEMA] + [MMA]) from 0% to 3%, 5%, and 7%, the corresponding $\overline{D}$ values of prepared microspheres are 38.0 μm (Figure 5.2a), 47.6 μm (Figure 5.2b), 57.5 μm (Figure 5.2c), and 53.6 μm (Figure 5.2d), respectively. It can be seen that the average size of microspheres is increased by adding PVP into the oil phase (Figure 5.3). The reason is that the addition of PVP increases the viscosity in oil phase and the interfacial tension between water phase and oil phase. The CV values only have a slight fluctuation around 3.5%, which indicates that the porous microspheres prepared using microfluidic technology have a good monodispersity. It is also worth noting that the surface morphology of porous microspheres is altered with increasing PVP content. The sphericity and mechanical strength of the microspheres became worse gradually with the increasing PVP content. Especially when the ratio of [PVP]/([HEMA] + [MMA]) is as high as 7%, some of the prepared microspheres are even deformed in deionized water (see Figure 5.2d). After the polymerization, PVP porogen disperses in the network of poly(HEMA-MMA) microspheres randomly. When porogen is removed, porous structures appear both on the surface and inside the microspheres. With the increasing content of PVP, the porosity of microspheres also increases. Finally, the microspheres prepared with higher PVP content have larger porosity, resulting in an easily deformed structure.

Porous structures in poly(HEMA-MMA) microspheres are visually observed by SEM and presented in Figure 5.4. When porogen is absent in the oil drops, the prepared poly(HEMA-MMA) microspheres have a smooth surface without any pores (Figure 5.4a,b). When PVP is added into the oil phase as porogen, the poly(HEMA-MMA) microspheres with a honeycomb porous structure are formed after removing PVP with water (Figure 5.4c–h). In addition, from the SEM images of microsphere cross-section, it is confirmed that both the surface and interior of microspheres have porous structures as shown in Figure 5.5. It can also be proved by SEM images (Figure 5.4) that with the increasing content of PVP the porosity of microspheres increases. The number of the pores in the microspheres increases with the increasing ratio of [PVP]/ ([HEMA] + [MMA]). Such microspheres with a honeycomb porous structure have a large specific surface area and would have a perfect performance in the fields of biological adsorption and chromatography. When the ratio of [PVP]/([HEMA] + [MMA]) reaches 7%, some of the porous microspheres exhibit deformed structures (Figure 5.4g,h), which are caused by the large quantity of pores in the network of porous microspheres. That is, the increase of porosity of the microspheres would cause a decrease in mechanical strength. To prepare microspheres with both porous structure and good mechanical strength, the ratio of [PVP]/([HEMA] + [MMA]) in the recipe should be less than 7%.

5.3 Microfluidic Fabrication of Porous PNIPAM Microparticles with Tunable Response Behaviors

Stimuli-sensitive hydrogel microspheres or microgels are polymeric particles consisting of cross-linked three-dimensional networks. They shrink or swell

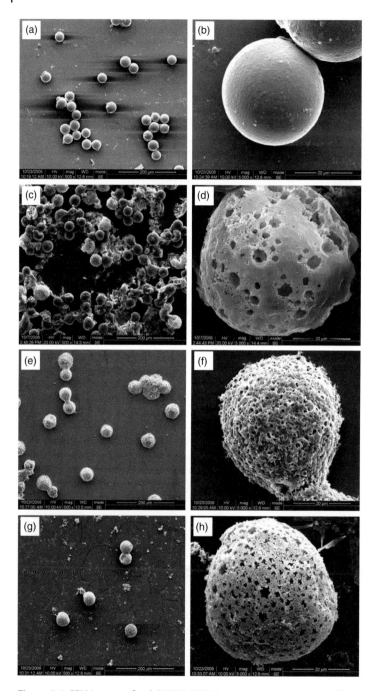

Figure 5.4 SEM images of poly(HEMA-MMA) porous microspheres. (a, b) M-P-I, (c, d) M-P-II, (e, f) M-P-III, and (g, h) M-P-IV. Scale bar in the left column is 50 μm and the right column is 20 μm. (Zhang *et al.* 2009 [5]. Reproduced with permission of Elsevier.)

Figure 5.5 SEM image of the cross-section of M-P-II porous microsphere. Scale bar is 20 μm. (Zhang et al. 2009 [5]. Reproduced with permission of Elsevier.)

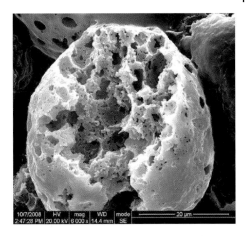

significantly by expelling or absorbing large amounts of water in response to external stimuli such as changes in temperature, pH, and electric or magnetic fields. The dramatic response and stimuli-specific behavior make these materials extremely valuable for numerous applications [2, 4, 9, 10], including drug delivery, chemical separations, sensors, catalysis, enzyme immobilization, and color-tunable crystals.

Several potential applications of these microgels, such as "smart" actuators, on–off switches, and pulse-release, require a short response time. Currently, numerous strategies are employed to speed up the response kinetics of typical thermo-responsive hydrogel microspheres such as PNIPAM microgels [2, 4, 6, 9]. One simple strategy is to use small size microgel particles. Since the characteristic time of gel swelling is proportional to the square of the linear dimension of the hydrogels [11], this results in a significant decrease in the response time. However, for certain applications such as actuators in a tube of fixed diameter, the size of microgels cannot be used as a control variable. Another well-known method is to chemically graft linear side chains of the same stimuli-sensitive polymer onto the cross-linked polymer network [12]. The linear side chains, with one free end each, respond to stimulus faster than the cross-linked network, and this leads to a shorter response time for the entire hydrogel as compared to a hydrogel with ungrafted polymer networks. However, the process of chemically grafting linear side chains onto a polymer network is complicated, limiting the widespread use of this technique. Another alternative is to fabricate hydrogels with heterogeneous internal microstructures [13] instead of a homogeneous net-like microstructure. The microgel-like particle clusters with numerous free ends inside the heterogeneous hydrogels can flex without any restriction [13] resulting in a faster response to stimuli as compared to microgels with a homogeneous microstructure. However, the heterogeneity of the microstructure may also cause some unwanted side effects; for example, the temperature-dependent equilibrium volume-deswelling ratio of PNIPAM hydrogels with a heterogeneous microstructure is much smaller than that of PNIPAM hydrogels with a homogeneous microstructure [13].

The author developed a microfluidic method to prepare thermo-sensitive PNIPAM porous microgels for controlling the volume-phase transition kinetics [2]. The kinetics of swelling and deswelling of stimuli-responsive polymeric hydrogels are typically governed by diffusion-limited transport of water in and out of the polymeric networks. Hence, if a facile way to control this transport of water is devised, the volume-phase transition kinetics of the microgels can be precisely tuned. This is accomplished by introducing different void structures inside the microgels [2]. An internal structure with voids offers less resistance to the transport of water compared to a voidless core. Thus, by varying the size and number of voids inside a microgel its swelling and shrinking dynamics can be effectively controlled. Here, we discuss the microfluidic approaches for fabrication of monodisperse porous PNIPAM microgels using solid nanoparticles as porogens [2].

5.3.1 Microfluidic Fabrication Strategy

The porous PNIPAM microparticles are prepared using a two-step method. First, monodisperse microgels with different number/sizes of solid polystyrene microspheres are synthesized in a microfluidic device. Second, the embedded microspheres are chemically dissolved to form spherical voids inside the microgels. The microfluidic device used to synthesize the microgels consists of cylindrical glass capillary tubes nested within square glass capillary tubes, as shown in Figure 5.6. It features a coaxial co-flow geometry ensured by matching the outside diameters of the cylindrical tubes to the inner dimensions of the square tubes. An aqueous suspension containing the monomer N-isopropylacrylamide (NIPAM), cross-linker N,N'-methylene-bis-acryamide (MBA), initiator ammonium persulfate (APS), and polystyrene microspheres is pumped into the left end of the left square tube. The continuous phase, kerosene containing a surfactant, PGPR, is pumped through the outer coaxial region between the left square tube and a tapered round microcapillary tube. The aqueous phase breaks into droplets at the entrance orifice of the tapered tube, forming monodisperse emulsion drops in the tube. The accelerator, N,N,N',N'-tetramethylethylenediamine (TEMED), dissolved in the continuous phase, is pumped through the outer coaxial region between the right square tube and the right end of the tapered round tube. When the accelerator meets the initiator, it starts a redox reaction that polymerizes the monomers, thus forming monodisperse microgels with embedded polystyrene microspheres. The addition of the reaction accelerator downstream of the other chemicals delays the polymerization process long enough to allow the formation of monodisperse emulsion drops of these chemicals in the continuous phase and eliminates the possibility of the entrance orifice of the round tube getting clogged by untimely polymerization of the monomers. Once the polymerization process is complete, the PNIPAM microgels are washed with isopropanol and subsequently immersed in xylene for a fixed period of time to dissolve the polystyrene beads leaving behind holes in the microgels. The microgels with spherical voids

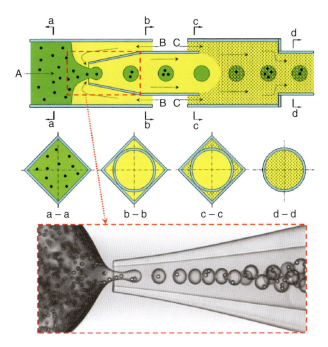

Figure 5.6 Microfluidic device for preparing monodisperse microgels with embedded solid particles. Fluid A is an aqueous suspension containing the monomer, initiator, cross-linker, and solid polystyrene particles; fluid B is an oil containing a surfactant; and fluid C is a mixture of an oil and a reaction accelerator. Illustrations a–a, b–b, c–c, and d–d are cross-section images of the capillary microfluidic device in relevant positions, which clearly show how the square capillary tubes and cylindrical tubes are assembled in the device. (Chu et al. 2007 [2]. Reproduced with permission of John Wiley & Sons.)

thus formed are again rinsed with isopropanol and are finally dispersed in water. Specific comparisons are made between the thermo-sensitive behavior of voidless microgels, without any voids, and those with hollow void structures, having either one large void of 75 µm diameter or different numbers of smaller voids.

Microgels containing different numbers of 25 µm diameter polystyrene beads and the microgels with spherical voids formed by dissolving these beads from similar microgels are shown in Figure 5.7. The optical micrographs suggest that the size and number of spherical voids in a microgel depend on the size and number of the encapsulated polystyrene beads. The precise circular periphery of the voids and unchanged size of the microgels also suggest that the chemical dissolution of the beads does not affect the structural integrity of the surrounding microgel structure.

5.3.2 Tunable Response Behaviors of Porous PNIPAM Microparticles

The dynamics of shrinking and swelling of the microgels are studied by heating them together with water from 23 to 47 °C and subsequently cooling them back to 23 °C in a transparent sample holder placed on a microscope-mounted

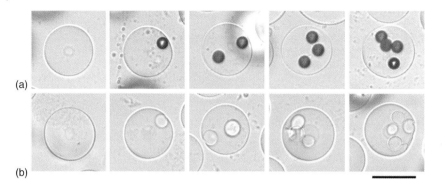

Figure 5.7 (a) Microgels with different number of embedded polystyrene beads; (b) microgels with spherical voids formed by dissolving the embedded beads from such microgels. Scale bar is 100 μm. (Chu et al. 2007 [2]. Reproduced with permission of John Wiley & Sons.)

thermal stage. The same sample holder and volume of water are used for microgels with different internal structures. Although the equilibrium volume change of the microgels is unaffected by the internal structure [2], their shrinking and swelling kinetics are affected by the induced void structure. When the temperature is increased from 23 to 47 °C, the microgel with multiple voids and the one with a hollow shell structure (one large void) shrink faster than the voidless microgel as shown in Figure 5.8a. For example, at $t = 49$ s, the diameter of the voidless microgel is the largest, while that of the one with multiple voids is the smallest. Similar behavior is observed when the microgels are cooled back to 23 °C: The microgel with multiple voids and the one with a hollow shell structure swell faster than the voidless microgel (Figure 5.8b). A quantitative comparison of the temporal evolution of sizes of the three microgels during heating and cooling is presented in Figure 5.9, where we have plotted the instantaneous diameters of the microgels as a function of time. The temperature of the microgels is shown in the lower panel, and $t = 0$ refers to the time when the temperature change is initiated. The different behavior for each sample is evidenced by the separation between the curves. The inset in Figure 5.9b reveals that the inception of swelling in the microgel with multivoids, indicated by a sudden change in its diameter, precedes that in the voidless microgel by as much as 10 s. The swelling of the microgels is perceptible only after a significant amount of water is transferred into the microgel. Thus, the result suggests that the transport of water in the microgels with voids occurs much before the voidless microgel.

The response rate of PNIPAM microgels to changes in temperature is governed by diffusion-limited transport of water in and out of the polymeric networks. Voids within a microgel offer much less resistance to the transport of water compared to the three-dimensional PNIPAM network; hence, the microgels with voids respond faster to changes in temperature than the voidless one. Furthermore, in the microgel with multivoids, the voids can form continuous channels from the core to the surface of the microgel, further expediting transport of water in and out of the microgel. This is presumably why the microgel with multivoids responds faster to temperature changes than even the microgel with the hollow shell structure.

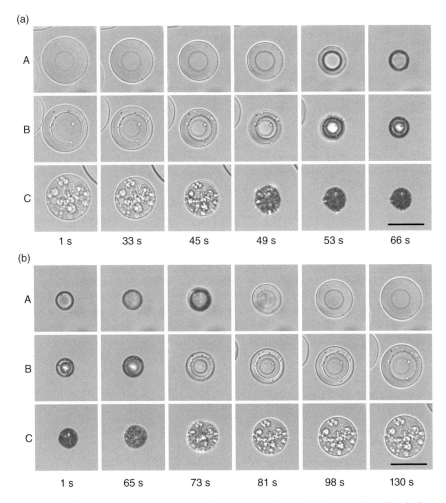

Figure 5.8 Effect of internal structure on the dynamic volume shrinking and swelling behavior of microgels upon (a) heating from 23 to 47 °C and (b) cooling from 47 to 23 °C. (A) Voidless microgel; (B) microgel with hollow shell structure; (C) microgel with multiple voids. Scale bar is 100 μm. (Chu et al. 2007 [2]. Reproduced with permission of John Wiley & Sons.)

To evaluate the effects of a progressive increase in the number of voids inside the microgels on their response kinetics, we prepared monodisperse PNIPAM microgels with one, two, three, and four voids of 25 μm diameter each. The dynamics of shrinking and swelling of these microgels are compared to that of a same-sized voidless microgel in Figure 5.10. When temperature is increased from 23 to 47 °C, the microgels with four voids shrink faster than all other samples, whereas the voidless microgel is the slowest to respond. The differences are more distinct during swelling. The inset in Figure 5.10b clearly reveals a systematic decrease in the time required for the inception of swelling with an increase in the number of voids. This suggests that the volume-phase transition kinetics of microgels can be tuned simply by varying the number of voids in them.

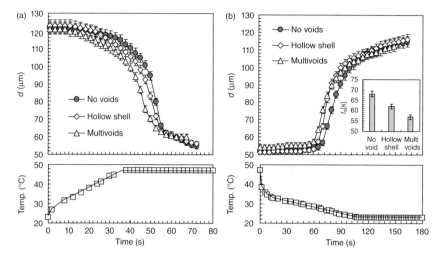

Figure 5.9 Time-dependent diameter of microgels with different internal structures upon (a) heating from 23 to 47 °C and (b) cooling from 47 to 23 °C; t_s is the time elapsed before the microgels begin to swell. (Chu et al. 2007 [2]. Reproduced with permission of John Wiley & Sons.)

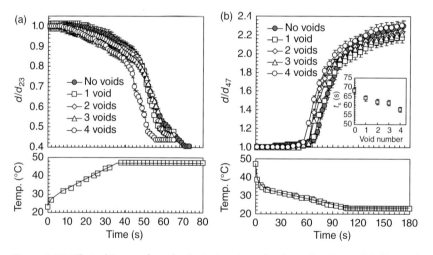

Figure 5.10 Effect of the number of spherical voids on the dynamic volume shrinking and swelling behavior of microgels. The diameter of spherical voids is 25 μm. The samples are (a) heated from 23 to 47 °C and (b) cooled from 47 to 23 °C; d_{23} and d_{47} are the microgel diameters at 23 and 47 °C, respectively, and t_s is the time elapsed before the microgels begin to swell. (Chu et al. 2007 [2]. Reproduced with permission of John Wiley & Sons.)

5.4 Microfluidic Fabrication of PNIPAM Microparticles with Open-Celled Porous Structure for Fast Response

As mentioned earlier, the improvement of the response rate of PNIPAM microgels is of significant importance for practical applications. Here, we introduce a simple method to prepare monodisperse and fast-responsive PNIPAM microgels with open-celled porous structure [6].

5.4.1 Microfluidic Fabrication Strategy

It has been reported that the hydrogel membranes of PNIPAM microcapsules can be ruptured upon stimuli-induced shrinking and squeezing out of the inner oil cores [14–16]. The shrinkage of the PNIPAM microcapsules can be induced by simply heating up or adding certain chemicals in the environment, and a broken hole can be generated in each capsule membrane upon squeezing out the inner oil core. Inspired by this phenomenon, we use fine oil droplets as porogens and subsequently squeeze the embedded oil droplets out from PNIPAM microgels to prepare open-celled porous PNIPAM microgels [6]. During the volume-phase transition processes of PNIPAM microgels, the open-celled porous structure in the microgels provides continuous channels for transferring water easily inward or outward convectively. Therefore, the response rate of the PNIPAM microgels with open-celled porous structure can be enhanced. Oil-in-water (O/W) primary emulsions prepared by homogeneous emulsification method is used as the inner phase, which is further emulsified into the outer oil phase in a capillary microfluidic device to form monodisperse (oil-in-water)-in-oil O/W/O emulsion templates. PNIPAM microgels with numerous embedded fine oil droplets are obtained by polymerizing the aqueous phase in the O/W/O emulsion templates. After adding isopropanol in the environment or heating up, the embedded fine oil droplets are squeezed out from the microgels, resulting in open-celled porous structure in the PNIPAM microgels.

Monodisperse oil-in-water-in-oil (O/W/O) emulsions are used as templates to prepare open-celled porous PNIPAM microgels (Figure 5.11). First, O/W primary emulsions are prepared by the homogeneous emulsification method. Colored oil phase (Figure 5.11a) is composed of benzyl benzoate, soybean oil, surfactant PGPR 90, and fluorescent dye Lumogen Red 300 (LR300). Aqueous phase (Figure 5.11b) is composed of deionized (DI) water, monomer NIPAM, cross-linker MBA, UV initiator 2,2′-azobis(2-methylpropionamidine)dihydrochloride (V-50), surfactant Pluronic F127, and glycerol for increasing the viscosity of the aqueous phase. After homogeneous emulsification at 12 000 rpm for 30 s (Figure 5.11c) by a homogenizer, O/W primary emulsions (Figure 5.11d) are obtained. And then, O/W primary emulsions as the inner phase and soybean oil containing PGPR 90 as the outer phase are further emulsified in a capillary microfluidic device to generate monodisperse O/W/O emulsion templates (Figure 5.11e). Two cylindrical capillary tubes are coaxially aligned within a square capillary tube by matching the outer diameters of the cylindrical tubes to the inner dimension of the square one. The inner diameters of the injection tube and collection tube are 550 and 350 µm, respectively, and the inner dimension of the square capillary tube is 1.0 mm. The end of injection tube is tapered by a micropuller, and the orifice dimension of tapered end is adjusted by a microforge to 120 µm. The prepared O/W/O emulsions (Figure 5.11f) are collected into soybean oil containing PGPR 90 and UV initiator BDK. For comparison, monodisperse W/O emulsions are also prepared as templates for preparing normal PNIPAM microgels. The inner phase pumped into the injection tube is replaced by aqueous phase used in the preparation of O/W primary emulsions. The flow rates of the inner phases in preparations of O/W/O and W/O emulsions are both fixed at 500 µl h^{-1}.

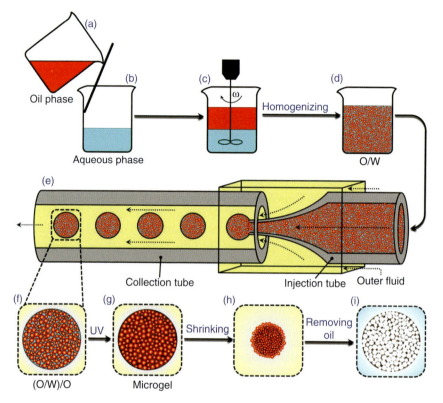

Figure 5.11 Schematic illustration of the preparation procedure of open-celled porous PNIPAM microgels. (a) Oil phase colored with LR300; (b) aqueous phase containing monomer and cross-linker. (c, d) Preparation of O/W primary emulsion through homogeneous emulsification method. (e, f) Preparation of O/W/O emulsion templates in a microfluidic device. (g) Formation of PNIPAM microgel embedded with numerous fine oil droplets via polymerization of middle aqueous phase of O/W/O emulsion template. (h, i) Preparation of PNIPAM microgel with open-celled porous structure through squeezing out the inside fine oil droplets upon stimuli-induced shrinkage of the PNIPAM microgel and removing the oil phase. (Mou et al. 2014 [6]. Reproduced with permission of American Chemical Society.)

By varying the flow rate of the outer oil phase, the outer diameters of O/W/O and W/O emulsions can be adjusted simply. To prepare open-celled porous PNIPAM microgels and normal PNIPAM microgels with different cross-linking degrees, two molar ratios of MBA cross-linker to NIPAM monomer in aqueous phases of W/O and O/W/O emulsions are varied as 2% and 4%.

Polymerization of the aqueous phase of O/W/O emulsions is initiated with UV irradiation for 20 min in an ice-water bath. After 12 h, O/W/O emulsions are converted into PNIPAM microgels embedded with numerous oil droplets (Figure 5.11g). The open-celled porous structure inside PNIPAM microgels is obtained through volume shrinkage of the microgels (Figure 5.11h,i) induced by two different methods. In the first method, isopropanol is added into the environment of PNIPAM microgels, and after 30 min the microgels are washed with isopropanol six times and subsequently washed with DI water six times to remove all the oils, unreacted chemicals, and surfactants. Open-celled porous PNIPAM microgels prepared by this method, with molar ratios of MBA to NIPAM being 2% and 4%, are coded as OPI-2 and OPI-4 microgels, respectively. In the second method, the PNIPAM microgels are put in a thermostatic bath at 25 °C for 30 min and then rapidly moved to another thermostatic bath at 50 °C for 30 min, and later it is moved back to the thermostatic bath at 25 °C for 30 min. After three cycles of heating up from 25 to 50 °C and cooling down from 50 to 25 °C, PNIPAM microgels are washed with isopropanol six times and subsequently washed with DI water six times to remove all the oils, unreacted chemicals, and surfactants. Open-celled porous PNIPAM microgels prepared in this way, with molar ratios of MBA to NIPAM being 2% and 4%, are coded as OPH-2 and OPH-4 microgels, respectively. Normal PNIPAM microgels are also obtained by polymerizing the aqueous phase of W/O emulsions with UV irradiation for 20 min in an ice-water bath. After 12 h, the prepared PNIPAM microgels are washed with isopropanol six times and subsequently washed with DI water six times to remove all the oils, unreacted chemicals, and surfactants. Normal PNIPAM microgels, with molar ratios of MBA to NIPAM being 2% and 4%, are coded as N-2 and N-4 microgels, respectively.

5.4.2 Morphologies and Microstructures of Porous PNIPAM Microparticles

The O/W primary emulsions prepared by homogeneous emulsification method are fine oil droplets dispersed in aqueous phase (Figure 5.12a). The diameters of the oil droplets are in the range of 2.9–18.9 µm (Figure 5.12b), and the average diameter is about 11.2 µm determined by measuring 500 emulsion droplets. The formation processes of O/W/O and W/O emulsions in the capillary microfluidic device are recorded by a high-speed camera (Figure 5.12c,d). The diameters of the generated emulsions are mainly influenced by the interface force between the inner and outer phases as well as the shearing force of the outer phase to the inner phase. When the flow rates of inner fluids are fixed at 500 µl h^{-1}, the outer diameters of O/W/O and W/O emulsion droplets decrease simply as the flow rates of outer phase fluids increase from 500 to 6000 µl h^{-1} (Figure 5.12e). The relationships between the emulsion diameters and the flow rates of the outer phase are almost the same in preparations of O/W/O and W/O emulsions. To prepare emulsions with the same outer diameter around 250 µm, the flow rates of the outer phase fluids in preparations of O/W/O and W/O emulsions are 3250 and 3300 µl h^{-1}, respectively.

The generated O/W/O emulsions are monodisperse, and numerous fine oil droplets are dispersed evenly inside the water droplets (Figure 5.13a–c). W/O emulsions are also monodisperse (Figure 5.13d). The monodispersity of emulsion templates or the resultant PNIPAM microgels is evaluated by an index called CV, which is defined as the ratio of the standard deviation of size distribution to its arithmetic mean. One hundred emulsion droplets or microgels are counted for obtaining each CV value. The calculated CV values of O/W/O and W/O emulsions are 1.6% and 1.8%, respectively, which indicate that the emulsion templates prepared by capillary microfluidic device are highly monodisperse.

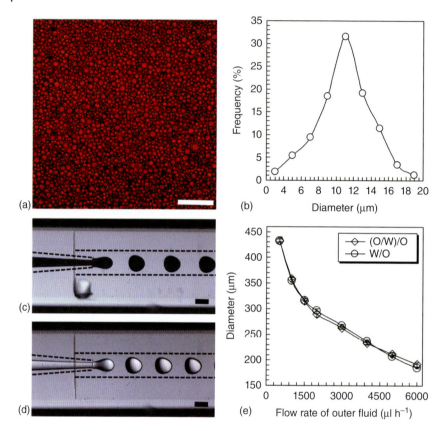

Figure 5.12 Preparation of O/W primary emulsions and both O/W/O and W/O emulsion templates. (a, b) Confocal laser scanning microscope (CLSM) micrograph (a) and size distribution (b) of O/W primary emulsions. Scale bar is 100 μm. (c, d) High-speed optical micrographs of the generation processes of O/W/O (c) and W/O (d) emulsions in capillary microfluidic device, in which the inner fluid flow rates are both fixed at 500 μl h^{-1} and the outer flow rates are 3250 μl h^{-1} (c) and 3300 μl h^{-1} (d), respectively. Scale bars are 200 μm. (e) Effects of the flow rates of outer oil fluids on the diameters of O/W/O and W/O emulsion droplets. The flow rates of inner fluids are fixed at 500 μl h^{-1}. (Mou et al. 2014 [6]. Reproduced with permission of American Chemical Society.)

After UV-irradiation-initiated polymerization, the emulsion templates are converted into PNIPAM microgels. Figure 5.14a is the CLSM (confocal laser scanning microscope) image of the prepared PNIPAM microgels containing numerous oil droplets in DI water at 25 °C. The red oil droplets are dispersed evenly in PNIPAM microgels, similar to those inside the emulsion templates. To fabricate PNIPAM microgels with open-celled porous structure, the shrinkage of PNIPAM microgels is induced by either adding isopropanol in the environment or heating up. Because the oil droplets in PNIPAM microgels based on O/W/O emulsion templates are very small and close to each other, it is hard to clearly observe the process of squeezing out the oil droplets from PNIPAM microgels. To clearly show the process of squeezing out oil cores from cross-linked

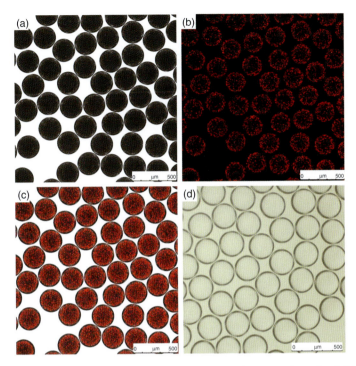

Figure 5.13 Micrographs of O/W/O and W/O emulsion templates. (a–c) CLSM micrographs of O/W/O emulsions, in which (a) shows the transmission channel image, (b) shows the red channel image, and (c) shows the overlay of red channel and transmission channel images. (d) Optical micrograph of W/O emulsions. Scale bars are 500 μm. (Mou et al. 2014 [6]. Reproduced with permission of American Chemical Society.)

PNIPAM microgels induced by either adding isopropanol in the environment or heating up, core–shell PNIPAM microcapsules each with a colored oil droplet core are prepared [16]. O/W/O emulsions are used as templates for the preparation of core–shell PNIPAM microcapsules. The composition of colored inner oil phase, middle water phase, and outer oil phase are all the same as that in the preparation of O/W/O emulsions for OPH-4 microgels. There exists a water layer between the inner oil core and PNIPAM hydrogel shell, and the surfactants mainly bind to the oil/water interface; therefore, the inner oil cores are all efficiently squeezed out from the core–sell PNIPAM microgels by either adding isopropanol in the environment or heating up from 25 to 50 °C. After squeezing out the inner oil core, a broken hole is generated in each PNIPAM microgel. Based on the same principle, PNIPAM microgels with open-celled porous structure can be obtained by squeezing out the inner fine oil droplets from the PNIPAM microgels. After totally removing the inner oil droplets, the microgels in DI water at 25 °C are opaque (Figure 5.14b,c), because of the heterogeneous open-celled porous structure inside microgels. On the contrary, normal PNIPAM microgels (Figure 5.14d) are transparent in DI water at 25 °C because of the homogeneous cross-linked structure inside microgels. Each kind of PNIPAM microgels has uniform morphology.

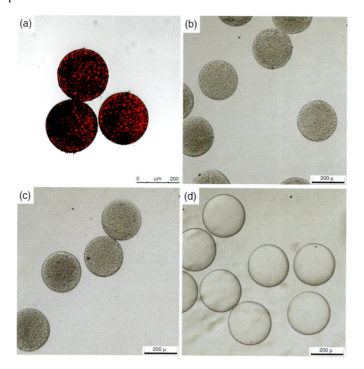

Figure 5.14 PNIPAM microgels with different structures in DI water at 25 °C. (a) CLSM micrograph of PNIPAM microgels embedded with numerous oil droplets. (b, c) Optical micrographs of OPI-4 (b) and OPH-4 (c) microgels with open-celled porous structures. (d) Optical micrograph of N-4 normal microgels. The molar ratio of MBA to NIPAM in the preparation recipe is 4%. Scale bars are 200 μm. (Mou et al. 2014 [6]. Reproduced with permission of American Chemical Society.)

All the PNIPAM microgels prepared by the microfluidic method are highly monodisperse. The size distribution of each kind of PNIPAM microgels is narrow and all the calculated CV values are smaller than 3%. The introduction of open-celled porous structure does not influence the size of the PNIPAM microgels at all. When the outer diameters of the W/O or O/W/O emulsion templates are almost the same, the dimensions of resultant PNIPAM microgels are almost the same too. However, the diameters of the resultant microgels are affected by the molar ratio of MBA to NIPAM monomer. With the increasing cross-linking degree of the PNIPAM network, the swelling degree of microgels in water is reduced. Therefore, the average size of microgels decreases as the cross-linking degree increases.

From the microstructures of freeze-dried PNIPAM microgels shown in SEM micrographs (Figure 5.15), it can be seen that the morphology of normal microgels is significantly different from that of open-celled porous microgels. The normal microgel exhibits a pinecone-like structure at one end and a tip-shaped tail structure at another end (N-4, Figure 5.15a1), which may be caused by the ice crystals with orientations in PNIPAM microgels when they are rapidly immersed into liquid nitrogen and the shrinkage of PNIPAM in the

5.4 Microfluidic Fabrication of PNIPAM Microparticles with Open-Celled Porous Structure | 97

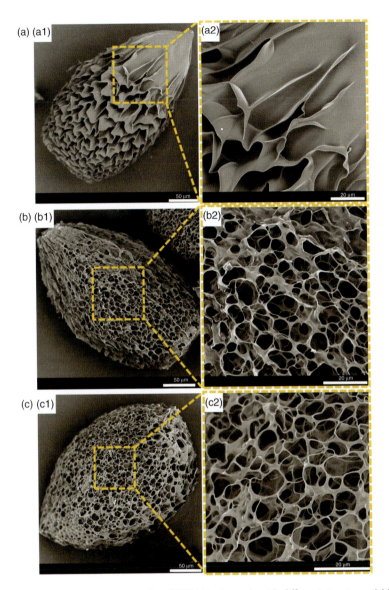

Figure 5.15 SEM micrographs of PNIPAM microgels with different structures. (a) N-4 normal microgels. (b, c) OPI-4 (b) and OPH-4 (c) microgels with open-celled porous structure. (a2), (b2), and (c2) are magnifications of the dashed areas in (a1), (b1), and (c1), respectively. The molar ratio of MBA to NIPAM in the preparation recipe is 4%. Scale bars are 50 μm in (a1), (b1), and (c1) and 20 μm in (a2), (b2), and (c2). (Mou *et al.* 2014 [6]. Reproduced with permission of American Chemical Society.)

dehydration process [17]. No interconnected pores are observed in the N-4 microgel (Figure 5.15a2). Completely different from the normal microgel N-4, open-celled porous microgels OPI-4 and OPH-4 show a sponge-like structure that is full of interconnected pores (Figure 5.15b,c). The sponge-like structure and the interconnected pores result from the open-celled porous structure inside the OPI-4 and OPH-4 microgels. Upon squeezing out numerous fine oil droplets from PNIPAM microgels, numerous interconnected hollow cavities are formed inside the microgels. There is no clear difference between the microstructures of OPI-4 and OPH-4 microgels (Figure 5.15b,c), which means that both the methods introduced in this study are effective for preparing PNIPAM microgels with open-celled porous structure.

5.4.3 Thermo-Responsive Volume Change Behaviors of PNIPAM Porous Microparticles

Figure 5.16 shows the snapshots of the dynamic thermo-responsive shrinking processes of PNIPAM microgels with different structures at different time intervals when the temperature increases from 25 to 50 °C quickly. The N-2 microgel shrinks slowly and subsequently some bubbles form on its surface, and the bubbles expand first, then collapse, and disappear at last (Figure 5.16a). These bubbles result from the dense skin layer effect of hydrogel, and similar phenomena are also described for macroscale PNIPAM hydrogels in the previous literature [18]. A dense skin layer, which will retard the outward diffusion of water, is formed at the outer surface of the N-2 microgel in initial shrinking process when the ambient temperature increases suddenly and exceeds the VPTT of PNIPAM. When the internal pressure increases to a degree that exceeds the strength of the dense skin layer, the bubbles appear on the surface. Water in the bubbles is squeezed out through the dense skin layer slowly and the bubbles gradually become smaller

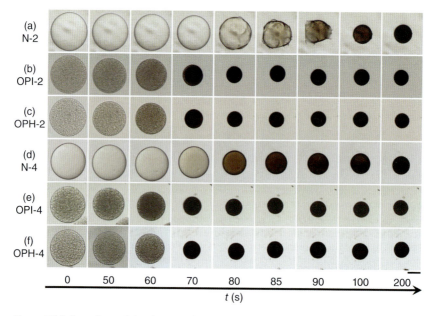

Figure 5.16 Snapshots of the dynamic thermo-responsive shrinking processes of PNIPAM microgels with different structures at different time intervals when the temperature increases from 25 to 50 °C quickly. Scale bar is 100 μm. (Mou et al. 2014 [6]. Reproduced with permission of American Chemical Society.)

and disappear at last. In the temperature-dependent equilibrium volume change study, the temperature is increased step by step and no pressure accumulation occurs; therefore, no bubbles form on the surface of the microgel. Remarkably different from the situation for N-2 microgels, no bubble appears on the surfaces of OPH-2 and OPI-2 microgels in the whole shrinking process. For OPH-2 and OPI-2 microgels, the interconnected open-celled pores from inside to the surfaces of PNIPAM microgels can serve as free channels for the outward transfer of water during the shrinking process. So, nearly no pressure accumulates inside OPH-2 and OPI-2 microgels, as a result no bubble appears on the surface of microgels. In other words, the introduction of the open-celled porous structure inside the PNIPAM microgels can effectively eliminate the skin layer effect in the shrinking process. As a result, the shrinking rates of OPH-2 and OPI-2 microgels are obviously faster than that of N-2 microgel. Before $t = 60$ s, OPH-2 and OPI-2 microgels are still in the swollen state. In the next 10 s (from $t = 60$ s to $t = 70$ s), both OPH-2 and OPI-2 microgels experience a very rapid shrinking process and show a shrunken state at $t = 70$ s. After $t = 80$ s, OPH-2 and OPI-2 microgels are close to the equilibrium state and their volumes nearly do not change anymore. However, N-2 microgel exhibits a swollen state before $t = 70$ s. From $t = 70$ s to $t = 100$ s, N-2 microgel shrinks slowly and it can reach the equilibrium state only after $t = 100$ s.

When the molar ratio of MBA to NIPAM is 4%, the PNIPAM microgels with open-celled porous structure (OPH-4 and OPI-4) also show remarkably faster response rate than the normal one (N-4) (Figure 5.16d–f). No bubble appears at the outer surface of N-4 microgel in its shrinking process. Because the strength of PNIPAM polymer network enhances as the cross-linking degree increases, the internal pressure is not high enough to form bubble through the skin layer.

The dynamic thermo-responsive shrinking ratio of PNIPAM microgels is defined as

$$(R_t)_{25 \to 50} = \frac{V_t}{V_{250}} = \left(\frac{D_t}{D_{250}}\right)^3 \tag{5.1}$$

where V_t and D_t are the volume (μm³) and diameter (μm) of PNIPAM microgels at t, and V_{250} and D_{250} are the volume (μm³) and diameter (μm) of PNIPAM microgels at $t = 250$ s, respectively. Figure 5.17 shows the dynamic changes in shrinking ratios of PNIPAM microgels with different structures. Fifteen microgels are measured at the same time to get an average value for a batch of microgels. The actual temperature around the PNIPAM microgels is shown in Figure 5.17c. Bubbles on the surface of N-2 microgels make it hard to measure the diameters exactly, so the error bars are relatively large during the process of bubble forming and disappearing. The OPI-2 and OPH-2 microgels start a sharp volume change at $t = 60$ s and they completely shrink within 15 s (Figure 5.17a). For the N-2 microgels, the sharp volume shrinking period ranges from $t = 70$ s to $t = 100$ s and the duration time is about 40 s. In other words, the PNIPAM microgels with open-celled porous structure shrink much faster than the normal ones. When the molar ratio of MBA to NIPAM increases to 4%, the PNIPAM microgels with open-celled porous structure also show obviously faster response rate than the normal ones (Figure 5.17b).

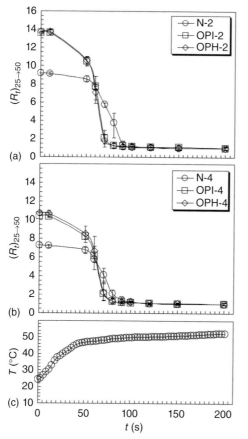

Figure 5.17 Dynamic thermo-responsive shrinking behaviors of PNIPAM microgels with different structures during the heating-up process from 25 to 50 °C. (a) The molar ratio of MBA to NIPAM in the microgel preparation recipe is 2%. (b) The molar ratio of MBA to NIPAM in the microgel preparation recipe is 4%. (c) Temperature change curve. (Mou et al. 2014 [6]. Reproduced with permission of American Chemical Society.)

The aforementioned results demonstrate that the introduction of the open-celled porous structure into PNIPAM microgels can effectively improve the response rates of PNIPAM microgels. The thermo-responsive shrinking rate of normal PNIPAM microgels is mainly governed by diffusion-limited transport of water through the polymeric network. The open-celled porous structure inside the microgels and the pores on their surfaces can effectively eliminate the skin layer effect and provide free channels for convectively transferring water. As a result, the PNIPAM microgels with open-celled porous structure can respond remarkably faster than the normal ones in the thermo-induced shrinking process.

The main factor that affects the dynamic thermo-responsive swelling behaviors of PNIPAM microgels is the resistance of water transferring from ambient into the microgels. Figure 5.18 shows the snapshots of the dynamic thermo-responsive swelling processes of PNIPAM microgels with different structures at different time intervals when the temperature decreases from 50 to 25 C quickly. All PNIPAM microgels show a full shrinking state after being equilibrated in DI water at 50 °C for 30 min. At $t=0$ s, which also refers to the time when the cooling process is initiated, N-2 microgel is slightly larger than OPI-2 and OPH-2 microgels. From $t=0$ s to $t=100$ s, N-2 microgel has nearly no

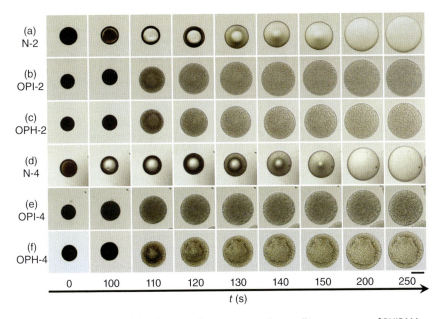

Figure 5.18 Snapshots of the dynamic thermo-responsive swelling processes of PNIPAM microgels with different structures at different time intervals upon quickly cooling from 50 to 25 °C. Scale bar is 100 μm. (Mou et al. 2014 [6]. Reproduced with permission of American Chemical Society.)

change in diameter, and from $t = 100$ s to $t = 250$ s it swells slowly. As expected, OPI-2 and OPH-2 microgels show very different thermo-responsive swelling characteristics. From $t = 0$ s to $t = 100$ s, OPI-2 and OPH-2 microgels only swell a little. However, OPI-2 and OPH-2 microgels experience a sharp volume change from $t = 100$ s to $t = 110$ s, and at this time their diameters are significantly larger than that of N-2 microgel. After $t = 140$ s, OPI-2 and OPH-2 microgels gradually reach the fully equilibrium swelling state. In other words, OPI-2 and OPH-2 microgels have much faster thermo-responsive swelling rate than N-2 microgel. The open-celled porous microgels with larger cross-linking degrees (OPI-4 and OPH-4) also swell much faster than the normal ones (Figure 5.18d–f).

The dynamic thermo-responsive swelling ratio of PNIPAM microgels is defined as

$$(R_t)_{50 \to 25} = \frac{V_t}{V_0} = \left(\frac{D_t}{D_0}\right)^3 \tag{5.2}$$

where V_t and D_t are the volume (μm³) and diameter (μm) of PNIPAM microgels at t, and V_0 and D_0 are the volume (μm³) and diameter (μm) of PNIPAM microgels at 0 s, respectively. The dynamic changes in thermo-responsive swelling ratios of PNIPAM microgels with different structures are illustrated in Figure 5.19. Fifteen microgels are measured at the same time to get an average value for a batch of microgels. The temperature around the PNIPAM microgels at different time intervals is shown in Figure 5.19c. Obviously, OPI-2 and OPH-2 microgels experience a sharp volume change at time ranging from $t = 100$ s

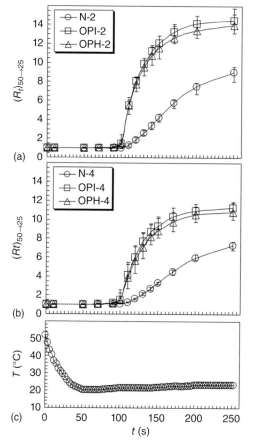

Figure 5.19 Dynamic thermo-responsive swelling behaviors of PNIPAM microgels with different structures during the cooling down process from 50 to 25 °C. (a) The molar ratio of MBA to NIPAM in the microgel preparation recipe is 2%. (b) The molar ratio of MBA to NIPAM in the microgel preparation recipe is 4%. (c) Temperature change curve. (Mou *et al.* 2014 [6]. Reproduced with permission of American Chemical Society.)

to $t=150$ s, and the volume expanding rate gradually decreases as they come close to an equilibrium state from $t=150$ s to $t=250$ s. Remarkably different from OPI-2 and OPH-2 microgels, N-2 microgels experience a slow and steady volume change from $t=100$ s to $t=250$ s (Figure 5.19a). After the beginning of swelling, the swelling volume ratios of OPI-2 and OPH-2 microgels are almost the same and both are larger than that of N-2 microgels. When the molar ratio of MBA to NIPAM in PNIPAM microgels is 4%, the thermo-responsive swelling volume ratios of open-celled porous PNIPAM microgels are also remarkably larger than the normal ones (Figure 5.19b). The fast thermo-responsive swelling behaviors of PNIPAM microgels are also due to the open-celled porous structure. Inside the conventional and normal PNIPAM microgels, the water transfer is governed by diffusion through cross-linked polymeric networks [11]. However, inside the open-celled porous microgels, water can convectively transfer from the environment into the PNIPAM microgels through the interconnected open-celled pores inside the microgels.

5.5 Summary

In this chapter, microfluidic strategies for controllable fabrication of monodisperse porous microparticles are introduced. The strategies are demonstrated by fabricating monodisperse porous poly(HEMA-MMA) microspheres using single emulsions as templates and PVP molecules as porogens, and fabricating monodisperse porous PNIPAM microgels using solid nanoparticles and fine oil droplets as porogens. The controllable porosity of the poly(HEMA-MMA) porous microspheres can be achieved by properly adjusting the porogen concentration in monomer solution. With the increasing PVP content, the porosity of poly(HEMA-MMA) microspheres increases. The monodisperse poly(HEMA-MMA) microspheres with controllable porous structures could be favorable in various applications, such as arterial embolization, immobilization of cells and enzymes, packing materials in chromatography, and controlled drug delivery. The response rate of thermo-sensitive PNIPAM microgels can be varied by fabricating different void structures inside monodisperse microgel particles. The more voids the microgel has, the faster is its response to changes in temperature. The PNIPAM microgels with open-celled porous structure demonstrate much faster response rates than the normal ones do in both thermo-responsive shrinking and swelling processes, because open-celled porous structure provides numerous interconnected free channels for water transferring outward and inward. The flexibility and control over the response kinetics should make the PNIPAM microgels more useful for a wide range of applications in pharmaceutics and cosmetics, and as sensors and actuators. In a word, the microfluidic approach provides new opportunities for design and fabrication of novel functional porous microparticles for different purposes.

References

1. Vianna-Soares, C.D., Kim, C.J., and Borenstein, M.R. (2003) Manufacture of porous cross-linked HEMA spheres for size exclusion packing material. *J. Porous Mater.*, **10**, 123–130.
2. Chu, L.Y., Kim, J.W., Shah, R.K., and Weitz, D.A. (2007) Monodisperse thermo-responsive microgels with tunable volume-phase transition kinetics. *Adv. Funct. Mater.*, **17**, 3499–3504.
3. Shiga, K., Muramatsu, N., and Kondo, T. (1996) Preparation of poly(D,L-lactide) and copoly(lactide-glycolide) microspheres of uniform size. *J. Pharm. Pharmacol.*, **48**, 891–895.
4. Xiao, X.C., Chu, L.Y., Chen, W.M., Wang, S., and Li, Y. (2003) Positively thermo-sensitive monodisperse core-shell microspheres. *Adv. Funct. Mater.*, **13**, 847–852.

5 Zhang, H., Ju, X.J., Xie, R., Cheng, C.J., Ren, P.W., and Chu, L.Y. (2009) A microfluidic approach to fabricate monodisperse hollow or porous poly(HEMA-MMA) microspheres using single emulsions as templates. *J. Colloid Interface Sci.*, **336**, 235–243.

6 Mou, C.L., Ju, X.J., Zhang, L., Xie, R., Wang, W., Deng, N.N., Wei, J., Chen, Q., and Chu, L.Y. (2014) Monodisperse and fast-responsive poly(N-isopropylacrylamide) microgels with open-celled porous structure. *Langmuir*, **30**, 1455–1464.

7 Dziubla, T.D., Torjman, M.C., Joseph, J.I., Murphy-Tatum, M., and Lowman, A.M. (2001) Evaluation of porous networks of poly(2-hydroxyethyl methacrylate) as interfacial drug delivery devices. *Biomaterials*, **22**, 2893–2899.

8 Utada, A.S., Chu, L.Y., Fernandez-Nieves, A., Link, D.R., Holtze, C., and Weitz, D.A. (2007) Dripping, jetting, drops, and wetting: the magic of microfluidics. *MRS Bull.*, **32**, 702–708.

9 Chu, L.Y., Xie, R., Ju, X.J., and Wang, W. (2013) *Smart Hydrogel Functional Materials*, Springer, Berlin, Heidelberg.

10 Pelton, R. (2000) Temperature-sensitive aqueous microgels. *Adv. Colloid Interface Sci.*, **85**, 1–33.

11 Tanaka, T. and Fillmore, D.J. (1979) The kinetics of swelling of gels. *J. Chem. Phys.*, **70**, 1214–1218.

12 Zhang, J., Chu, L.Y., Cheng, C.J., Mi, D.F., Zhou, M.Y., and Ju, X.J. (2008) Graft-type poly(N-isopropylacrylamide-co-acrylic acid) microgels exhibiting rapid thermo- and pH-responsive properties. *Polymer*, **49**, 2595–2603.

13 Ju, X.J., Chu, L.Y., Zhu, X.L., Hu, L., Song, H., and Chen, W.M. (2006) Effects of internal microstructures of poly(N-isopropylacrylamide) hydrogels on thermo-responsive volume phase-transition and controlled-release characteristics. *Smart Mater. Struct.*, **15**, 1767–1774.

14 Chu, L.Y., Utada, A.S., Shah, R.K., Kim, J.W., and Weitz, D.A. (2007) Controllable monodisperse multiple emulsions. *Angew. Chem. Int. Ed.*, **46**, 8970–8974.

15 Liu, L., Wang, W., Ju, X.J., Xie, R., and Chu, L.Y. (2010) Smart thermo-triggered squirting capsules for nanoparticle delivery. *Soft Matter*, **6**, 3759–3763.

16 Mou, C.L., He, X.H., Ju, X.J., Xie, R., Liu, Z., Liu, L., Zhang, Z., and Chu, L.Y. (2012) Change in size and structure of monodisperse poly(N-isopropylacrylamide) microcapsules in response to varying temperature and ethyl gallate concentration. *Chem. Eng. J.*, **210**, 212–219.

17 Cheng, C.J., Chu, L.Y., Zhang, J., Wang, H.D., and Wei, G. (2008) Effect of freeze-drying and rehydrating treatment on the thermo-responsive characteristics of poly(N-isopropylacrylamide) microspheres. *Colloid Polym. Sci.*, **286**, 571–577.

18 Kaneko, Y., Sakai, K., Kikuchi, A., Yoshida, R., Sakurai, Y., and Okano, T. (1995) Influence of freely mobile grafted chain length on dynamic properties of comb-type grafted poly(N-isopropylacrylamide) hydrogels. *Macromolecules*, **28**, 7717–7723.

6

Microfluidic Fabrication of Uniform Hierarchical Porous Microparticles

6.1 Introduction

The natural hierarchical porous materials, such as biominerals and trabecular bones, provide fantastic inspiration for the development of materials with hierarchical pores from nanometer to micrometer scale for various applications [1–3]. Especially, porous polymeric microparticles have attracted wide interests in myriad fields, such as adsorption [4, 5], drug delivery [6, 7], tissue engineering [8], sensing/detecting [9, 10], and chromatography [11]. Elaborate integration of controllable highly interconnected hierarchical porous structures containing micrometer- and nanometer-sized pores within the polymeric microparticles enables combined advantages of pores at both micro- and nanoscale to achieve fascinating properties for enhanced applications. Generally, the nanometer-sized pores with high interconnectivity and functionality can provide large functional surface area for interaction with molecules, while the micrometer-sized pores can offer easier access with low resistance for the molecules especially biomacromolecules through the porous matrix. Combination of the nanometer-sized pores with micrometer-sized pores allows synergetic advantages of the large functional surface area and the enhanced mass transport to achieve advanced overall performance. Control of the porosities of both nanometer- and micrometer-sized pores plays an important role in determining the loading capacity and mass transfer profile. Meanwhile, the uniform size and shape of microparticles are crucial for regulating their assembly behaviors [12, 13], drug release kinetics [14, 15], and packing performance [16–18]. Thus, development of uniform microparticles containing controllable highly interconnected hierarchical porous structures, with tunable pore size, porosity, functionality, and particle shape, is imperative for enhancing their broad applications.

Typically, porous structures in materials can be created by template-directed synthesis [19]. Pores with sizes ranging from nanometer to micrometer scale can be produced by templates such as assembled surfactants [20, 21] or copolymers [4, 22–24], colloids [25–29], emulsion drops [30, 31], bubbles [32], phase-separated domains [33], and bacterials [34]. By combining multiple pore-templating strategies, hierarchical porous materials can be developed [35–39]. Further manipulation of the macroscopic interfaces of the porous matrix in particle shape can produce porous particles, but they are usually with polydisperse sizes. With excellent control of microdrop interfaces [40, 41], microfluidic techniques provide a powerful platform for engineering controllable porous polymeric microparticles with monodisperse size. Based on microfluidics, most of the pore-templating strategies can be adapted into the controllable microfluidic drops for pore generation [19]. Thus, uniform porous polymeric microparticles can be fabricated by using monodisperse emulsion drops from microfluidics to shape the microparticle [42], and using pore templates within the emulsions such as droplets [43, 44], bubbles [45], assembled copolymers [46] as well as packed colloids [10], to engineer the pores. However, these strategies that employ pore templates at one size scale usually produce porous polymeric microparticles containing pores with single size scale. Hierarchical porous inorganic microparticles have been developed by assembling silica nanoparticles on the inner drops of microfluidic W/O/W emulsions via evaporation of the middle solvent layer [47]. The inner drops create micrometer-sized pores, while the voids between the packed silica nanoparticles create nanometer-sized pores in the inorganic microparticles; however, the nanoporosity remains difficult to be flexibly adjusted. Moreover, it requires posttreatment for further functionalization of these inorganic microparticles. Thus, for hierarchical porous polymeric microparticles, it is still challenging to simultaneously achieve accurate and independent control of their hierarchical porous structures with tunable pore size, porosity, functionality, as well as the particle shape in a single step. Therefore, techniques to realize such controls for developing controllable functional polymeric microparticles with highly interconnected hierarchical porous structures are highly desired.

Recently, the author's group developed a simple and versatile microfluidic approach for controllable fabrication of uniform polymeric microparticles containing highly interconnected hierarchical porous structures [48]. The pore size, porosity, functionality, and particle shape of the hierarchical porous microparticles can be individually and flexibly manipulated in one step. These microparticles elaborately combine the advantages of enhanced mass transfer, large functional surface area, and flexibly tunable functionalities from the hierarchical porous structures, thus providing an efficient strategy to physically and chemically achieve enhanced synergetic performances for extensive applications. In this chapter, the design, fabrication, and performance of these uniform hierarchical porous microparticles are introduced.

6.2 Microfluidic Strategy for Fabrication of Uniform Hierarchical Porous Microparticles

The fabrication procedure of the controllable highly interconnected hierarchical porous microparticles is schematically illustrated in Figure 6.1. Monodisperse water-in-oil-in-water (W/O/W) emulsions, with partially water-soluble organic monomer solution containing cross-linker and surfactant as the oil phase, are generated from microfluidics for the microparticle fabrication (Figure 6.1a).

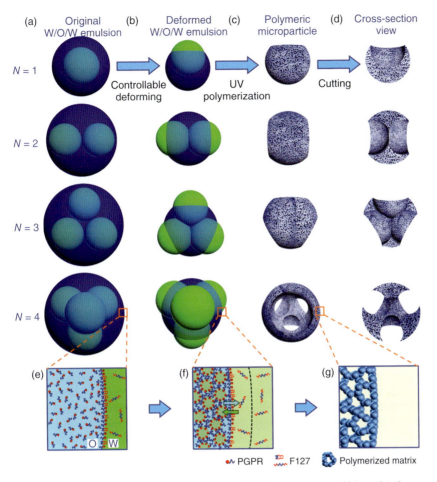

Figure 6.1 Strategy for fabrication of controllable highly interconnected hierarchical porous microparticles. (a–d) Fabrication of microparticles with controllable micrometer-sized pore structures and shapes from controllably deformed W/O/W emulsions containing oil phase partially miscible with the aqueous phases. (e–g) Formation of nanoporous structures by employing mass-transfer-induced formation of aqueous nanodrops in oil phase as templates. (Zhang et al. 2015 [48]. Reproduced with permission of American Chemical Society.)

In the W/O/W emulsions, the middle oil phase is partially miscible with the inner aqueous microdrop and outer aqueous continuous phase. This can lead to deformation of the emulsions into controllable shapes depending on the confined packing structure of the inner microdrops (Figure 6.1a,b). This process also facilitates the mass exchange between the oil phase and aqueous phase, resulting in diffusion of water molecules into the oil phase to form aqueous nanodrops due to the presence of excess surfactant (Figure 6.1e,f). By UV-polymerizing the deformed W/O/W emulsions, micrometer- and nanometer-sized pores can be, respectively, templated from the inner microdrops and nanodrops to form uniform hierarchically engineered microparticles with highly interconnected hierarchical porous structures and controllable shapes (Figure 6.1c,d,g). For these hierarchical porous microparticles, the nanoporous structure of the polymeric matrix determines the specific surface area, while the micrometer-sized pores determine the shape of the nanoporous polymeric matrix. The porosity of the micrometer- and nanometer-sized pores can be separately tuned by simply changing the size and number (N) of the inner microdrops and the amount of the surfactant in the oil phase. Meanwhile, these hierarchical porous microparticles can be flexibly functionalized by adding functional nanoparticles and monomers in the middle oil phase of W/O/W emulsions in the fabrication. Therefore, this approach allows controllable fabrication of uniform polymeric microparticles containing highly interconnected hierarchical porous structures, with pore size, porosity, functionality, and particle shape individually and flexibly manipulated in a single step.

6.3 Controllable Microfluidic Fabrication of Uniform Hierarchical Porous Microparticles

6.3.1 Preparation of Hierarchical Porous Microparticles

The fabrication strategy for producing hierarchical porous microparticles is demonstrated by preparing hierarchically engineered poly(methyl methacrylate-*co*-ethylene glycol dimethacrylate) (poly(MMA-*co*-EGDMA)) microparticles with controllable hierarchical porous structures and shapes. A two-stage glass-capillary microfluidic device [49] is used to generate W/O/W emulsions for fabricating the hierarchical porous poly(MMA-*co*-EGDMA) microparticles. In the microfluidic device, the inner diameters of the injection tube, transition tube, and collection tube are, respectively, 550, 150, and 300 μm. Typically, 4 ml methyl methacrylate (MMA) containing cross-linker ethylene glycol dimethacrylate (EGDMA) (0.2 ml, 5 % (v/v)), surfactant polyglycerol polyricinoleate (PGPR) (0.2 g, 5% (w/v)), and photoinitiator 2-hydroxy-2-methyl-1-phenyl-1-propanone (HMPP) (20 μl, 0.5% (v/v)) is used as the middle fluid and 50 ml deionized water containing glycerin (2.5 g, 5% (w/v)) and Pluronic F127 (0.5 g, 1% (w/v)) is used as the inner and outer aqueous fluids. By separately pumping the inner, middle, and outer fluids into the injection, transition, and collection tubes of microfluidic device, monodisperse W/O/W emulsions are generated. The generated emulsions are collected in a glass holder and kept for 20 min. This allows mass exchange between the oil phase and the aqueous phases for deformation of the W/O/W emulsions and formation of aqueous nanodrops in the oil phase. Then, the controllably deformed W/O/W emulsions are converted into hierarchical porous microparticles by UV-initiated polymerization for 20 min. At last, the obtained microparticles are washed with ethanol and deionized water for further use.

6.3.2 Hierarchical Porous Microparticles with Micrometer-Sized Pores from Deformed W/O/W Emulsions

Figure 6.2a schematically illustrates the two-stage glass-capillary microfluidic device for generating monodisperse W/O/W emulsions as templates for

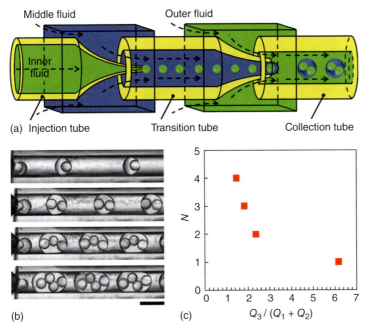

Figure 6.2 Controllable generation of W/O/W emulsion templates from microfluidics. (a) Schematic illustration of microfluidic device for controllably generating monodisperse W/O/W emulsions. (b) High-speed optical micrographs showing the generation process of controllable W/O/W emulsions in the microfluidic device. (c) Effect of flow rates on the number (N) of inner microdrops in the W/O/W emulsions. The flow rates of the inner fluid (Q_1) and middle fluid (Q_2) are 400 and 1000 µl h^{-1}, respectively. The flow rate of the outer fluid (Q_3) is adjusted to tune the N values. Scale bar is 400 µm. (Zhang et al. 2015 [48]. Reproduced with permission of American Chemical Society.)

synthesis of uniform hierarchical porous microparticles. Based on the high controllability of microfluidics, precise control of the emulsion drop size [49] and the inner drop number (N) (Figure 6.2b,c) can be achieved by simply tuning flow rates. Therefore, monodisperse W/O/W emulsions can be generated from microfluidics (Figure 6.3a), with controllable inner microdrops as pore templates to create porous structures with tunable size and porosity of micrometer-sized pores.

Because the middle oil phase is partially miscible with the inner and outer aqueous phases, the W/O/W emulsions containing inner microdrops with controlled numbers can controllably deform into desired shapes. The deformed shapes depend on the confined packing structure of the inner microdrops (Figure 6.3a,b). Moreover, such deformation process can lead to partial dewetting of the middle oil phase on the inner microdrops, resulting in a thin film consisting of surfactant bilayers between the inner aqueous microdrop and outer aqueous phase (Figure 6.3c). Meanwhile, this deformation process also squeezes the inner microdrops together, forming a thin oil film between the inner and outer aqueous phases. The thin films can be ruptured after UV-polymerizing the emulsion templates; thus, they are crucial for creating highly interconnected

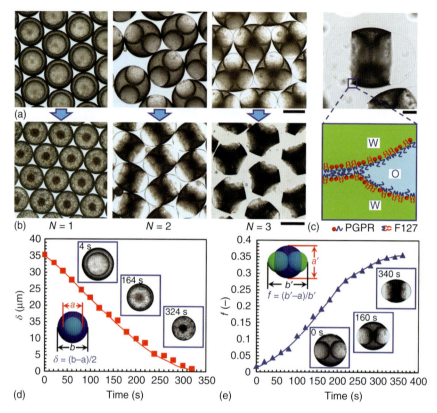

Figure 6.3 Controllable deformation of W/O/W emulsion templates. (a, b) Optical micrographs of W/O/W emulsions containing inner microdrops with tunable number (N) before (a) and after (b) controllable deformation. (c) Optical micrograph of a deformed W/O/W emulsion containing two inner microdrops, with oil shell partially dewetting on the inner microdrop to form a surfactant bilayer. (d, e) Time-dependent change in the shell thickness (δ) of W/O/W emulsions with one inner microdrop (d), and in the ellipticity (f) of W/O/W emulsions with two inner microdrops (e). Scale bars are 200 μm in (a) and (b), and 50 μm in (c). (Zhang et al. 2015 [48]. Reproduced with permission of American Chemical Society.)

open micrometer-sized pores. Besides the packing structure of the inner microdrops, the processing time also influences the deformed structure of the W/O/W emulsions, such as their shell thickness and shapes (Figure 6.3d,e). For example, as shown in Figure 6.3d, for W/O/W emulsions with one inner microdrop, after processing for 6 min, the emulsions remain spherical but the oil shell thickness decreases from tens of micrometers to several micrometers. For W/O/W emulsions with two inner microdrops, the emulsion shape changes from sphere into ellipsoid with passage of time (Figure 6.3e). Therefore, combination of the controllable inner microdrops and the processing time allows good control of the emulsion structure as well as the resultant microparticle shape and the micrometer-sized pore structure. Based on this, after further UV polymerization of the controllably deformed emulsions containing inner microdrops with

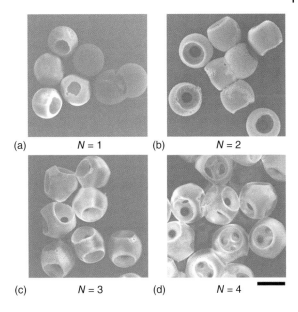

Figure 6.4 SEM images of poly(MMA-*co*-EGDMA) microparticles with tunable number ($N = 1–4$) of highly interconnected micrometer-sized pores and controllable shapes. Scale bar is 200 µm. (Zhang *et al.* 2015 [48]. Reproduced with permission of American Chemical Society.)

tunable numbers, uniform microparticles with controllable shapes and highly interconnected micrometer-sized open pores can be produced (Figure 6.4).

6.3.3 Integration of Nanometer- and Micrometer-Sized Pores for Creating Hierarchical Porous Microparticles

Integration of controllable nanometer-sized pores in the microparticles containing micrometer-sized pores is achieved by using mass-transfer-induced formation of aqueous nanodrops in the oil shell of W/O/W emulsions as templates for the nanometer-sized pores. As discussed in Section 6.3.2, during the abovementioned emulsion deformation process, the partially miscible property of oil phase with aqueous phase facilitates the mass exchange between the oil and aqueous phases. Therefore, water molecules diffuse into the oil shell to form aqueous nanodrops, which can be stabilized by the excess amount of surfactant PGPR in the oil phase. The nanodrop formation is confirmed by adding EGDMA ($\rho = 1.051\,\mathrm{g\,cm^{-3}}$) microdrops containing 10% (w/v) PGPR, with density larger than water, into the continuous aqueous phase (Figure 6.5a). Fluorescent rhodamine B is dissolved in the aqueous phase to trace the diffused water in the oil microdrop during the formation of aqueous nanodrops. As monitored by confocal laser scanning microscope (CLSM), on increasing the processing time, increasing numbers of red aqueous nanodrops are observed inside the EGDMA microdrop, resulting in an increased fluorescence intensity (Figure 6.5a). Since the nanodrop formation induced by mass exchange is mainly due to the presence of excess surfactant PGPR, change of the surfactant amount in the oil phase allows adjustment of the nanodrop formation; thus, the nanoporosity can be tuned. This is demonstrated by fabricating poly(EGDMA) (PEGDMA) microparticles with different PGPR contents for control of their nanoporous structures. Such PEGDMA microparticles are fabricated from single emulsion

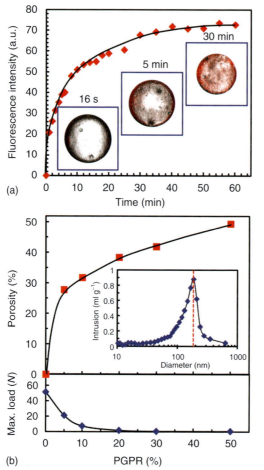

Figure 6.5 Mass-transfer-induced creation of nanoporous structures. (a) Time-dependent change in the fluorescent intensity of EGDMA drops with 10% (w/v) PGPR in aqueous phase. (b) Porosity and mechanical strength of PEGDMA microparticles fabricated with different PGPR contents. The inset shows the size distribution of the nanometer-sized pores of PEGDMA microparticles with 10% (w/v) PGPR. The "Max. load" in the Y-axis presents the maximum load at fracture upon compression. (Zhang et al. 2015 [48]. Reproduced with permission of American Chemical Society.)

drops of EGDMA solutions containing 0% (w/v), 5% (w/v), 10% (w/v), 20% (w/v), 30% (w/v), and 50% (w/v) of PGPR. As a result, the nanoporosity of PEGDMA microparticles synthesized from the EGDMA microdrops increases from 0% to 49.4% on increasing the PGPR contents from 0% (w/v) to 50% (w/v) (Figure 6.5b). With the increasing PGPR content, the mechanical strength of the PEGDMA microparticle decreases due to the increased porosity (Figure 6.5b). As observed in the SEM images (Figure 6.6), the PEGDMA microparticles with 0% (w/v) PGPR show nonporous structures (Figure 6.6a) due to the absence of nanodrops in the oil phase, while the PEGDMA microparticles prepared from EGDMA drops with 10% (w/v) PGPR content (Figure 6.5a) exhibit highly interconnected nanoporous structures (Figure 6.6e). It is worth noting that, by gradually increasing the PGPR content from 0% to 10% (w/v), the PEGDMA microparticles exhibit a structure transition from nonporous structure (Figure 6.6a), isolated porous structure (Figure 6.6b), to highly interconnected porous structure (Figure 6.6c–e), depending on the PGPR content as well as the fraction of aqueous nanodrops in the oil phase. Moreover, by further increasing the PGPR

Figure 6.6 SEM images of PEGDMA microparticles with 0% (a1), 5% (b1), 7% (c1), 9% (d1), 10% (e1), and 50% (f1) PGPR contents, and their outer surfaces (a2–f2) and cross-sections (a3–f3). Scale bars are 50 μm in (a1)–(f1), and 20 μm in the rest. (Zhang *et al.* 2015 [48]. Reproduced with permission of American Chemical Society.)

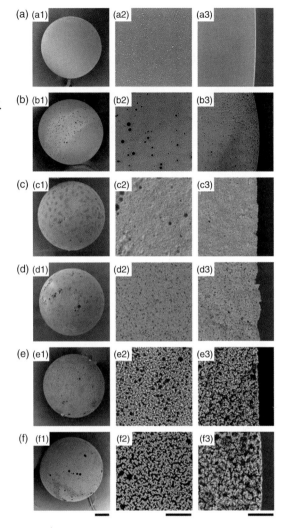

content from 10% to 50% (w/v), microparticles with highly interconnected porous structures of increased porosity can be obtained (Figures 6.6f and 6.5b). As a typical example, the nanoporous PEGDMA microparticles prepared with 10% (w/v) PGPR show nanoporous structures with average pore size of ∼180 nm (inset in Figure 6.5b), as measured by mercury intrusion porosimetry.

Elaborate combination of the highly interconnected nanometer- and the micrometer-sized pore structures enables the fabrication of highly interconnected hierarchical porous microparticles (Figure 6.7). The SEM images of the resultant hierarchical porous poly(MMA-*co*-EGDMA) microparticles clearly show the single micrometer-sized pore in the microparticle, and the highly interconnected nanoporous structures on the outer surface and the cross-section (Figure 6.7a). Moreover, the emulsion-templated synthesis process of the hierarchical porous microparticles enables flexible functionalization of

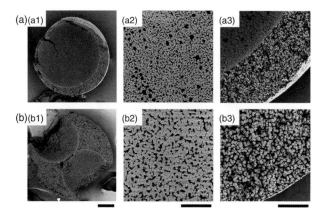

Figure 6.7 Integration of nanometer- and micrometer-sized pores for fabricating controllable hierarchical porous microparticles. SEM images of ruptured hierarchical porous poly(MMA-*co*-EGDMA) microparticles with one micrometer-sized pore (a1) and magnetic hierarchical porous poly(MMA-*co*-EGDMA-*co*-GMA) microparticles with two micrometer-sized pores (b1), as well as their magnified outer surfaces (a2–b2) and cross-sections (a3–b3). Scale bars are 50 μm in (a1)–(b1), and 20 μm in the rest. (Zhang *et al.* 2015 [48]. Reproduced with permission of American Chemical Society.)

the microparticles by incorporating functional components into the oil phase of the emulsion templates. This advantage is demonstrated by adding glycidyl methacrylate (GMA) and magnetic nanoparticles in the oil phase of emulsion templates to produce magnetic hierarchical porous poly(MMA-*co*-EGDMA-*co*-GMA) microparticles containing two micrometer-sized pores (Figure 6.7b). This fabrication approach enables controllable formation of highly interconnected porous microparticles with both micrometer- and nanometer-sized pores, tunable particle shapes, and flexible functionalities in a single step, thus providing a facile and versatile strategy for fabricating functional hierarchical porous microparticles.

6.4 Hierarchical Porous Microparticles for Oil Removal

6.4.1 Concept of the Hierarchical Porous Microparticles for Oil Removal

The hierarchical porous poly(MMA-*co*-EGDMA) microparticles provide porous matrix with hydrophobic properties for the adsorption of oil drops from water. When using microparticles with only nanoporous structure for oil adsorption, the hydrophobic surface of their nanometer-sized pores allows wetting of oil drop on the nanoporous matrix for oil loading, while their nanoporosity determines the capacity for loading oil (Figure 6.8a). With the integration of micrometer-sized pores in the nanoporous matrixes, the hierarchical porous microparticles can provide extra micrometer-sized pores as microcontainers for oil loading, thus resulting in a larger capacity for capturing oil (Figure 6.8b). Moreover, by further incorporating the hierarchical porous microparticles with magnetic nanoparticles, the microparticles can be modified

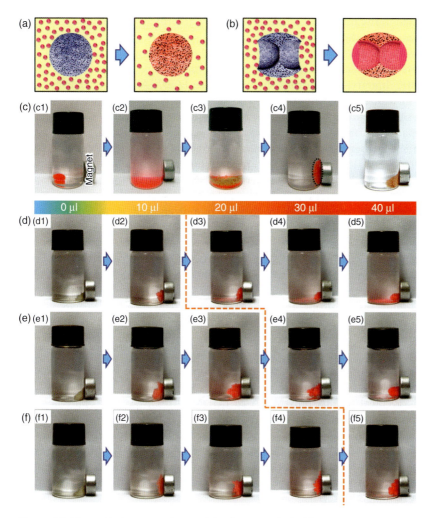

Figure 6.8 Hierarchical porous microparticles for removal of oil from water. (a, b) Schematic illustration of oil capture with nanoporous (a) and hierarchical porous (b) microparticles. (c) Magnetic hierarchical porous poly(MMA-*co*-EGDMA) microparticles for magnetic-guided removal of EGDMA microdrops from water. A large EGDMA drop (c1) is shaken into microdrops (c2), and then microparticles are added (c3); next, the oil-adsorbed microparticles are separated by a magnet (c4) and then washed with ethanol for recycle (c5). (d–f) Nanoporous (d) and hierarchical porous poly(MMA-*co*-EGDMA-*co*-GMA) microparticles with two (e) and three (f) micrometer-sized pores for quantitative oil removal. (Zhang *et al.* 2015 [48]. Reproduced with permission of American Chemical Society.)

with magnetic-responsive property for removing oil drops from water via magnetic manipulation.

6.4.2 Hierarchical Porous Microparticles for Magnetic-Guided Oil Removal

For magnetic functionalization of the poly(MMA-*co*-EGDMA) microparticles for magnetic-guided oil removal, 0.5% (w/v) magnetic nanoparticles are dispersed in the middle oil phase by ultrasonic treatment. The magnetic-guided removal of oil drops from water is demonstrated in Figure 6.8c. EGDMA dyed with LR300 (red color), which is used as the sample oil, is added into water (Figure 6.8c1) and then shaken into microdrops (Figure 6.8c2). Then, magnetic hierarchical porous poly(MMA-co-EGDMA) microparticles with three micrometer-sized pores are added (Figure 6.8c3) and mixed with the EGDMA microdrops by shaking for oil adsorption. After that, the oil-adsorbed microparticles can be easily separated by a magnet for oil removal (Figure 6.8c4) and can be further recycled by washing off the adsorbed oil with ethanol for reuse (Figure 6.8c5). These hierarchical porous microparticles can also be employed for magnetic-guided removal of other oil that can spread on their porous matrix.

Next, the effect of hierarchical porous structures on the oil capturing capacity is quantitatively investigated by using nanoporous and hierarchical porous poly(MMA-co-EGDMA-co-GMA) microparticles for magnetic-guided removal of EGDMA oil drops (Figure 6.8d–f). For fabrication of the nanoporous and hierarchical porous poly(MMA-co-EGDMA-co-GMA) microparticles, GMA, MMA, EGDMA, and PGPR with volume ratio of $25:25:50:10$ are added in the oil phase of their emulsion templates. As shown in Figure 6.8d–f, by gradually adding the oil drops, the poly(MMA-co-EGDMA-co-GMA) nanoporous microparticles (Figure 6.8d) exhibit maximum oil capture upon total addition of ~20 μl oil drops (Figure 6.8d3), while those with hierarchical porous structures containing two (Figure 6.8e) and three (Figure 6.8f) micrometer-sized pores, respectively, exhibit maximum oil capture at total addition of ~30 μl (Figure 6.8e4) and ~40 μl (Figure 6.8f5) oil drops. All the results show that the hierarchical porous poly(MMA-co-EGDMA-co-GMA) microparticles can capture larger amount of oil than the ones with only nanoporous structures, and the maximum capacity for oil capture increases with the increasing number of micrometer-sized pores. Therefore, as compared with the homogeneous nanoporous structures, the hierarchical porous structures provide opportunities for creating microparticles with larger oil capturing capacity from the same mass quality of matrix materials.

6.5 Hierarchical Porous Microparticles for Protein Adsorption

6.5.1 Concept of Hierarchical Porous Microparticles for Protein Adsorption

The hierarchical porous microparticles enable enhanced synergetic performance for improved protein adsorption. As compared with microparticles with only nanoporous structures (Figure 6.9a1), the microparticles with hierarchical porous structures provide highly interconnected micrometer-sized pores as easier access for the protein molecules to diffuse into the porous matrix (Figure 6.9a2). Meanwhile, these micrometer-sized pores also provide the nanoporous matrix with larger interfacial area between the microparticle and the bulk solution of protein, including the outer surface of the microparticle and the internal surface of the micrometer-sized pores, for the protein molecules

Figure 6.9 Hierarchical porous microparticles for protein adsorption. (a) Schematic illustration of dynamic processes of protein adsorption with nanoporous (a1) and hierarchical porous (a2) microparticles. (b) BSA adsorption profiles of nonporous PEGDMA microparticles, nanoporous PEGDMA microparticles, magnetic nanoporous poly(MMA-*co*-EGDMA-*co*-GMA) microparticles (PMEG), and magnetic hierarchical porous poly(MMA-*co*-EGDMA-*co*-GMA) microparticles with two micrometer-sized pores (PMEG-2). (Zhang et al. 2015 [48]. Reproduced with permission of American Chemical Society.)

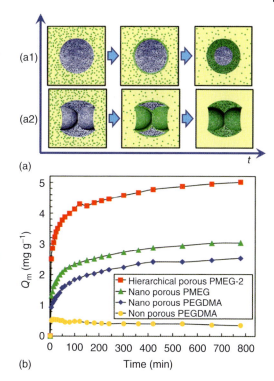

to diffuse into the nanoporous matrix. Therefore, faster adsorption of protein molecules can be achieved by using the hierarchical porous microparticles.

6.5.2 Hierarchical Porous Microparticles for Enhanced Protein Adsorption

The enhanced synergetic performance of hierarchical porous microparticles for protein adsorption is quantitatively investigated by using bovine serum albumin (BSA) as the model protein. Four different types of microparticles, including nonporous PEGDMA microparticles, nanoporous PEGDMA microparticles, magnetic nanoporous poly(MMA-*co*-EGDMA-*co*-GMA) microparticles, and magnetic hierarchical porous poly(MMA-*co*-EGDMA-*co*-GMA) microparticles containing two micrometer-sized pores, are employed. During the adsorption experiment, the microparticles are immersed in BSA solutions to allow free diffusion of the BSA molecules into the porous matrix of the microparticles. Due to the highly interconnected nanoporous structures for adsorbing BSA molecules, the nanoporous PEGDMA microparticles with specific surface area of $16.37 \, m^2 \, g^{-1}$ show a faster adsorption rate with higher adsorption capacity than the nonporous PEGDMA microparticles (Figure 6.9b). By further using magnetic nanoporous poly(MMA-*co*-EGDMA-*co*-GMA) microparticles with specific surface area of $16.83 \, m^2 \, g^{-1}$ as adsorbents (Figure 6.9b), improved performance for BSA adsorption can be achieved. Their nanoporous structures with GMA-modified interfaces provide better interactions with BSA molecules for enhanced adsorption. Moreover, the presence of magnetic nanoparticles in the microparticles facilitates the remote separation via simple magnetic manipulation. By creating highly interconnected micrometer-sized pores in the microparticles for enhanced mass-transfer kinetics, the BSA adsorption of such functional nanoporous poly(MMA-*co*-EGDMA-*co*-GMA) microparticles can be further improved. As shown in Figure 6.9b, with hierarchical porous structures, the hierarchical porous poly(MMA-*co*-EGDMA-*co*-GMA) microparticles containing two micrometer-sized pores exhibit a much faster adsorption rate (3.88 mg g^{-1} at $t = 60$ min) than the nanoporous poly(MMA-*co*-EGDMA-*co*-GMA) ones with only nanoporous structures (2.17 mg g^{-1} at $t = 60$ min).

In the adsorption experiments, poly(MMA-*co*-EGDMA-*co*-GMA) microparticles containing nanoporous and hierarchical porous structures with the same total mass quality are used as adsorbents, and their polymeric matrixes are both prepared from oil phase with the same composition. Therefore, both types of microparticles exhibit polymeric matrixes with the same nanoporous structures and the same total mass quality. This means that both types of these poly(MMA-*co*-EGDMA-*co*-GMA) microparticles exhibit the same specific surface areas, which are provided by the nanoporous polymeric matrixes, for maximum BSA adsorption. Therefore, the faster adsorption performance of the hierarchical porous poly(MMA-*co*-EGDMA-*co*-GMA) microparticles containing two micrometer-sized pores mainly results from their highly interconnected micrometer-sized pores. All the results indicate that the functional hierarchical porous microparticles with combined advantages of enhanced mass-transfer kinetics and flexibly tunable functionalities are promising as adsorbents for enhanced protein adsorption.

6.6 Summary

In summary, this chapter describes a facile and flexible microfluidic approach for one-step fabrication of uniform controllable polymeric microparticles containing highly interconnected hierarchical porous structures. The controllably deformed W/O/W emulsions from microfluidics enable the production of uniform hierarchical porous microparticles with precisely and individually controlled pore size, porosity, functionality, and particle shape by using the inner microdrops and the nanodrops as templates for, respectively, creating micrometer- and nanometer-sized pores. The hierarchical porous microparticles with controllable physical structures and chemical properties combine the advantages of porous structures at both micro- and nanoscales as well as the tunable functionalities to achieve enhanced synergetic performances for broad applications.

References

1 Parlett, C.M.A., Wilson, K., and Lee, A.F. (2013) Hierarchical porous materials: catalytic applications. *Chem. Soc. Rev.*, **42**, 3876–3893.
2 Li, Y., Fu, Z.Y., and Su, B.L. (2012) Hierarchically structured porous materials for energy conversion and Storage. *Adv. Funct. Mater.*, **22**, 4634–4667.

3 Yuan, Z.Y. and Su, B.L. (2006) Insights into hierarchically meso-macroporous structured materials. *J. Mater. Chem.*, **16**, 663–677.
4 Yu, H., Qiu, X., Nunes, S.P., and Peinemann, K.V. (2014) Biomimetic block copolymer particles with gated nanopores and ultrahigh protein sorption capacity. *Nat. Commun.*, **5**, 4110.
5 Abbaspourrad, A., Carroll, N.J., Kim, S.H., and Weitz, D.A. (2013) Surface functionalized hydrophobic porous particles toward water treatment application. *Adv. Mater.*, **25**, 3215–3221.
6 Liu, X.H., Jin, X.B., and Ma, P.X. (2011) Nanofibrous hollow microspheres self-assembled from star-shaped polymers as injectable cell carriers for knee repair. *Nat. Mater.*, **10**, 398–406.
7 Edwards, D.A., Hanes, J., Caponetti, G., Hrkach, J., BenJebria, A., Eskew, M.L., Mintzes, J., Deaver, D., Lotan, N., and Langer, R. (1997) Large porous particles for pulmonary drug delivery. *Science*, **276**, 1868–1871.
8 Chung, H.J. and Park, T.G. (2007) Surface engineered and drug releasing pre-fabricated scaffolds for tissue engineering. *Adv. Drug Delivery Rev.*, **59**, 249–262.
9 Nakayama, D., Takeoka, Y., Watanabe, M., and Kataoka, K. (2003) Simple and precise preparation of a porous gel for a colorimetric glucose sensor by a templating technique. *Angew. Chem. Int. Ed.*, **42**, 4197–4200.
10 Zhao, X.W., Cao, Y., Ito, F., Chen, H.H., Nagai, K., Zhao, Y.H., and Gu, Z.Z. (2006) Colloidal crystal beads as supports for biomolecular screening. *Angew. Chem. Int. Ed.*, **45**, 6835–6838.
11 Qu, J.B., Wan, X.Z., Zhai, Y.Q., Zhou, W.Q., Su, Z.G., and Ma, G.H. (2009) A novel stationary phase derivatized from hydrophilic gigaporous polystyrene-based microspheres for high-speed protein chromatography. *J. Chromatogr. A*, **1216**, 6511–6516.
12 Sacanna, S., Irvine, W.T.M., Chaikin, P.M., and Pine, D.J. (2010) Lock and key colloids. *Nature*, **464**, 575–578.
13 Damasceno, P.F., Engel, M., and Glotzer, S.C. (2012) Predictive self-assembly of polyhedra into complex structures. *Science*, **337**, 453–457.
14 De La Vega, J.C., Elischer, P., Schneider, T., and Hafeli, U.O. (2013) Uniform polymer microspheres: monodispersity criteria, methods of formation and applications. *Nanomedicine*, **8**, 265–285.
15 Xu, Q., Hashimoto, M., Dang, T.T., Hoare, T., Kohane, D.S., Whitesides, G.M., Langer, R., and Anderson, D.G. (2009) Preparation of monodisperse biodegradable polymer microparticles using a microfluidic flow-focusing device for controlled drug delivery. *Small*, **5**, 1575–1581.
16 Weitz, D.A. (2004) Packing in the spheres. *Science*, **303**, 968–969.
17 Donev, A., Cisse, I., Sachs, D., Variano, E., Stillinger, F.H., Connelly, R., Torquato, S., and Chaikin, P.M. (2004) Improving the density of jammed disordered packings using ellipsoids. *Science*, **303**, 990–993.
18 John, U., Lennart, S., Arvid, B., and Jan, B. (1983) Monodisperse polymer particles – a step forward for chromatography. *Nature*, **303**, 95–96.
19 Gokmen, M.T. and Du Prez, F.E. (2012) Porous polymer particles-a comprehensive guide to synthesis, characterization, functionalization and applications. *Prog. Polym. Sci.*, **37**, 365–405.

20 Lu, Y.F., Fan, H.Y., Stump, A., Ward, T.L., Rieker, T., and Brinker, C.J. (1999) Aerosol-assisted self-assembly of mesostructured spherical nanoparticles. *Nature*, **398**, 223–226.

21 Kresge, C.T., Leonowicz, M.E., Roth, W.J., Vartuli, J.C., and Beck, J.S. (1992) Ordered mesoporous molecular sieves synthesized by a liquid-crystal template mechanism. *Nature*, **359**, 710–712.

22 Zhao, D.Y., Feng, J.L., Huo, Q.S., Melosh, N., Fredrickson, G.H., Chmelka, B.F., and Stucky, G.D. (1998) Triblock copolymer syntheses of mesoporous silica with periodic 50 to 300 angstrom pores. *Science*, **279**, 548–552.

23 Warren, S.C., Messina, L.C., Slaughter, L.S., Kamperman, M., Zhou, Q., Gruner, S.M., DiSalvo, F.J., and Wiesner, U. (2008) Ordered mesoporous materials from metal nanoparticle-block copolymer self-assembly. *Science*, **320**, 1748–1752.

24 Yang, P.D., Zhao, D.Y., Margolese, D.I., Chmelka, B.F., and Stucky, G.D. (1998) Generalized syntheses of large-pore mesoporous metal oxides with semicrystalline frameworks. *Nature*, **396**, 152–155.

25 Dinsmore, A.D., Hsu, M.F., Nikolaides, M.G., Marquez, M., Bausch, A.R., and Weitz, D.A. (2002) Colloidosomes: selectively permeable capsules composed of colloidal particles. *Science*, **298**, 1006–1009.

26 Velev, O.D., Lenhoff, A.M., and Kaler, E.W. (2000) A class of microstructured particles through colloidal crystallization. *Science*, **287**, 2240–2243.

27 Zakhidov, A.A., Baughman, R.H., Iqbal, Z., Cui, C.X., Khayrullin, I., Dantas, S.O., Marti, I., and Ralchenko, V.G. (1998) Carbon structures with three-dimensional periodicity at optical wavelengths. *Science*, **282**, 897–901.

28 Velev, O.D., Jede, T.A., Lobo, R.F., and Lenhoff, A.M. (1997) Porous silica via colloidal crystallization. *Nature*, **389**, 447–448.

29 Holland, B.T., Blanford, C.F., and Stein, A. (1998) Synthesis of macroporous minerals with highly ordered three-dimensional arrays of spheroidal voids. *Science*, **281**, 538–540.

30 Imhof, A. and Pine, D.J. (1997) Ordered macroporous materials by emulsion templating. *Nature*, **389**, 948–951.

31 Imhof, A. and Pine, D.J. (1998) Uniform macroporous ceramics and plastics by emulsion templating. *Adv. Mater.*, **10**, 697–700.

32 Kim, T.K., Yoon, J.J., Lee, D.S., and Park, T.G. (2006) Gas foamed open porous biodegradable polymeric microspheres. *Biomaterials*, **27**, 152–159.

33 Nakanishi, K. and Tanaka, N. (2007) Sol-gel with phase separation. Hierarchically porous materials optimized for high-performance liquid chromatography separations. *Acc. Chem. Res.*, **40**, 863–873.

34 Davis, S.A., Burkett, S.L., Mendelson, N.H., and Mann, S. (1997) Bacterial templating of ordered macrostructures in silica and silica-surfactant mesophases. *Nature*, **385**, 420–423.

35 Sai, H., Tan, K.W., Hur, K., Asenath-Smith, E., Hovden, R., Jiang, Y., Riccio, M., Muller, D.A., Elser, V., Estroff, L.A., Gruner, S.M., and Wiesner, U. (2013) Hierarchical porous polymer scaffolds from block copolymers. *Science*, **341**, 530–534.

36 Stein, A., Li, F., and Denny, N.R. (2008) Morphological control in colloidal crystal templating of inverse opals, hierarchical structures, and shaped particles. *Chem. Mater.*, **20**, 649–666.

37 Fan, J.B., Huang, C., Jiang, L., and Wang, S. (2013) Nanoporous microspheres: from controllable synthesis to healthcare applications. *J. Mater. Chem. B*, **1**, 2222–2235.

38 Wu, D., Xu, F., Sun, B., Fu, R., He, H., and Matyjaszewski, K. (2012) Design and preparation of porous polymers. *Chem. Rev.*, **112**, 3959–4015.

39 Yang, P.D., Deng, T., Zhao, D.Y., Feng, P.Y., Pine, D., Chmelka, B.F., Whitesides, G.M., and Stucky, G.D. (1998) Hierarchically ordered oxides. *Science*, **282**, 2244–2246.

40 Atencia, J. and Beebe, D.J. (2005) Controlled microfluidic interfaces. *Nature*, **437**, 648–655.

41 Wang, W., Zhang, M.J., and Chu, L.Y. (2014) Functional polymeric microparticles engineered from controllable microfluidic emulsions. *Acc. Chem. Res.*, **47**, 373–384.

42 Xu, S.Q., Nie, Z.H., Seo, M., Lewis, P., Kumacheva, E., Stone, H.A., Garstecki, P., Weibel, D.B., Gitlin, I., and Whitesides, G.M. (2005) Generation of monodisperse particles by using microfluidics: control over size, shape, and composition. *Angew. Chem. Int. Ed.*, **44**, 724–728.

43 Wang, W., Zhang, M.J., Xie, R., Ju, X.J., Yang, C., Mou, C.L., Weitz, D.A., and Chu, L.Y. (2013) Hole–shell microparticles from controllably evolved double emulsions. *Angew. Chem. Int. Ed.*, **52**, 8084–8087.

44 Choi, S.-W., Zhang, Y., and Xia, Y. (2009) Fabrication of microbeads with a controllable hollow interior and porous wall using a capillary fluidic device. *Adv. Funct. Mater.*, **19**, 2943–2949.

45 Wan, J., Bick, A., Sullivan, M., and Stone, H.A. (2008) Controllable microfluidic production of microbubbles in water-in-oil emulsions and the formation of porous microparticles. *Adv. Mater.*, **20**, 3314–3318.

46 Zhang, J., Coulston, R.J., Jones, S.T., Geng, J., Scherman, O.A., and Abell, C. (2012) One-step fabrication of supramolecular microcapsules from microfluidic droplets. *Science*, **335**, 690–694.

47 Lee, D. and Weitz, D.A. (2009) Nonspherical colloidosomes with multiple compartments from double emulsions. *Small*, **5**, 1932–1935.

48 Zhang, M.-J., Wang, W., Yang, X.-L., Ma, B., Liu, Y.-M., Xie, R., Ju, X.-J., Liu, Z., and Chu, L.-Y. (2015) Uniform microparticles with controllable highly interconnected hierarchical porous structures. *ACS Appl. Mater. Interfaces*, **7**, 13758–13767.

49 Chu, L.Y., Utada, A.S., Shah, R.K., Kim, J.W., and Weitz, D.A. (2007) Controllable monodisperse multiple emulsions. *Angew. Chem. Int. Ed.*, **46**, 8970–8974.

7

Microfluidic Fabrication of Monodisperse Hollow Microcapsules

7.1 Introduction

Hollow microcapsules can encapsulate various chemical or biological substances in their inner spaces, and it is thus possible to attain a controlled permeation of chemical or biological molecules by using appropriate microcapsules. Because of their characteristics such as small size, huge total surface area, large inner volume, and stable membrane, hollow microcapsules have found many applications in various fields from drug delivery to the textile, foods, pesticide, energy storage, and chemical and environmental industries.

Uniform particle size is important for hollow microcapsules to improve their performances in various applications, especially in the field of drug delivery systems. The distribution of the microcapsules *in vivo*, and the interaction with biological cells, is greatly affected by the particle size [1, 2]. Monodisperse microcapsules can increase the bioavailability and decrease the side effects. In addition, the practical and theoretical evaluation such as drug release kinetics will become simple and precise if the size distribution of microcapsules is narrow [1–4].

The authors' group developed several microfluidic strategies for fabrication of monodisperse hollow microcapsules, including monodisperse ethyl cellulose (EC) hollow microcapsules with a simple method that combines microfluidic double emulsification and solvent diffusion [5], monodisperse hollow calcium alginate microcapsules via internal gelation in microfluidic-generated double emulsions [6], monodisperse glucose-responsive microcapsules via polymerization in microfluidic-generated double emulsions [7], and monodisperse multi-stimuli-responsive microcapsules via combining cross-linked chitosan acting as pH-responsive capsule membrane, embedded magnetic nanoparticles realizing "site-specific targeting" and embedded temperature-responsive submicrospheres serving as "microvalves" in microfluidic-generated double emulsions [4]. In this chapter, microfluidic strategies are introduced to controllably prepare monodisperse hollow microcapsules with microfluidic-generated double emulsions as templates.

7.2 Microfluidic Fabrication of Monodisperse Ethyl Cellulose Hollow Microcapsules

7.2.1 Microfluidic Fabrication Strategy

Using microfluidic devices, monodisperse water-in-oil-in-water (W/O/W) double emulsions are obtained and used as templates to fabricate monodisperse EC microcapsules. Microfluidic device for generation of monodisperse W/O/W double emulsions is shown in Figure 7.1. The microfluidic device is fabricated by assembling glass capillary tubes on glass slides. To assemble the capillaries, a cylindrical capillary is inserted into and coaxially aligned with a square capillary tube by matching the outer diameter of the cylindrical tube to the inner dimension of the square one. The inner diameters of the injection, transition, and collection tubes are 580, 200, and 580 μm, respectively. A micropuller is used to taper the end of cylindrical capillaries, and the orifice dimensions of tapered ends are adjusted by a microforge. The inner diameters of the tapered ends of the injection and transition tubes are 40 and 200 μm, respectively.

To prepare the W/O/W double emulsions, the inner aqueous fluid, middle organic fluid, and outer aqueous fluid are separately pumped into the injection, transition, and collection tubes. Due to the coaxial co-flow geometry, monodisperse W/O single emulsions are generated in the transition tube and monodisperse W/O/W double emulsions are generated in the collection tube (Figure 7.1). EC microcapsules with different size and morphology can be achieved by using different inner and outer fluids. The flow rates are adjusted to achieve optimum emulsification conditions for generating double emulsions. Recipes for preparing EC microcapsules (No. MC-W-W, MC-W-CS, MC-NaCl-CS, and MC-CS-CS) and the viscosity coefficients of solutions are listed in Table 7.1. EC polymer is dissolved in a nontoxic solvent, ethyl acetate (EA). Both inner and outer fluids are presaturated by 8% (v/v) EA. The viscosity coefficients of solutions are determined by a rotational viscometer at 20 °C.

The obtained double emulsions are collected in a container and immersed in excess DI water. EC microcapsules are fabricated via precipitation and

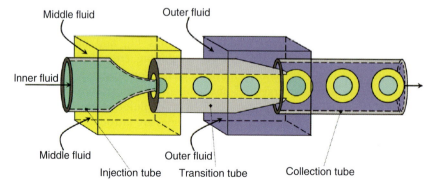

Figure 7.1 Schematic illustration of the microfluidic device for fabricating monodisperse W/O/W double emulsions. (Liu et al. 2009 [5]. Reproduced with permission of Elsevier.)

Table 7.1 Recipes for using microfluidic device to prepare EC microcapsules.

Microcapsule code	Inner fluid[a]					Outer fluid[a]			
	F127[b] (%) (w/v)	NaCl (g l^{-1})	HAc (%) (v/v)	CS (%) (w/v)	Viscosity (mPa s)	F127 (%) (w/v)	HAc (%) (v/v)	CS (%) (w/v)	Viscosity (mPa s)
MC-W-W[c]	1	—	—	—	1.35	1	—	—	1.35
MC-W-CS	1	—	—	—	1.35	1	1	1	13.0
MC-NaCl-CS	1	5.1	—	—	1.35	1	1	1	13.0
MC-CS-CS	1	—	1	1	13.0	1	1	1	13.0

a) Inner and outer fluids are presaturated with EA (8%, v/v). Middle fluid is the EA solution containing 5%w/v EC and 2%w/v Span 20. The viscosity of middle fluid is 8 mPa s.
b) *Abbreviations*: F127: Pluronic F127; HAc: acetic acid; CS: chitosan.
c) MC-W-W is solidified in mixture of ethanol and water (1 : 4 v/v).

solidification of EC polymers when EA diffuses into external water, as shown in Figure 7.2. Ethanol is added into external water when MC-W-W is collected. After solidification completes, EC microcapsules are washed with DI water until there is no residual EA.

7.2.2 Morphologies and Structures of Ethyl Cellulose Hollow Microcapsules

The size and size distribution of microcapsules are determined by measuring the diameters of 100 microcapsules in optical micrographs with analysis software on the basis of optical micrographs. In the analysis, the size distribution is expressed by a coefficient of variation (CV), which is defined as the ratio of the standard deviation of size distribution to its arithmetic mean, as shown in Equations 4.1 and 4.2.

The optical and SEM micrographs of EC microcapsules prepared using water as both inner and outer fluids (No. MC-W-W) are shown in Figure 7.3a–c. During the solidification process in pure water, the double emulsion droplets tend to float on the top surface of water due to the lower density of EA. The consequent evaporation of EA at the interface between water and air results in irregular shapes of

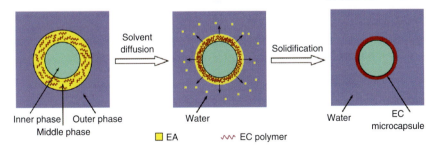

Figure 7.2 Schematic illustration of the EC microcapsule preparation via solvent diffusion using the W/O/W double emulsion as template. (Liu *et al.* 2009 [5]. Reproduced with permission of Elsevier.)

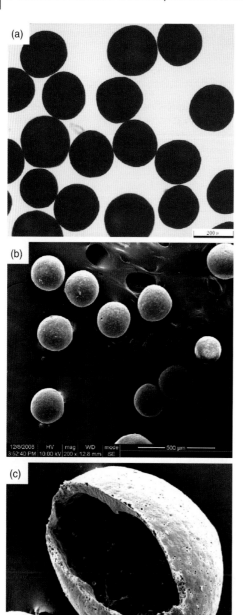

Figure 7.3 Optical micrograph (a) and SEM images (b, c) of MC-W-W microcapsules. The flow rates of inner, middle, and outer fluids for preparation of MC-W-W are 150, 500, and 25 000 µl h^{-1}, respectively. Scale bars are (a) 200 µm, (b) 500 µm, (c) 100 µm. (Liu et al. 2009 [5]. Reproduced with permission of Elsevier.)

microcapsules. Thus, ethanol is added into the water to accelerate solidification, so that the emulsion droplets are solidified and precipitated completely before they reach the top surface of the water. It can be clearly seen that the prepared MC-W-W EC microcapsules are monodisperse and have hollow inner structures. The optical micrograph in Figure 7.3a shows that the sphericity of MC-W-W EC microcapsules is not good. It has been reported that solvent removal at a slower rate from the embryonic microcapsules could cause polymer inward shrinkage [8]. The insufficiently spherical shapes of MC-W-W EC microcapsules shown in Figure 7.3a might have resulted from the uneven shrinkage of EC matrix during solidification.

When using bulk emulsification, viscosity is also an important parameter that could affect the size of obtained emulsions. Decreasing the viscosity of dispersed phase or increasing the viscosity of continuous phase could also help fabricate smaller emulsion droplets [8]. Since EC solution is quite viscous, it is hard to reduce its viscosity unless the EC concentration is lowered, whereas lower concentration of EC would cause loose matrix structure of the resulting microcapsules [9]. Because of its good biocompatibility, nontoxicity, and biodegradability, CS is widely used in the pharmaceutical field. Owing to its high molecular weight, CS in HAc solution exhibits considerable viscosity even at low concentrations. In this work, the viscosity coefficient of dilute HAc solution containing 1% CS is 13.0 mPa s, which is much higher than that of the solution containing only F127 (1.35 mPa s). Thus, instead of water, the viscous CS/HAc solution is used as outer fluid to obtain smaller double emulsion droplets for the preparation of EC microcapsules.

Figure 7.4 shows the optical and SEM micrographs of MC-W-CS EC microcapsules, which are prepared with the W/O/W double emulsion droplets that are generated by employing water as inner fluid and CS/HAc solution as outer fluid. The EC microcapsules with hollow structure exhibit obvious collapse morphology. It is worth noting that the collapse of microcapsules occurs in water (Figure 7.4a), which indicates that the collapsed structure of microcapsules is not caused by sample freeze-drying or cutting for SEM observation. Such collapse may be caused by the osmotic pressure gradient between the inner and outer aqueous fluids during solidification due to the introduction of CS/HAc into the outer fluid. Because EC shell is a kind of rigid matrix after complete solidification, the prepared EC microcapsules could not swell when they are immersed into pure water. To eliminate the influence of osmotic pressure on the microcapsule structure, the inner osmotic pressure is enhanced by adding NaCl into inner fluid. The ionic concentration of NaCl added into the inner fluid is calculated to be iso-osmotic with HAc and CS in outer fluid. With this recipe, MC-NaCl-CS EC microcapsules are prepared, and their optical and SEM micrographs are presented in Figure 7.5. The morphology of EC microcapsules MC-NaCl-CS is similar to that of MC-W-CS microcapsules, both of which have collapsed morphology in water according to the optical micrographs (Figures 7.3a and 7.4a). The results indicate that NaCl solute fails to balance the osmotic pressure, possibly because of its small size, which enables it to penetrate through the EC matrix during solidification.

To obtain absolutely equal osmotic pressures between inner and outer aqueous fluids, the same solutions are employed as both the inner and outer fluids.

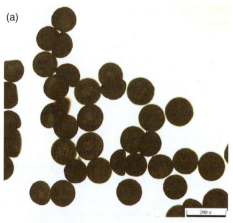

Figure 7.4 Optical micrograph (a) and SEM images (b, c) of MC-W-CS microcapsules. The flow rates of inner, middle, and outer fluids for preparation of MC-W-CS are 80, 200, and 6000 μl h^{-1}, respectively. Scale bars are (a) 200 μm, (b) 500 μm, (c) 100 μm. (Liu *et al.* 2009 [5]. Reproduced with permission of Elsevier.)

Figure 7.5 Optical micrograph (a) and SEM images (b, c) of MC-NaCl-CS microcapsules. The flow rates of inner, middle, and outer fluids for preparation of MC-NaCl-CS are 200, 500, and 10 000 μl h^{-1}, respectively. Scale bars are (a) 200 μm, (b) 500 μm, (c) 100 μm. (Liu et al. 2009 [5]. Reproduced with permission of Elsevier.)

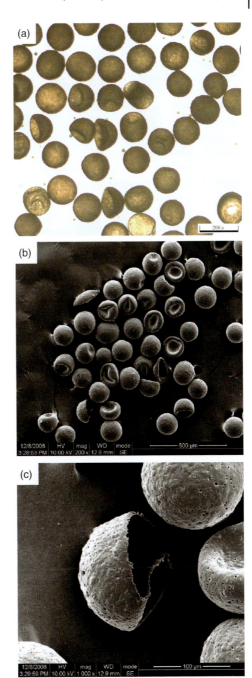

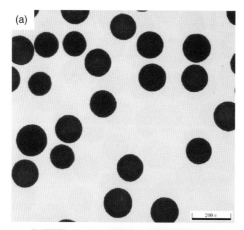
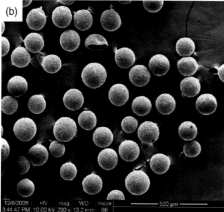
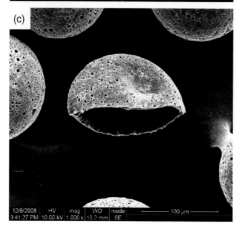

Figure 7.6 Optical micrograph (a) and SEM images (b, c) of MC-CS-CS microcapsules. The flow rates of inner, middle, and outer fluids for preparation of MC-CS-CS are 200, 500, and 12 000 µl h^{-1}, respectively. Scale bars are (a) 200 µm, (b) 500 µm, (c) 100 µm. (Liu *et al.* 2009 [5]. Reproduced with permission of Elsevier.)

Figure 7.7 Size distributions of MC-W-W and MC-CS-CS microcapsules dispersed in water. (Liu et al. 2009 [5]. Reproduced with permission of Elsevier.)

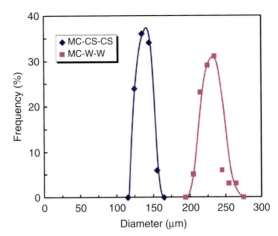

Figure 7.6 shows the optical and SEM micrographs of MC-CS-CS EC microcapsules, which are prepared by using CS/HAc solution at same concentration as both inner and outer fluids. All the resulted microcapsules show no collapsed structure under optical microscope and are of perfect sphericity (Figure 7.6). The results also imply that CS is not able to permeate through EC matrix during solidification due to its considerably large molecular weight, unlike small molecules such as NaCl or HAc, which could penetrate through the EC matrix freely. To fabricate EC microcapsules with satisfactory sphericity, the osmotic pressure gradient between the inner and outer fluids of the template double emulsions should be avoided.

By using the more viscous CS/HAc solution as the inner and outer fluids instead of water, a dramatic decrease is observed in the size of prepared EC microcapsules, as shown in Figure 7.7. The average diameter of MC-CS-CS EC microcapsules is $137 \pm 17\,\mu m$, whereas that of MC-W-W EC microcapsules is $228 \pm 39\,\mu m$. Generally, low flow rate of outer fluid usually favors the preparation of large microcapsules. However, although the flow rate of outer fluid for MC-CS-CS microcapsule preparation is only $12\,ml\,h^{-1}$ and that for MC-W-W microcapsule preparation is as large as $25\,ml\,h^{-1}$, the fabricated MC-W-W microcapsules are still much larger than MC-CS-CS microcapsules. The results imply that for the preparation system in this study, the influence of fluid viscosity on the size of prepared microcapsules is greater than that of the flow rate. The CV values for MC-CS-CS and MC-W-W microcapsules are 6.26% and 5.40%, respectively, which indicate that both types of these prepared EC microcapsules show good monodispersity.

7.3 Microfluidic Fabrication of Monodisperse Calcium Alginate Hollow Microcapsules

Alginates have been one of the most popular biomaterials because of the convenient sources, nontoxicity, excellent biocompatibility, and biodegradability [10]. Alginates are anionic polysaccharides composed of β-D-mannuronic acid and α-L-guluronic acid, and they form hydrogels with multivalent cations such as Ca^{2+}, Ba^{2+}, or Fe^{3+}. The preparation process of alginate microspheres or microcapsules involves two steps, that is, droplet generation and ionic cross-linking. Most of the attempts to prepare alginate microcapsules using single emulsions as templates usually result in solid microspheres. Although calcium alginate microcapsules with lower cross-linking degree inside can be obtained via external gelation, the inner cavities of those microcapsules are not well defined, because the precise control of cross-linking process dependent on the gelation time and cross-linker concentration is still difficult. Here we show the preparation of micron-sized monodisperse calcium alginate microcapsules by combining the microfluidic emulsification with internal gelation [6].

7.3.1 Microfluidic Fabrication Strategy

The capillary microfluidic device is fabricated by assembling glass capillary tubes on glass slides, as shown in Figure 7.1. Oil-in-water-in-oil (O/W/O) double emulsions containing Na-alginate in the middle aqueous layer are generated using a capillary microfluidic device and serve as templates for synthesis of calcium alginate hollow microcapsules, as shown in Figure 7.8. A mixture of soybean oil and benzyl benzoate is used as inner oil phase. The middle aqueous phase contains surfactant Pluronic F127, $CaCO_3$ nanoparticles, sodium alginate, and photoacid generator diphenyliodonium nitrate (PAG). The outer oil phase is soybean oil with surfactant polyglycerol polyricinoleate (PGPR). To prepare monodisperse O/W/O double emulsions as templates, inner oil phase, middle water phase, and outer oil phase solutions are separately pumped into the injection, transition, and collection tubes through polyethylene tubing attached to disposable syringes. The fluid flows are driven by syringe pumps. Based on the coaxial co-flow

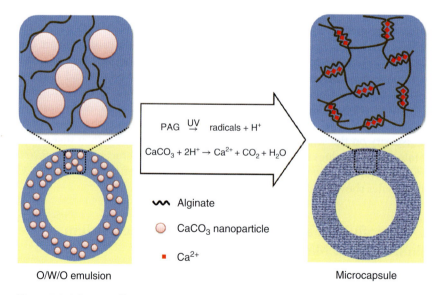

Figure 7.8 Schematic illustration of fabrication of calcium alginate microcapsule via internal gelation in O/W/O double emulsion. "PAG" stands for diphenyliodonium nitrate. (Liu *et al.* 2013 [6]. Reproduced with permission of Elsevier.)

geometry, monodisperse O/W single emulsions are generated in the transition tube and monodisperse O/W/O double emulsions are generated in the collection tube. The flow rates of the inner, middle, and outer fluids are $Q_1 = 300\,\mu l\,h^{-1}$, $Q_2 = 400\,\mu l\,h^{-1}$, and $Q_3 = 4000\,\mu l\,h^{-1}$, respectively. The obtained O/W/O double emulsions are collected in a petri dish and are subjected to UV irradiation in ice bath for 30 min. Upon UV irradiation, the photoacid generator releases protons, and as a result the pH value decreases. Therefore, calcium ions are released from calcium carbonate and cross-link the alginate polymers (Figure 7.8). To obtain hollow microcapsules, the inner and outer oil phases of the products are washed with isopropanol, and the microcapsules are finally dispersed in water.

7.3.2 Morphologies and Structures of Calcium Alginate Hollow Microcapsules

Monodisperse microcapsules can be obtained on the premise of the monodisperse O/W/O double emulsion templates. Microfluidic emulsification allows precise control of both the emulsion size and the number of inner droplets [11]. The prepared O/W/O double emulsions are with uniform structure with one oil droplet in one water droplet, as shown in Figure 7.9a. Based on these

Figure 7.9 Optical micrograph (a) and size distribution (b) of monodisperse O/W/O emulsions containing Na-alginate in the middle layer. (Liu et al. 2013 [6]. Reproduced with permission of Elsevier.)

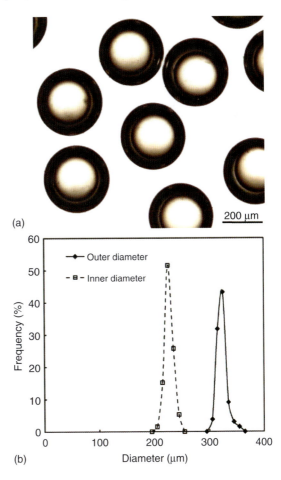

monodisperse double emulsions, microcapsules with well-defined cavity can be obtained. The average outer and inner diameters of the O/W/O double emulsions prepared with microfluidics are 323 and 227 μm, respectively, as shown in Figure 7.9b. The CV, which is defined as the ratio of the standard deviation of the size distribution to its arithmetic mean, is used to characterize the size monodispersity of particles. The CV values of the outer and inner diameters of double emulsions are 2.5% and 3.1%, respectively, which illustrate highly monodisperse property.

Figure 7.10a is the confocal laser scanning microscope (CLSM) image on the transmission channel of the hollow calcium alginate microcapsules. It shows that monodisperse calcium alginate microcapsules are successfully made through internal gelation of alginate polymers by Ca^{2+}. The Ca-alginate microcapsule membrane is quite transparent because of the hydrophilicity of the Ca-alginate polymers and thus the high water content in the swollen microcapsule membrane. The average outer and inner diameters of the prepared

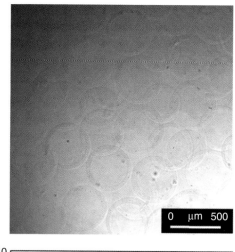
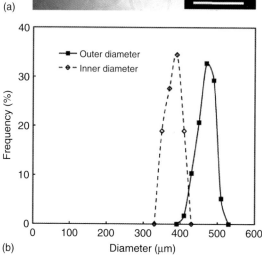

Figure 7.10 CLSM image on the transmission channel (a) and size distribution (b) of monodisperse calcium alginate microcapsules. The inner oil phase has been removed after UV curing to obtain hollow calcium alginate microcapsules. (Liu *et al.* 2013 [6]. Reproduced with permission of Elsevier.)

microcapsules are 467 and 383 μm, respectively (Figure 7.10b), which are both much larger than those of their O/W/O emulsion templates due to the swelling of the microcapsule membrane in deionized water. The CV values of the outer and inner diameters of the Ca-alginate microcapsules are 4.8% and 5.0%, respectively, which indicates that the Ca-alginate microcapsules still remain highly monodisperse after swelling. The average thickness of the microcapsule membranes is 42 μm with a CV value of 4.2%.

Alginate hydrogel materials have been widely used for immobilization of protein, enzyme, and microbes. Because divalent cation cross-linked alginate hydrogels have large pore sizes in the hydrogel networks, protein molecules can diffuse into the Ca-alginate microcapsules. The Ca-alginate microcapsules are left to soak in the fluorescein isothiocyanate (FITC)-labeled BSA (bovine serum albumin) (BSA-FITC) solution overnight. The BSA-FITC-soaked microcapsules are washed with deionized water repeatedly and observed under CLSM. As shown in Figure 7.11a, although the microcapsule membrane shows

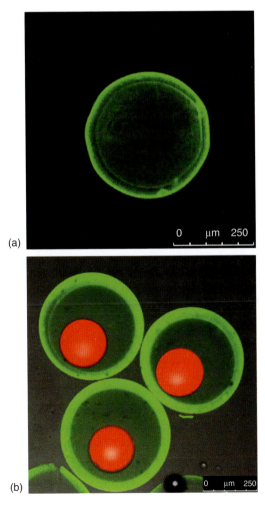

Figure 7.11 CLSM images of monodisperse hollow (a) and core–shell (b) Ca-alginate microcapsules loaded with BSA-FITC molecules. (Liu *et al.* 2013 [6]. Reproduced with permission of Elsevier.)

stronger fluorescence, the fluorescence can be obviously observed inside the microcapsule. The results indicate that BSA-FITC molecules that have fluorescence can diffuse through the Ca-alginate membranes and stay inside the microcapsules. Similar results are obtained in the core–shell microcapsules with oil core, as shown in Figure 7.11b. The results demonstrate that the as-prepared Ca-alginate microcapsules are able to encapsulate the protein molecules after the capsule preparation, which provide an alternative route for loading active drugs or chemicals into carriers to avoid the inactivation in the carrier preparation.

7.4 Microfluidic Fabrication of Monodisperse Glucose-Responsive Hollow Microcapsules

Over the past decades, sugar-responsive systems have been widely investigated in various fields such as sensors, drug delivery systems, bioseparations, and microreactors. As one of the simple sugars, glucose is very important as the target molecule for these systems, due to the key role of glucose in biological functions. For example, diabetes mellitus, one of the most widely spread diseases, is a disorder of glucose regulation, which usually results in glucose accumulation in blood. The tumor cells also accumulate glucose faster than normal cells [12], where the high glucose concentration is believed to have strong influence on the activity of tumor hexokinase type II promoter [13]. Glucose-responsive systems that can change their volumes or other properties in response to glucose concentration show great potential as drug delivery carriers for controlled release of insulin for diabetes therapy [14] or targeted delivery of anticancer drugs for cancer treatment [13]. Such glucose-responsive systems enable self-regulated drug delivery while constantly monitoring the blood glucose concentration for diabetes [15] and cancer, which reduces the pain of patients from insulin injections and chemotherapy.

Microfluidic technology with precise manipulation of emulsion droplets and high encapsulation efficiency has already shown great potential in the fabrication of monodisperse microcapsules for encapsulation and delivery. Based on microfluidics, here we show a simple emulsion-template approach for fabricating monodisperse hydrogel-based microcapsules with repeated glucose response under physiological temperature and glucose concentration conditions [7].

7.4.1 Microfluidic Fabrication Strategy

To construct a totally synthetic microcapsule with long-term stability and repeated glucose response, glucose-responsive 3-acrylamidophenylboronic acid (AAPBA) and thermo-responsive poly(N-isopropylacrylamide) (PNIPAM) are respectively employed as the glucose sensor and actuator for constructing the microcapsule shell. The sensor AAPBA can reversibly form complex with cis-diol such as glucose [16]. In aqueous solution, PBA derivatives exist in equilibrium between an uncharged form and a charged form, both of which can react reversibly with glucose. Especially, only the charged form can stably complex with glucose through reversible covalent bonding, whereas the uncharged form is highly susceptible to hydrolysis [17]. The actuator PNIPAM is a famous thermo-responsive material that can reversibly switch between a swollen and a shrunken state via temperature changes, exhibiting a volume-phase transition temperature (VPTT) (~32 °C) close to the physiological temperature (37 °C). However, incorporation of hydrophobic PBA moiety and PNIPAM into microcapsule shell makes its VPTT lower than 32 °C. So certain amount of hydrophilic acrylic acid (AAc) is used for VPTT adjustment to make the microcapsule achieve a maximum swelling/shrinking volume change in response to glucose concentration change at 37 °C.

The concept of the glucose-responsive hollow microcapsule is schematically illustrated in Figure 7.12. In an environment with pH value close to the pK_a of AAPBA moiety ($pK_a = 8.6$), where the PBA is supposed to be present in both the uncharged and the charged forms, the glucose-responsive microcapsule is initially shrunken at 37 °C (Figure 7.12a,c). When the glucose concentration increases, the charged form of PBA in the hydrogel shell forms stable complex with glucose through reversible covalent bonding. The complex consumes charged PBA forms and shifts the dissociation equilibrium of PBA, which converts more hydrophobic and uncharged PBA groups into hydrophilic and charged phenylborate ions [14]. This makes the VPTT of the microcapsule shift to a higher temperature and builds up a Donnan potential, resulting in a glucose-induced swelling of the microcapsule at 37 °C (Figure 7.12b,c). During the glucose-response process, the PNIPAM-based VPTT shift can exhibit a thermal-phase transition amplification of the glucose-induced swelling behavior [16]. Similarly, decrease in glucose concentration causes a

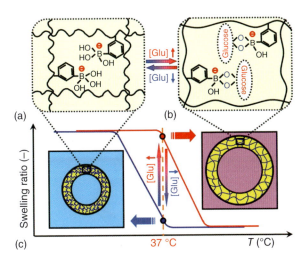

Figure 7.12 Schematic illustration of the glucose-responsive microcapsule with reversible glucose-induced swelling/shrinking behavior. Based on the complex of AAPBA with glucose, the glucose-responsive microcapsule can reversibly transit between a shrunken state (a) and a swollen state (b) by changing the glucose concentration. (c) Formation/decomposition of AAPBA–glucose complex induces a VPTT shift of the microcapsules, which causes reversible glucose-induced swelling/shrinking behaviors of the microcapsule. (Zhang et al. 2013 [7]. Reproduced with permission of The Royal Society of Chemistry.)

glucose-induced shrinking of the microcapsule, due to the decomposition of the PBA–glucose complex. The microcapsules with reversible glucose-responsive swelling/shrinking behaviors under physiological conditions show great potential as glucose sensors and self-regulated delivery systems for diabetes and cancer.

Monodisperse double emulsions from microfluidics are used as templates for the synthesis of glucose-responsive microcapsules (Figure 7.13). The microfluidic device (Figure 7.13b) is the same as that shown in Figure 7.1. Typically, deionized water containing monomer N-isopropylacrylamide (NIPAM), AAPBA, and cross-linker N,N'-methylene-bis-acrylamide (MBA), initiator 2,2′-azobis(2-amidinopropane dihydrochloride) (V50), glycerin, and Pluronic F127 are used as the middle aqueous fluids to construct the glucose-responsive poly(NIPAM-co-AAPBA) (PNA) shell of the microcapsules. The molar ratio of NIPAM to AAPBA is kept at 9:1. To adjust the volume-phase transition behavior of the microcapsule, different amounts of AAc are added into the middle fluid to fabricate the poly(NIPAM-co-AAPBA-co-AAc) (PNAA) microcapsule with tunable VPTT (Figure 7.13a). Soybean oil containing PGPR and $CaCO_3$ nanoparticles for emulsion stabilization, and soybean oil containing PGPR and photoinitiator 2,2-dimethoxy-2-phenylacetophenone (BDK) are used as the outer and inner fluids, respectively. O/W/O double emulsions are generated by separately pumping the inner, middle, and outer fluids into the injection, transition, and collection tubes of the microfluidic device and then collected in

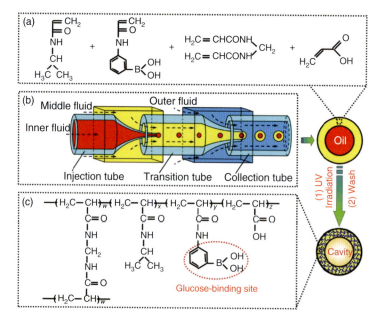

Figure 7.13 Schematic illustration of the fabrication process of glucose-responsive microcapsule. (a) Middle aqueous phase containing NIPAM, AAPBA, AAc monomers, and MBA cross-linker for constructing microcapsule shell. (b) Capillary microfluidic device used to generate monodisperse O/W/O double emulsions. (c) The chemical structure of the microcapsule shell after UV-initiated polymerization. (Zhang et al. 2013 [7]. Reproduced with permission of The Royal Society of Chemistry.)

a container. The flow rates of the inner, middle, and outer fluids are, respectively, 500, 600, and 3000 µl h^{-1}. The inner oil core of the double emulsions is labeled with Sudan Red to emphasize the core–shell structure. The collected O/W/O double emulsions are converted into microcapsules by polymerization under UV irradiation for 20 min in an ice-water bath. Under UV light, monomers contained in the middle aqueous phase of the double emulsions are polymerized to build the glucose-responsive hydrogel shell of the microcapsule (Figure 7.13c). A 250 W UV lamp with an illuminance spectrum of 250–450 nm is employed to produce UV light. After washing with isopropanol and deionized water for several times, the microcapsules are obtained.

Figure 7.14 presents the representative optical micrographs of O/W/O double emulsions and the resulted glucose-responsive microcapsules in isopropanol at 25 °C, respectively. The double emulsions show clear core/shell structures with oil core containing Sudan Red (Figure 7.14b) and uniform size with narrow size distribution (Figure 7.14d). The calculated CV values for the inner diameters (ID) and outer diameters (OD) of O/W/O double emulsions are 1.25% and 1.38%,

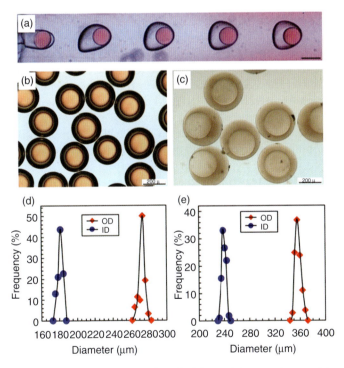

Figure 7.14 Characterization of the double emulsion templates and the resulted glucose-responsive microcapsules. (a) High-speed optical micrograph showing the formation of O/W/O double emulsions in microfluidic device. (b, c) Optical micrographs of the monodisperse double emulsions (b) and the glucose-responsive microcapsules in isopropanol at 25 °C (c). (d, e) Size distributions of the outer diameter (OD) and inner diameter (ID) for double emulsions (d) and glucose-responsive microcapsules in isopropanol at 25 °C (e). Scale bars are 200 µm. (Zhang *et al.* 2013 [7]. Reproduced with permission of The Royal Society of Chemistry.)

respectively, indicating good monodispersity of the emulsion droplets. Using these uniform double emulsions as templates, glucose-responsive microcapsules with hollow structures can be prepared by UV-initiated polymerization of the monomer-containing middle aqueous layer of the double emulsions and washing off the oil core (Figure 7.14c). The microcapsules also show narrow size distributions and slightly increased ID and OD values that resulted from the swollen hydrogel shell in isopropanol at 25 °C (Figure 7.14e). The CV values for the ID and OD of the microcapsules are 1.23% and 1.38%, respectively, also showing highly monodisperse structures. The eccentric hollow cavity results from the density difference between the inner oil core and the middle aqueous layer of emulsion templates. More uniform shell could be obtained by matching the densities of inner fluid and middle fluid [18]. The results show that an effective and promising microfluidic strategy is developed to fabricate monodisperse glucose-responsive hydrogel microcapsules with hollow structures.

7.4.2 Glucose-Responsive Behaviors of Microcapsules

The AAPBA moiety in the hydrogel shell of the PNA microcapsules can work as glucose sensor to make the PNA microcapsules glucose-responsive. The effect of glucose concentration on the equilibrium swelling/shrinking behaviors of PNA microcapsules is shown in Figure 7.15. In all cases, the samples display decreasing volumes with increasing temperature, due to the thermo-responsive volume-phase transition of the PNIPAM networks in the microcapsule shell. Incorporation of hydrophobic AAPBA into PNIPAM networks makes the VPTT of PNA microcapsules in pure water shift to a lower temperature (~22.5 °C). When placed in $0\,g\,l^{-1}$ GBS with pH close to the pK_a of AAPBA, more AAPBA moieties become hydrophilic and charged form due to ionization, which causes a higher VPTT (~27.2 °C) and more swollen state of the microcapsules than those in pure water. Upon stepwise increase of the glucose concentration of GBS from 0 to $3.0\,g\,l^{-1}$, the PNA microcapsules exhibit a glucose-induced swelling behavior that is in direct proportion to the glucose concentration, due to the complexation of AAPBA moieties with glucose. These results indicate both the good thermo-responsive property and glucose sensitivity of the PNA microcapsules.

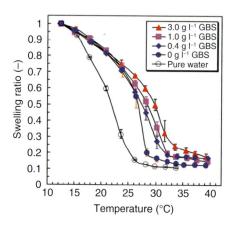

Figure 7.15 Temperature-dependent volume changes of PNA microcapsules in GBS containing different glucose concentrations. (Zhang et al. 2013 [7]. Reproduced with permission of The Royal Society of Chemistry.)

However, although the complex of AAPBA with 0.4–3.0 g l^{-1} glucose results in higher VPTT and more swelling state of the PNA microcapsule, it is still difficult to make their T_{opt} locate at 37 °C because all PNA microcapsules exhibit shrinking at 37 °C, as shown in Figure 7.15. To solve this problem, hydrophilic AAc is introduced into the PNA microcapsule for further VPTT adjustment.

PNAA microcapsules with 2.4 mol% AAc are fabricated to achieve a good glucose-responsive swelling/shrinking volume change at physiological temperature [7]. Figure 7.16 shows the glucose-induced swelling (Figure 7.16a) and shrinking (Figure 7.16b) behaviors of PNAA microcapsules in response to glucose concentration changes between 0.4 and 3.0 g l^{-1} at 37 °C. At 37 °C, the PNAA microcapsule is initially in a shrunken state (Figure 7.16a1). When the glucose concentration is increased suddenly from 0.4 to 3.0 g l^{-1}, the PNAA microcapsule changes from the shrunken state to a swollen state dramatically (Figure 7.16a2–a4), because more AAPBA moieties complex with glucose. Contrarily, when the glucose concentration is decreased suddenly from 3.0 back to 0.4 g l^{-1}, the PNAA microcapsule returns to the shrunken state again (Figure 7.16b1–b4). In both of the swelling (Figure 7.16a1,a2) and shrinking (Figure 7.16b1,b2) process, the microcapsule exhibits an obvious volume change within $t = 0.75$ min after glucose concentration change, indicating fast glucose-responsive swelling/shrinking behaviors. Moreover, the reversible

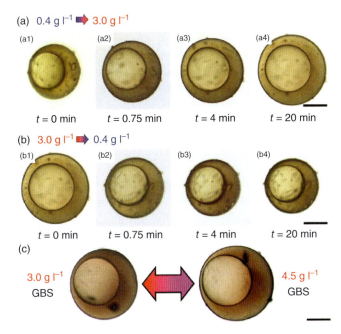

Figure 7.16 Optical micrographs of the reversible glucose-induced swelling/shrinking behaviors of PNAA microcapsules with 2.4 mol% AAc in response to glucose concentration change at 37 °C. (a, b) Reversible swelling/shrinking volume change behaviors from 0.4 to 3.0 g l^{-1} GBS (a) and from 3.0 to 0.4 g l^{-1} GBS (b). (c) Reversible swelling/shrinking volume change of PNAA microcapsules equilibrated in 3.0 and 4.5 g l^{-1} GBS at 37 °C. Scale bars are 100 μm. (Zhang et al. 2013 [7]. Reproduced with permission of The Royal Society of Chemistry.)

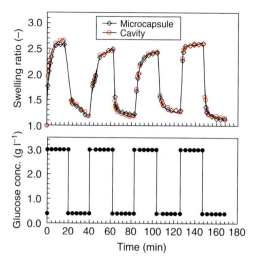

Figure 7.17 The repeated and reversible swelling/shrinking behaviors of PNAA microcapsules with 2.4 mol% AAc in response to repeated changes in glucose concentration between 0.4 and 3.0 g l^{-1} at 37 °C. (Zhang et al. 2013 [7]. Reproduced with permission of The Royal Society of Chemistry.)

volume changes of the PNAA microcapsules that, respectively, equilibrated in 3.0 and 4.5 g l^{-1} at 37 °C (Figure 7.16c) exhibit a more swollen microcapsule in 4.5 g l^{-1} than that in 3.0 g l^{-1}, showing a ∼1.3 times volume change. The reversible volume increase of the PNAA microcapsules from 0 to 3.0 and to 4.5 g l^{-1} GBS indicates a good glucose-responsive volume change behavior in proportion to the glucose concentration. These results show the good and reversible swelling/shrinking response of the PNAA microcapsules to changes in the glucose concentration at physiological temperature.

To investigate the ability of the PNAA microcapsules for repeated glucose response, the glucose-responsive swelling/shrinking behaviors are studied by repeatedly transferring the microcapsules between 0.4 and 3.0 g l^{-1} GBS at 37 °C (Figure 7.17). The sizes of the microcapsules and the hollow cavities increase with suddenly increasing glucose concentration from 0.4 to 3.0 g l^{-1} and decrease with suddenly decreasing glucose concentration from 3.0 to 0.4 g l^{-1}, exhibiting reversible and repeated glucose-induced swelling/shrinking behaviors with ∼2.6 times volume change. The glucose-induced swelling/shrinking behaviors enable the microcapsules as potential self-regulated delivery systems for controlled drug release.

7.4.3 Glucose-Responsive Drug Release Behaviors of Microcapsules

To demonstrate the potential application of the PNAA microcapsules for drug delivery, rhodamine B and FITC-labeled insulin (FITC-insulin) that are used as model molecules and model drugs are, respectively, encapsulated into PNAA microcapsules with 2.4 mol% AAc for glucose-responsive release. Figure 7.18 shows the glucose-responsive dynamic release behavior of rhodamine B from the PNAA microcapsules at 37 °C. In the drug release experiments, the microcapsules are first equilibrated in 0.4 g l^{-1} GBS containing high concentration of rhodamine B for drug loading. After washing with 0.4 g l^{-1} GBS solution, the microcapsules are quickly transferred into 0.4 g l^{-1} GBS without rhodamine B and then moved into the observation field of CLSM for monitoring the drug

Figure 7.18 The glucose-responsive release behaviors of rhodamine B from PNAA microcapsules with 2.4 mol% AAc at 37 °C. (a) Glucose-responsive swelling; (b) glucose-responsive release, in which the fluorescence intensity is measured in the region that covers the direct neighborhood of PNAA microcapsules. (Zhang et al. 2013 [7]. Reproduced with permission of The Royal Society of Chemistry.)

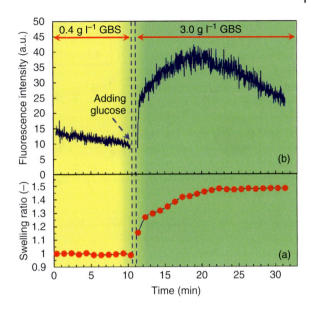

release process. This operation process takes ∼20 s, so the initial drug release behavior that causes the increase of fluorescence intensity at the very beginning is missed. During the first 10 min, the fluorescence intensity that reflects the amount of rhodamine B released from the microcapsule keeps at a low level, indicating a sustained release with slow rate in 0.4 g l^{-1} GBS. The slight decrease of the fluorescence intensity with time is mainly due to the diffusion of the released rhodamine B from the monitoring region into surrounding enviroment; because this diffusion rate is faster than the release rate of rhodamine B across the cross-linked hydrogel shell. Meanwhile, the slightly decreased concentration of rhodamine B inside the microcapsule during the release process also causes a slightly decreased release rate. After suddenly increasing the glucose concentration from 0.4 to 3.0 g l^{-1}, the PNAA microcapsule undergoes a glucose-induced swelling volume change, which leads to a rapid increase of fluorescence intensity, indicating a suddenly enhanced release of rhodamine B with a faster rate. The increased release is attributed to the enlarged space size delineated by the polymer networks in the cross-linked shell caused by the glucose-induced swelling volume change. The enlarged spaces delineated by the polymer networks provide larger channels with less resistance for faster release of rhodamine B molecules. After the PNAA microcapsules reach equilibrium swollen state in 3.0 g l^{-1} GBS, sustained release of rhodamine B with a relatively faster rate than that in 0.4 g l^{-1} GBS is observed. The results show that the glucose-responsive swelling/shrinking behaviors of PNAA microcapsules enable control of shell permeability for glucose-responsive controlled drug release.

Figure 7.19 shows the release behaviors of FITC-insulin from the PNAA microcapsules that are immersed in 0.4 and 3.0 g l^{-1} GBS at 37 °C. For the PNAA microcapsule with more swollen shell placed in 3.0 g l^{-1}, the fluorescence intensity that reflects the amount of FITC-insulin that remained in the microcapsule shows a faster decrease than that in 0.4 g l^{-1} GBS, indicating a faster release

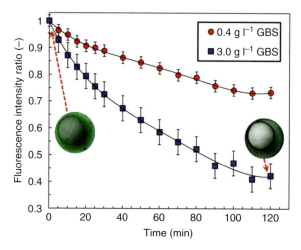

Figure 7.19 The glucose-responsive release behaviors of FITC-insulin from PNAA microcapsules with 2.4 mol% AAc at 37 °C. The fluorescence intensity is measured in the region that covers the PNAA microcapsules. (Zhang et al. 2013 [7]. Reproduced with permission of The Royal Society of Chemistry.)

of FITC-insulin in $3.0\,g\,l^{-1}$. Since the PNAA microcapsules exhibit more swollen state in $4.5\,g\,l^{-1}$ GBS than those in $3.0\,g\,l^{-1}$ GBS, the PNAA microcapsules that enable drug release in response to glucose concentration change from 0.4 to $3.0\,g\,l^{-1}$, can also be used for glucose-responsive drug release in response to glucose concentration change from 0.4 to $4.5\,g\,l^{-1}$. The PNAA microcapsules with glucose-responsive drug release behaviors show great potential as self-regulated delivery systems for diabetes therapy and cancer treatment.

7.5 Microfluidic Fabrication of Monodisperse Multi-Stimuli-Responsive Hollow Microcapsules

Smart microcapsules, which can control the release of their encapsulated contents according to various environmental stimuli, have attracted great interests from various fields in recent years. Because of their relatively faster response rate and other advantages such as small size, large inner volume, huge total surface area, and stable capsule membrane, these environmental stimuli-responsive microcapsules are considered to be the most ideally intelligent drug delivery systems. By encapsulated inside these microcapsules, drugs or chemicals can be released at a desired rate only when and/or where the release is needed [19]. A considerable amount of researches have been carried out on temperature-, pH-, magnetic-, and other stimuli-responsive microcapsules in recent years. However, in many cases, different environmental changes may occur at the same time, thus single stimulus-responsive microcapsules are insufficient for practical applications. Therefore, it is much more favorable that microcapsules possess multiple stimuli-responsive properties simultaneously. More importantly, the patients' conditions are usually complex and diverse. To achieve best effects and reduce side effects of drugs, it is necessary to regulate the release dosage and the release rate in time according to patients' individual differences.

Up to now, the stimuli-responsive microcapsules used as drug delivery systems are mainly developed based on the "on–off" mechanism, which can be divided into two categories. The first one is prepared by grafting stimuli-responsive materials into the pores of microcapsules [1, 15, 19] or by incorporating stimuli-responsive particles into the microcapsule membranes [20]. The microcapsule substrates have no environmental response, and the controlled-release properties only depend on the abrupt deswelling/swelling properties of the graft chains or the embedded particles, which can regulate the transmembrane diffusional permeation of drug molecules. Another one is fabricated by directly utilizing the stimuli-responsive materials as the microcapsule membrane. The release mechanisms of these microcapsules mainly rely on the deswelling/swelling properties of the capsule membrane themselves [7], or the squeezing action from sudden shrinking of the capsule membranes [11], or the decomposition of capsule membranes [18]. Generally, due to the "on–off" state of microcapsules, the drug controlled-release is often in an extreme "on" or "off" state. Their drug release rate cannot be flexibly adjusted according to patients' individual differences; therefore, smart microcapsules with adjustable controlled-release rate are more rational for drug administration. Besides, pH is an important stimulus for stimuli-responsive controlled-release systems, because pH differentiation exists at many specific and pathological sites in human body. Temperature is also an important factor that can be easily controlled, and the site-specific targeting drug delivery can be achieved easily by magnetic targeting. Consequently, the design and preparation of microcapsules with pH-, temperature-, and magnetic-responsive synergistic effects and adjustable controlled-release rate are of great potential in drug delivery.

Here, we introduce the microfluidic fabrication of multi-stimuli-responsive microcapsules with adjustable controlled-release by embedding temperature-responsive submicrospheres as "microvalves" into the magnetic- and pH-responsive microcapsule membrane [4]. The proposed microcapsules can simultaneously achieve targeted delivery by applying an external magnetic field, self-regulated drug release according to pH difference at pathological sites, and adjustable controlled-release relying on temperature regulation. That is, this kind of microcapsule can achieve targeting aggregation at specific pathological sites and effectively adjustable controlled-release according to patients' individual differences, which is of great importance for realizing more rational drug administration.

7.5.1 Microfluidic Fabrication Strategy

The fabrication procedure and the controlled-release mechanism of the multi-stimuli-responsive microcapsules are schematically illustrated in Figure 7.20. The microcapsule is composed of biocompatible and pH-responsive chitosan capsule membrane embedded with magnetic nanoparticles and temperature-responsive submicrospheres, which are respectively prepared by chemical co-precipitation method [21] and precipitation polymerization [22]. For drug delivery systems, microcapsules with narrow size distribution are preferable, since the drug loading levels and release kinetics are directly affected by the size distribution of microcapsules. Microfluidic techniques, which have been developed for generating highly monodisperse emulsions, are used for preparing microcapsules with uniform size. As shown in Figure 7.20a, the chitosan aqueous solution containing magnetic-responsive nanoparticles and

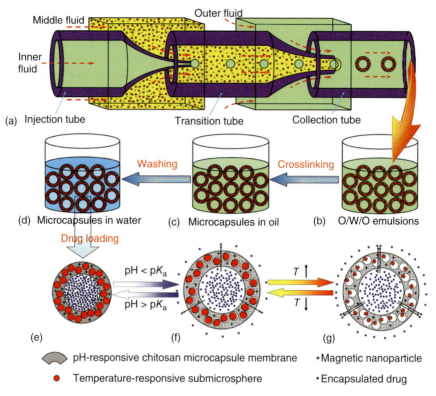

Figure 7.20 Schematic illustration of fabrication process (a–d) and controlled-release mechanism (e–g) of the proposed multi-stimuli-responsive microcapsules with adjustable controlled-release. The capillary microfluidic device (a) is used for generating O/W/O double emulsions (b), and microcapsules are prepared by using these emulsions as templates via cross-linking reaction (c, d). When pH > pK_a, the microcapsule is in a shrunken state; therefore, the release rate is low (e), while a high release rate results due to the swollen state of the microcapsule when pH < pK_a (f). By increasing the environmental temperature, the interspace size between the capsule membrane and submicrospheres becomes larger due to the shrinkage of submicrospheres, so the drug release rate is further increased (g). The pH-/thermo-responsive behaviors are reversible. (Wei et al. 2014 [4]. Reproduced with permission of John Wiley & Sons.)

temperature-responsive submicrospheres is used as the middle fluid (MF), while the oil phase containing cross-linker glutaraldehyde (GA) is used as the inner fluid (IF) and outer fluid (OF). The obtained oil-in-water-in-oil (O/W/O) emulsions (Figure 7.20b) are used as templates for the preparation of multi-stimuli-responsive microcapsules (Figure 7.20c,d).

By embedding magnetic nanoparticles into the microcapsule membrane, site-specific targeting can be achieved under an external magnetic field. Cross-linked chitosan have typical cationic pH-responsive properties. At specific sites where the environmental pH is higher than the pK_a value of chitosan (about 6.2–7.0) [18], such as in the normal tissue (the body physiological pH is about 7.4), the microcapsule membrane is in the compact state due to the pH-responsive shrinking, which results in a low release rate of drugs from the microcapsule (Figure 7.20e). On the other hand, when the environmental pH is lower than the pK_a value of chitosan, such as at some chronic wound sites whose pH values are as low as 5.45, the microcapsule membrane is in the loose state due to the pH-responsive swelling, which leads to a high release rate (Figure 7.20f). That is, the smart microcapsules can realize self-regulated drug release according to pH difference at pathological sites. More importantly, through the addition of temperature-responsive submicrospheres as "microvalves," drug release rate can be effectively adjusted through regulating external temperature via local heating/cooling. The interspace size between the microcapsule membrane and submicrospheres can be flexibly regulated through the temperature-dependent volume change of submicrospheres, which results in adjustable controlled-release properties. When ambient temperature is increased, the interspaces become lager due to the shrinking of submicrospheres, and therefore drug release rate is higher (Figure 7.20g); on the contrary, when the temperature is decreased, the interspaces become smaller due to the swelling of submicrospheres, and then a lower release rate is resulted (Figure 7.20f).

Chitosan microcapsules (CS), chitosan microcapsules with magnetic sensitivity (CS-M), and chitosan microcapsules with both magnetic- and thermosensitivity (CS-M-T) are all prepared using O/W/O double emulsions as cross-linking templates by a microfluidic technique. The capillary microfluidic device is the same as that shown in Figure 7.1. The compositions of different fluids for preparation of microcapsules are listed in Table 7.2. The generated O/W/O emulsions

Table 7.2 The recipes of different fluids for fabricating multi-stimuli-responsive microcapsules.

Code	IF (inner fluid)	MF (middle fluid)	OF (outer fluid)
1#	Oil($V_{\text{GA-S-BB}} : V_{\text{SO}} = 1:20$)[a)] + 3%w/v PGPR[b)]	Pure water + 4%w/v CS + 0.25%w/v HEC + 0.5%w/v F127	Oil($V_{\text{GA-S-BB}} : V_{\text{SO}} = 1:50$) + 8%w/v PGPR
2#	Oil($V_{\text{GA-S-BB}} : V_{\text{SO}} = 1:20$) + 3%w/v PGPR	Pure water + 4%w/v CS + 0.25%w/v HEC + 0.5%w/v F127	Oil($V_{\text{GA-S-BB}} : V_{\text{SO}} = 1:75$) + 8%w/v PGPR
3#	Oil($V_{\text{GA-S-BB}} : V_{\text{SO}} = 1:50$) + 3%w/v PGPR	Pure water + 4%w/v CS + 0.25%w/v HEC + 0.5%w/v F127	Oil($V_{\text{GA-S-BB}} : V_{\text{SO}} = 1:50$) + 8%w/v PGPR
4#	Oil($V_{\text{GA-S-BB}} : V_{\text{SO}} = 1:50$) + 3%w/v PGPR	Pure water + 4%w/v CS + 0.5%w/v F127	Oil($V_{\text{GA-S-BB}} : V_{\text{SO}} = 1:50$) + 8%w/v PGPR
5#	Oil($V_{\text{GA-S-BB}} : V_{\text{SO}} = 1:50$) + 3%w/v PGPR	Pure water + 4%w/v CS + 0.5%w/v F127 + 0.15%w/v magnetic nanoparticles	Oil($V_{\text{GA-S-BB}} : V_{\text{SO}} = 1:50$) + 8%w/v PGPR
6#	Oil($V_{\text{GA-S-BB}} : V_{\text{SO}} = 1:50$) + 3%w/v PGPR	Pure water + 4%w/v CS + 0.5%w/v F127 + 0.15%w/v magnetic nanoparticles + 1%w/v submicrospheres	Oil($V_{\text{GA-S-BB}} : V_{\text{SO}} = 1:50$) + 8%w/v PGPR

a) Note: "$V_{\text{GA-S-BB}}$" represents the volume of glutaraldehyde-saturated benzyl benzoate; "V_{SO}" represents the volume of soybean oil.
b) Abbreviations: PGPR: polyglycerol polyricinoleate; CS: chitosan; HEC: hydroxyethyl cellulose; F127: Pluronic F127.

are collected in a container, and then left to stand for 24 h to make sure the chitosan in the water phase is completely cross-linked. The obtained microcapsules are washed using a mixture of ethyl acetate and isopropanol to remove the inner and outer oil solutions, and finally the microcapsules are dispersed into water for further use.

The morphology of the magnetic nanoparticles is confirmed by transmission electron microscopy (TEM). As shown in Figure 7.21a, the size of prepared magnetic nanoparticles is around 20 nm, and there is no obvious aggregation, which indicates that the nanoparticles are well dispersed in water. Magnetic property of the nanoparticles is measured with a vibrating sample magnetometer (VSM). As shown in Figure 7.21b, the saturation magnetization (M_s) of the magnetic nanoparticles is 72.32 emu g^{-1}, and hysteresis and coercivity are almost undetectable, which suggests that the superparamagnetic property of the prepared nanoparticles is satisfactory. The temperature-responsive

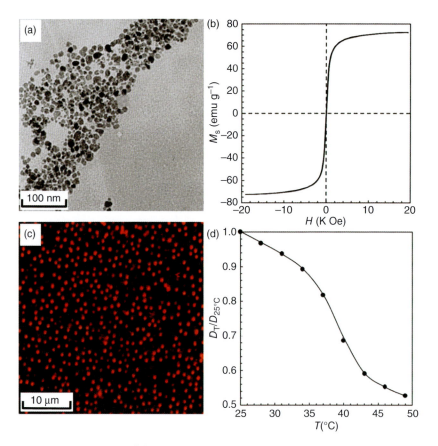

Figure 7.21 TEM image of the prepared magnetic nanoparticles (a), magnetization curve of magnetic nanoparticles at room temperature (b), CLSM image of the temperature-responsive submicrospheres in water at room temperature on red fluorescent channel (c), and temperature-responsive property of the hydrodynamic diameters of submicrospheres (d). (Wei et al. 2014 [4]. Reproduced with permission of John Wiley & Sons.)

submicrospheres are labeled with red fluorescence dye to make them easy to be observed. Moreover, in order to make the temperature-responsive submicrospheres have obvious volume-phase transition near human body temperature (37 °C), acrylamide (AAm) is used as a hydrophilic co-monomer to modulate the thermosensitivity of the submicrospheres. Figure 7.21c shows the CLSM image of obtained poly(*N*-isopropylacrylamide-*co*-acrylamide) (P(NIPAM-*co*-AAm)) submicrospheres in water at room temperature on the red fluorescent channel excited at 543 nm. The submicrospheres show red fluorescence and exhibit good monodispersity, and are well dispersed in water. The equilibrium deswelling ratio ($D_T/D_{25°C}$) of P(NIPAM-*co*-AAm) submicrospheres as a function of temperature is shown in Figure 7.21d. The results show that these temperature-responsive submicrospheres deswell gradually with increasing temperature, and the equilibrium deswelling ratio shows significant change near body temperature (37 °C). This kind of temperature-dependent volume-phase transition provides feasibility for subsequent adjustable controlled-release near body temperature.

The optical and fluorescent microscope images of O/W/O double emulsions generated by the microfluidic method are shown in Figure 7.22. The emulsions, which act as templates for fabricating microcapsules via cross-linking reaction, are highly monodisperse and quite stable (Figure 7.22a′–c′). Because only temperature-responsive submicrospheres are labeled with red fluorescence,

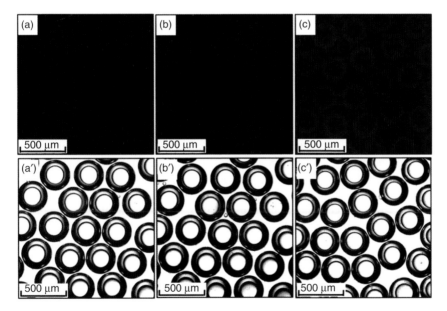

Figure 7.22 CLSM images of different O/W/O double emulsion templates on red fluorescent channel (a–c); and optical microscope images of different emulsion templates (a′–c′). (a) and (a′) represent emulsions prepared with only chitosan in MF (4#), (b) and (b′) represent emulsions prepared with both chitosan and magnetic nanoparticles in MF (5#), (c) and (c′) represent emulsions prepared with chitosan, magnetic nanoparticles, and temperature-responsive submicrospheres in MF (6#). Scale bars are 500 μm. (Wei *et al*. 2014 [4]. Reproduced with permission of John Wiley & Sons.)

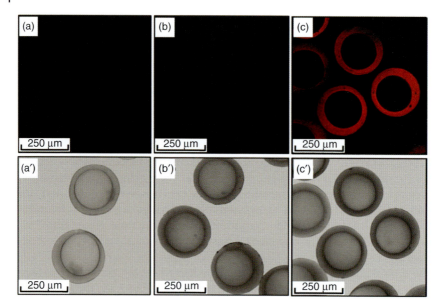

Figure 7.23 CLSM images of different microcapsules on red fluorescent channel (a–c); and optical microscope images of different microcapsules (a′–c′). (a) and (a′) represent CS microcapsules, (b) and (b′) represent CS-M microcapsules, (c) and (c′) represent CS-M-T microcapsules. The microcapsules are dispersed in buffer solution of pH 7.4 at 37 °C. Scale bars are 250 μm. (Wei et al. 2014 [4]. Reproduced with permission of John Wiley & Sons.)

emulsions with only chitosan in middle water phase (Figure 7.22a) or with both chitosan and magnetic nanoparticles in middle water phase (Figure 7.22b) exhibit no fluorescence, while emulsions containing chitosan, magnetic nanoparticles, and submicrospheres (Figure 7.22c) show clearly red fluorescence.

Figure 7.23 shows CLSM images and optical microscope images of different microcapsules cross-linked from different double emulsion templates. The microcapsules are dispersed in buffer solution of pH 7.4 at 37 °C, which simulate the body physiological pH and temperature. Under the same condition, only chitosan microcapsules containing both magnetic nanoparticles and temperature-responsive submicrospheres (CS-M-T, Figure 7.23c) show red fluorescent, while the chitosan microcapsules (CS, Figure 7.23a) and the magnetic nanoparticles embedded chitosan microcapsules (CS-M, Figure 7.23b) show no fluorescence. The red fluorescence from the capsule membranes of CS-M-T microcapsules indicates the successful embedding of temperature-responsive submicrospheres into the microcapsules. The optical microscope images of different microcapsules (Figure 7.23a′–c′) look almost the same, which all show that the obtained microcapsules are with good sphericity and monodispersity.

7.5.2 Stimuli-Responsive Behaviors of Microcapsules

Microcapsules with pH sensitivity are studied widely, because pH difference exists at many physiological, biological, and/or chemical systems. Chronic wounds have been reported to have pH values between 8.65 and 5.45, and cancer tissue is also reported to be acidic extracellularly. The cationic pH-responsive microcapsules, which have acid-responsive swelling property due to protonation, are suitable for rate-controlled release and sustained drug release in acidic conditions via self-regulated adjustment of molecular diffusion permeation. Chitosan is a well-known cationic polysaccharide with excellent biological activity, good biocompatibility, and biodegradability, so microcapsules based on chitosan are attracting more and more interests in various applications. Microcapsules made from cross-linked chitosan exhibit typical cationic pH-responsive characteristics. When the environmental pH is lower than the pK_a value of chitosan, the microcapsules swell due to the protonation of amino groups, and when the pH is higher than its pK_a value, the deprotonation of amino groups results in the deswelling of the microcapsules. Swelling ratios ($OD_{pH}/OD_{pH7.4}$) of microcapsule outer diameter at certain pH to that at body physiological pH 7.4 are used to characterize the acid-induced swelling change of these prepared microcapsules.

The influences of the preparation conditions on the pH sensitivities of microcapsules are investigated, and the recipes of different fluids are listed in Table 7.2. As shown in Figure 7.24, all microcapsules exhibit good pH sensitivity,

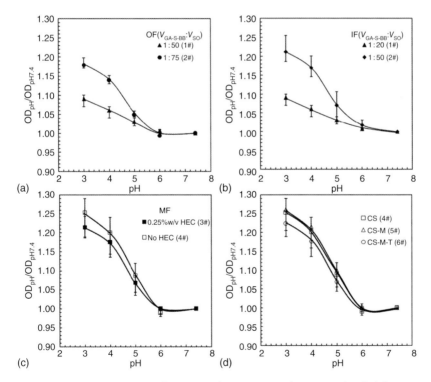

Figure 7.24 pH-responsive swelling ratios of CS microcapsules prepared with different concentrations of GA in OF (a), with different concentrations of GA in IF (b), with and without HEC addition in MF (c), and pH-responsive swelling ratios of CS, CS-M, and CS-M-T microcapsules (d). $T = 37\,°C$. (Wei et al. 2014 [4]. Reproduced with permission of John Wiley & Sons.)

and the swelling ratios increase with the decrease in external pH values. From Figure 7.24a it can be seen that, with the same IF and MF, the chitosan microcapsules prepared with different GA concentrations in OF show different swelling extents. The chitosan microcapsules prepared with lower GA concentration exhibit higher swelling ratio compared with that prepared with higher GA concentration. The similar results are also observed from microcapsules prepared with different GA concentrations in IF, as shown in Figure 7.24b. GA is the cross-linker for chitosan microcapsules. With the decrease of GA concentration, the cross-linking density of microcapsules decreases, which results in the increase in swelling ratios of microcapsules. Microcapsules prepared with too low cross-linking density are easy to be broken, so GA concentrations of IF and OF are chosen to be 1/50 (v/v).

For the preparation of chitosan microcapsules, hydroxyethyl cellulose (HEC) is added into MF to increase fluid viscosity. The chitosan microcapsules prepared with HEC addition are much stable than those prepared without HEC addition. The pH-responsive behaviors of microcapsules prepared with and without addition of HEC are also studied. Microcapsules prepared without HEC addition show slightly larger swelling ratio than that prepared with HEC addition at a same pH condition (Figure 7.24c). The reason is that HEC is a macromolecular compound, which has long polymer chains. The entanglements between HEC and chitosan molecular chains restrict the swelling of microcapsules, resulting in lower swelling ratio of microcapsules prepared with the addition of HEC. Considering the addition of magnetic nanoparticles and submicrospheres could also increase the viscosity of MF, and to make particles well dispersed in MF, the chitosan solution without addition of HEC is used in the subsequent studies.

By adding magnetic nanoparticles into MF, chitosan microcapsules with magnetic sensitivity (CS-M) are obtained. With the addition of magnetic nanoparticles and thermosensitive submicrospheres into MF, the generated chitosan microcapsules possess both magnetic and thermosensitivity (CS-M-T). The pH sensitivities of these microcapsules are also investigated, as shown in Figure 7.24d. The addition of magnetic nanoparticles has nearly no influence on the pH sensitivity of chitosan microcapsules, but the addition of submicrospheres has slight influence on the pH sensitivity of microcapsules. The magnetic nanoparticles are very small (about 20 nm) and the amount of nanoparticles added into the microcapsules is relatively low (0.15%w/v), so the pH sensitivity of microcapsules is nearly not influenced by the addition of magnetic nanoparticles. However, the temperature-responsive submicrospheres are relatively large (about 1 μm in water at room temperature) and the content of submicrospheres in the microcapsules is relatively high (1%w/v). So, CS-M-T microcapsules exhibit slightly lower pH-responsive swelling ratios than CS and CS-M microcapsules at the same condition.

Drug release at specific pathological sites can effectively reduce the side effects. Incorporation of superparamagnetic nanoparticles into the chitosan capsule membrane enables the microcapsules to realize magnetic-guided targeting delivery. The CS-M (Figure 7.25a,a′,a″) and CS-M-T (Figure 7.25b,b′,b″) microcapsules both exhibit satisfactory magnetic properties. Without the

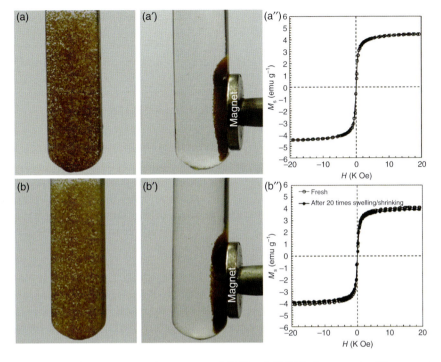

Figure 7.25 Magnetic-responsive property of CS-M and CS-M-T microcapsules. Microcapsules in buffer solution of pH 7.4 at room temperature without magnet (a, b) and with an external magnet (a', b'), and magnetization curves of microcapsules at room temperature (a", b"). (a), (a'), and (a") represent CS-M microcapsules, and (b), (b'), and (b") represent CS-M-T microcapsules. (Wei et al. 2014 [4]. Reproduced with permission of John Wiley & Sons.)

external magnetic field, the microcapsules are randomly dispersed in water (Figure 7.25a,b), while after applying the external magnetic field, the microcapsules are aggregated at the sites where the magnet is placed (Figure 7.25a',b'). A VSM is also used to measure the magnetic properties of CS-M and CS-M-T microcapsules, and their magnetization curves are displayed in Figure 7.25a",b". Their hysteresis and coercivity are almost undetectable, which suggests that the CS-M and CS-M-T microcapsules also have superparamagnetic properties. The M_s of CS-M and CS-M-T microcapsules are 4.47 emu g^{-1} and 4.11 emu g^{-1}, respectively, which are much lower than that of pure magnetic nanoparticles (72.32 emu g^{-1}). This is attributed to the low content of nanoparticles in the microcapsule membrane. To confirm the nanoparticles can be stably kept inside the capsule membranes, the magnetic properties of nanoparticle-embedded microcapsules before and after repeatedly swelling/shrinking for 20 times are compared. The results show that M_s values of CS-M-T microcapsules before and after repeatedly swelling/shrinking for 20 times are almost the same (Figure 7.25b"), which indicates that the nanoparticles embedded in the capsule membranes do not diffuse out even when the membranes undergo repeated swelling. Before being added in the chitosan solution, the nanoparticles are negatively charged because they have been modified with trisodium citrate.

While the chitosan is positively charged at acidic conditions (pH < pK_a of chitosan). Although the capsule membranes are swollen in the acidic conditions, the nanoparticles can be kept stably inside the membrane due to electrostatic attraction between the negatively charged nanoparticles and the positively charged chitosan networks.

7.5.3 Controlled-Release Characteristics of Multi-Stimuli-Responsive Microcapsules

The pH-responsive controlled-release behaviors of vitamin B12 (VB12) as model drug from CS-M-T microcapsules are shown in Figure 7.26. The permeability coefficient of VB12 molecules (P_{VB12}) from microcapsules exhibits obvious pH-dependent characteristics. When the environmental pH is lower than the body physiological value (pH 7.4), the value of P_{VB12} increases with the decreasing pH value in external circumstances. The capsule membrane is in the loose state in acidic conditions due to the microcapsule swelling, resulting in a high release rate. Another parameter defined as $P_{pH}/P_{pH7.4}$ is also introduced to characterize the degree of permeability change compared to that at the body physiological pH. It can be seen that the $P_{pH}/P_{pH7.4}$ value also increases with decreasing pH value. As expected, the pH-responsive controlled-release behaviors of the CS-M-T microcapsules are consistent with their pH-dependent swelling results.

The adjustable drug controlled-release of microcapsules is achieved by regulating the interspace size between the microcapsule membrane and submicrospheres, while the interspace size is determined by the size change of temperature-responsive submicrospheres embedded in the capsule membrane. To estimate the adjustable controlled-release behaviors induced by temperature change, the pH value of external circumstances should be fixed. Very low colonic pH values have been found in some severe active ulcerative colitis (the lowest pH values are 2.3–3.4) and some chronic wounds have been reported to have pH value as low as 5.45 [23]. To reflect the adjustable controlled-release under pathologically acidic conditions, two pH values (pH 3 and pH 5) are chosen.

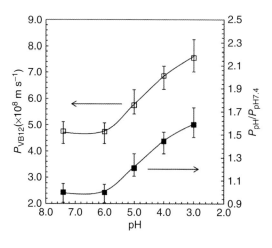

Figure 7.26 pH-responsive release of model drug VB12 from CS-M-T microcapsules at 37 °C. (Wei et al. 2014 [4]. Reproduced with permission of John Wiley & Sons.)

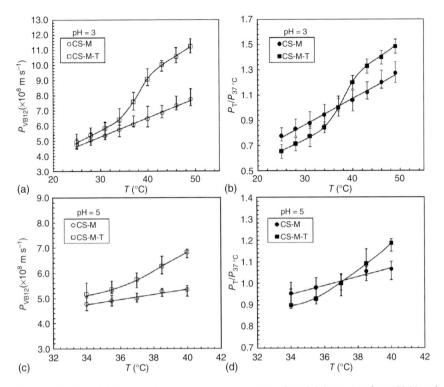

Figure 7.27 Adjustable controlled-release characteristics of model drug VB12 from CS-M and CS-M-T microcapsules in buffer solutions with different pH values and at different temperatures. (Wei *et al.* 2014 [4]. Reproduced with permission of John Wiley & Sons.)

The adjustable controlled-release of VB12 from CS-M-T microcapsules at pH 3 in response to temperature is exhibited in Figure 7.27a,b. The CS-M microcapsules are used as the control group. It is desirable that the P_{VB12} value of CS-M-T microcapsules increases with the increase of temperature from 25 to 49 °C (Figure 7.27a). In particular, the P_{VB12} value of CS-M-T microcapsules increases significantly in the temperature range of 34–43 °C, while it increases relatively slowly in the temperature ranges of 25–34 °C and 43–49 °C. The temperature-dependent P_{VB12} change of CS-M-T microcapsules corresponds well with the temperature sensitivity of the submicrospheres shown in Figure 7.21d. The temperature-responsive submicrospheres are used as the "microvalves" to adjust the release rate of microcapsules. Interspace size between the capsule membrane and submicrospheres, which determines the VB12 release rate from the microcapsules, can be adjusted by temperature-dependent volume change of submicrospheres. Therefore, the changing trend of VB12 permeability coefficients is consistent with the temperature-responsive volume change of submicrospheres. When the temperature is increased, the submicrospheres embedded into the capsule membrane shrink, which causes the interspaces between the capsule membrane and submicrospheres for VB12 passing through become larger. As a result, the value of P_{VB12} becomes higher with increasing temperature. When the temperature is decreased, the interspaces become smaller due to the swelling of submicrospheres, thereby resulting in a lower value of P_{VB12}. In contrast, the permeability coefficients of VB12 from CS-M microcapsules change slightly in the whole temperature range (25–49 °C), and there is no obvious change in the range of 34–43 °C under the same condition. This slight change of P_{VB12} is due to the accelerated diffusivity of VB12 with increasing temperature. From practical application point of view, the adjustable controlled-release should be operated near body temperature (37 °C). A parameter defined as $P_T/P_{37°C}$ is introduced to evaluate the adjustable controlled-release characteristics with respect to VB12 release near body temperature, in which P_T is the permeability coefficient of VB12 from microcapsules at certain temperature T, while $P_{37°C}$ is the permeability coefficient of VB12 at 37 °C. From Figure 7.27b, it can be seen that the CS-M-T microcapsules can achieve a better and more effective adjustment of drug release in the temperature range near 37 °C than CS-M microcapsules through controlling the temperature due to the embedded temperature-responsive submicrospheres. The adjustable release characteristics of VB12 at pH 5 are also studied, as shown in Figure 7.27c,d. Just like the release behaviors at pH 3, the permeability coefficient of VB12 from CS-M-T microcapsules increases significantly with the increase of temperature in the range of 34–40 °C, while the P_{VB12} of CS-M microcapsules changes slightly under the same circumstances. $P_T/P_{37°C}$, as a function of temperature, also reflects the effective adjustment of drug controlled-release for CS-M-T microcapsules at pH 5 in the temperature range near 37 °C. That is, the CS-M-T microcapsules can also achieve an effective adjustable drug delivery than CS-M microcapsules at pH 5.

The temperature-controlled-release behaviors of microcapsules at normal body physiological pH (about 7.4) are shown in Figure 7.28. At pH 7.4, which is higher than the pK_a value of chitosan, the chitosan membranes are in the compacted state. FITC-dextran is selected as the model drug. The permeability of the capsule membrane can also be modulated by changing the temperature even when chitosan is in the compacted state at pH 7.4. The release rate of FITC-dextran from CS-M-T microcapsules also increases significantly with the

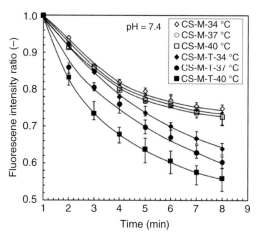

Figure 7.28 Release behaviors of FITC-dextran from CS-M and CS-M-T microcapsules in buffer solution of pH 7.4 at different temperatures. The fluorescence intensity is measured in the region that covers the microcapsules. (Wei et al. 2014 [4]. Reproduced with permission of John Wiley & Sons.)

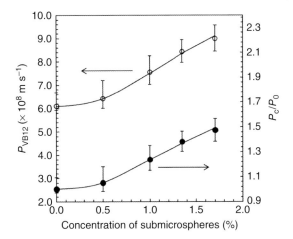

Figure 7.29 Controlled-release characteristics of model drug VB12 from CS-M-T microcapsules prepared with different concentrations of temperature-responsive submicrospheres in buffer solution of pH 3 at 37 °C. (Wei et al. 2014 [4]. Reproduced with permission of John Wiley & Sons.)

increase of temperature. In contrast, there is no obvious change in the release rate of FITC-dextran from CS-M microcapsules when the temperature is varied in the same range. That is, even when chitosan membranes are in the compacted state at pH 7.4, the embedded temperature-responsive submicrospheres are still efficient as "microvalves."

The influences of the added amount of submicrospheres into the capsule membrane on the controlled-release behaviors are shown in Figure 7.29. The controlled-release characteristics of VB12 from CS-M-T microcapsules containing different contents of submicrospheres at temperature of 37 °C and pH of 3 are investigated. With increasing the added amount of submicrospheres, the value of P_{VB12} increases obviously. The number of interspaces between the capsule membrane and submicrospheres is directly determined by the added amount of submicrospheres. Consequently, with the increase of submicrospheres amount, the value of P_{VB12} increases due to the increase of the number of interspaces for VB12 passing through. When the added amount of submicrospheres is low, the submicrospheres are well dispersed in the capsule membrane and thus they do not form percolated networks. However, when a large amount of submicrospheres is added in the capsule membrane, the submicrospheres may be arrayed closely one by one although they are well dispersed, so that some connected porous networks may be formed through the shrinkage of submicrospheres at 37 °C. Therefore, when a large amount of submicrospheres is added, the permeability of CS-M-T microcapsules at 37 °C increases a lot. A parameter defined as P_C/P_0 is also used to reflect the degree of influence of the addition amount of submicrospheres on the VB12 controlled-release, in which P_C is the permeability coefficient of VB12 from CS-M-T microcapsules prepared with certain concentration C of submicrospheres and P_0 is that from chitosan microcapsules prepared without submicrospheres. With an increase in the amount of embedded submicrospheres in the capsule membrane, the degree of influence on the VB12 controlled-release also increases.

7.6 Summary

In this chapter, microfluidic strategies for controllable fabrication of monodisperse hollow microcapsules are introduced. The strategies are demonstrated by fabricating monodisperse EC hollow microcapsules with a simple method that combines microfluidic double emulsification and solvent diffusion, monodisperse hollow calcium alginate microcapsules via internal gelation in microfluidic-generated double emulsions, monodisperse glucose-responsive microcapsules via polymerization in microfluidic-generated double emulsions, and monodisperse multi-stimuli-responsive microcapsules via combining cross-linked chitosan acting as pH-responsive capsule membrane, embedded magnetic nanoparticles realizing "site-specific targeting" and embedded temperature-responsive submicrospheres serving as "microvalves" in microfluidic-generated double emulsions. EC microcapsules with different size and morphology are prepared by using different inner and outer aqueous fluids. With the internal gelation strategy, the shell thickness and the internal structure of alginate hollow microcapsules can be efficiently and precisely controlled with the virtue of microfluidics for generating controllable multiple emulsions as templates. The PNAA hollow microcapsules with glucose-responsive P(NIPAM-*co*-AAPBA-*co*-AAc) shell exhibit reversible and repeated swelling/shrinking responses to glucose concentration changes within the physiological blood glucose concentration range at 37 °C. The multi-stimuli-responsive hollow microcapsules are promising to achieve a more rational drug delivery and controlled-release according to patients' individual differences. In a word, the microfluidic approach provides new opportunities for design and fabrication of novel functional hollow microcapsules for myriad applications, including drug, chemical or nutrient encapsulation and delivery, cell or enzyme encapsulation and immobilization, and so on.

References

1. Chu, L.Y., Park, S.H., Yamaguchi, T., and Nakao, S. (2002) Preparation of micron-sized monodispersed thermo-responsive core-shell microcapsules. *Langmuir*, **18**, 1856–1864.
2. Shiga, K., Muramatsu, N., and Kondo, T. (1996) Preparation of poly(D,L-lactide) and copoly(lactide-glycolide) microspheres of uniform size. *J. Pharm. Pharmacol.*, **48**, 891–895.
3. Chu, L.Y., Kim, J.W., Shah, R.K., and Weitz, D.A. (2007) Monodisperse thermo-responsive microgels with tunable volume-phase transition kinetics. *Adv. Funct. Mater.*, **17**, 3499–3504.
4. Wei, J., Ju, X.J., Zou, X.Y., Xie, R., Wang, W., Liu, Y.M., and Chu, L.Y. (2014) Multi-stimuli-responsive microcapsules for adjustable controlled-release. *Adv. Funct. Mater.*, **24**, 3312–3323.
5. Liu, L., Yang, J.P., Ju, X.J., Xie, R., Yang, L., Liang, B., and Chu, L.Y. (2009) Microfluidic preparation of monodisperse ethyl cellulose hollow microcapsules with non-toxic solvent. *J. Colloid Interface Sci.*, **336**, 100–106.

6 Liu, L., Wu, F., Ju, X.J., Xie, R., Wang, W., Niu, C.H., and Chu, L.Y. (2013) Preparation of monodisperse calcium alginate microcapsules via internal gelation in microfluidic-generated double emulsions. *J. Colloid Interface Sci.*, **404**, 85–90.

7 Zhang, M.J., Wang, W., Xie, R., Ju, X.J., Liu, L., Gu, Y.Y., and Chu, L.Y. (2013) Microfluidic fabrication of monodisperse microcapsules for glucose-response at physiological temperature. *Soft Matter*, **9**, 4150–4159.

8 Soppimath, K.S. and Aminabhavi, T.M. (2002) Ethyl acetate as a dispersing solvent in the production of poly(DL-lactide-*co*-glycolide) microspheres: effect of process parameters and polymer type. *J. Microencapsulation*, **19**, 281–292.

9 Yang, C.Y., Tsay, S.Y., and Tsiang, R.C.C. (2000) An enhanced process for encapsulating aspirin in ethyl cellulose microcapsules by solvent evaporation in an O/W emulsion. *J. Microencapsulation*, **17**, 269–277.

10 Lin, Y.H., Liang, H.F., Chung, C.K., Chen, M.C., and Sung, H.W. (2005) Physically crosslinked alginate/N,O-carboxymethyl chitosan hydrogels with calcium for oral delivery of protein drugs. *Biomaterials*, **26**, 2105–2113.

11 Chu, L.Y., Utada, A.S., Shah, R.K., Kim, J.W., and Weitz, D.A. (2007) Controllable monodisperse multiple emulsions. *Angew. Chem. Int. Ed.*, **46**, 8970–8974.

12 Ganapathy, V., Thangaraju, M., and Prasad, P.D. (2009) Nutrient transporters in cancer: relevance to Warburg hypothesis and beyond. *Pharmacol. Ther.*, **121**, 29–40.

13 Manna, U. and Patil, S. (2010) Glucose-triggered drug delivery from borate mediated layer-by-layer self-assembly. *ACS Appl. Mater. Interfaces*, **2**, 1521–1527.

14 Kataoka, K., Miyazaki, H., Bunya, M., Okano, T., and Sakurai, Y. (1998) Totally synthetic polymer gels responding to external glucose concentration: their preparation and application to on-off regulation of insulin release. *J. Am. Chem. Soc.*, **120**, 12694–12695.

15 Chu, L.Y., Liang, Y.J., Chen, W.M., Ju, X.J., and Wang, H.D. (2004) Preparation of glucose-sensitive microcapsules with a porous membrane and functional gates. *Colloids Surf. B*, **37**, 9–14.

16 Hoare, T. and Pelton, R. (2007) Engineering glucose swelling responses in poly(N-isopropylacrylamide)-based microgels. *Macromolecules*, **40**, 670–678.

17 Lapeyre, V., Gosse, I., Chevreux, S., and Ravaine, V. (2006) Monodispersed glucose-responsive microgels operating at physiological salinity. *Biomacromolecules*, **7**, 3356–3363.

18 Liu, L., Yang, J.P., Ju, X.J., Xie, R., Liu, Y.M., Wang, W., Zhang, J.J., Niu, C.H., and Chu, L.Y. (2011) Monodisperse core-shell chitosan microcapsules for pH-responsive burst release of hydrophobic drugs. *Soft Matter*, **7**, 4821–4827.

19 Chu, L.Y., Yamaguchi, T., and Nakao, S. (2002) A molecular recognition microcapsule for environmental stimuli-responsive controlled-release. *Adv. Mater.*, **14**, 386–389.

20 Yu, Y.L., Zhang, M.J., Xie, R., Ju, X.J., Wang, J.Y., Pi, S.W., and Chu, L.Y. (2012) Thermo-responsive monodisperse core-shell microspheres with PNIPAM core and biocompatible porous ethyl cellulose shell embedded with PNIPAM gates. *J. Colloid Interface Sci.*, **376**, 97–106.

21 Wang, W., Liu, L., Ju, X.J., Zerrouki, D., Xie, R., Yang, L., and Chu, L.Y. (2009) A novel thermo-induced self-bursting microcapsule with magnetic-targeting property. *ChemPhysChem*, **10**, 2405–2409.
22 Pelton, R. (2000) Temperature-sensitive aqueous microgels. *Adv. Colloid Interface Sci.*, **85**, 1–33.
23 Fallingborg, J., Christensen, L.A., Jacobsen, B.A., and Rasmussen, S.N. (1993) Very low intraluminal colonic pH in patients with active ulcerative colitis. *Dig. Dis. Sci.*, **38**, 1989–1993.

8

Microfluidic Fabrication of Monodisperse Core–Shell Microcapsules

8.1 Introduction

Considerable efforts have been devoted to the design and fabrication of multifunctional microcapsules due to their potential applications in numerous fields, such as controlled release [1, 2], actives protection [3, 4], and confined microreaction [5, 6]. Due to the large internal space, core–shell microcapsules can provide their core compartment for encapsulation with enhanced loading efficiency, as compared with homogeneous microspheres. Typically, traditional methods for fabrication of core–shell microparticles usually require harsh conditions to remove the sacrificial core materials, which are usually solid organic or inorganic particles. Meanwhile, it still remains challenging for the traditional methods to flexibly and precisely engineer the structures of the shell and the core. With nested droplet-in-droplet structures, double emulsions provide ideal templates for creating microparticles with core–shell structures [7–11]. The inner and outer droplets in the double emulsions can be used as templates to construct the core compartment and the outer shell of the microparticles, respectively. With excellent control over microdroplets [12], microfluidic technique allows accurate manipulation of the sizes of the inner and outer droplets to finely tune the size of the core compartment and the shell. Moreover, the formation process of droplet with microfluidics enables elaborate incorporation of various functional materials in both the inner and outer droplets to flexibly and individually tailor the structures and functions of the core and the shell [13]. This microfluidic strategy allows controllable fabrication of core–shell microparticles with diverse shells and cores for broad encapsulation and delivery applications.

Recently, the authors' group developed core–shell microparticles with an oil core and a shell consisting of various stimuli-responsive hydrogels for controllable encapsulation and triggered release. In this chapter, the microfluidic fabrication of core–shell microcapsules with a shell consisting of thermo-responsive, alcohol-responsive, ion-responsive, and pH-responsive hydrogels and their performances for encapsulation and triggered release are introduced.

8.2 Microfluidic Strategy for Fabrication of Monodisperse Core–Shell Microcapsules

Typically, microcapsules with core–shell structures can be produced using the inner droplet of double emulsions as the inner core compartment and the outer drop as the outer shell. As compared with the hollow microcapsules with water-filled inner compartments introduced in Chapter 7, oil-filled compartments inside the core–shell microcapsules can provide preferable interiors for encapsulating oil-soluble components. Generally, core–shell microcapsules with oil-filled compartments can be produced by converting the monomer-containing middle aqueous layer of O/W/O double emulsions into a polymeric shell. The detailed strategies for fabrication of core–shell microcapsules with different functional materials are introduced in the following sections.

8.3 Smart Core–Shell Microcapsules for Thermo-Triggered Burst Release

8.3.1 Fabrication of Core–Shell Microcapsules for Thermo-Triggered Burst Release of Oil-Soluble Substances

The concept of the core–shell microcapsules for thermo-triggered burst release of oil-soluble substances is schematically illustrated in Figure 8.1. Each of the microcapsules possess a core/shell structure, with an oil core and a thermo-responsive shell composed of poly(N-isopropylacrylamide) (PNIPAM) and homogeneously embedded superparamagnetic Fe_3O_4 nanoparticles. The oil core can be used to encapsulate oil-soluble substances. The PNIPAM shell with embedded Fe_3O_4 nanoparticles combines the magnetic- and thermo-responsive properties. The magnetic-responsive property enables the transport of the microcapsules to a desired site under external magnetic guide. The thermo-responsive PNIPAM shell undergoes hydrophilic-swelling/hydrophobic-shrinking phase changes when temperature changes above its volume-phase transition temperature (VPTT); thus, when increasing the temperature above the VPTT, the PNIPAM shell shrinks rapidly and finally squeezes the oil core out together with the oil-soluble substances for burst release. As a result, the microcapsules can be delivered to the desired site via magnetic guide first and then burst-release their encapsulated oil-soluble substances by a thermo-trigger. To construct such

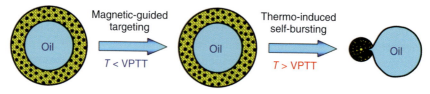

Figure 8.1 Schematic illustration of the concept of the core–shell microcapsules for thermo-triggered burst release of oil-soluble substances. (Wang et al. 2009 [11]. Reproduced with permission of John Wiley & Sons.)

microcapsules, typically, 0.452 g monomer N-isopropylacrylamide (NIPAM), 0.0308 g cross-linker N,N′-methylene-bis-acrylamide (MBA), 0.024 g initiator ammonium persulfate (APS), and 0.04 g surfactant Pluronic F127 were added into an aqueous solution composed of 3 ml ferrofluid and 1 ml deionized water, and used as the middle fluid in the microfluidic device shown in Figure 2.5a. Meanwhile, soybean oil and soybean oil with 5% (w/v) surfactant polyglycerol polyricinoleate (PGPR 90) were, respectively, used as the inner and outer fluids. Monodisperse O/W/O double emulsions generated from the microfluidic device were collected in collection solution, which was soybean oil with 5% (w/v) surfactant PGPR 90 and 0.2% (w/v) photoinitiator 2,2-dimethoxy-2-phenylacetophenone (BDK). Then, the collected double emulsions were converted into microcapsules by exposure to UV irradiation for 8 min and further reaction for 1 h at temperature below the VPTT. Under UV light, the activated BDK diffuses to the interface between the outer oil phase and middle aqueous phase, where it initiates the copolymerization of NIPAM and MBA in the middle aqueous phase to build the hydrogel shell.

Figure 8.2a shows the Fe_3O_4 nanoparticles with diameter of 10 nm, which are used in the middle fluid to provide magnetic-responsive property to the shell. Figure 8.2b,c shows monodisperse O/W/O double emulsions and the resultant core–shell microcapsules in water at 20 °C, respectively. The microcapsules are featured with oil-core/hydrogel-shell structure, with their size similar to those of the O/W/O double emulsions. The PNIPAM shells of the microcapsules exhibit light brown color because of the homogeneously embedded Fe_3O_4 nanoparticles in their networks. Sudan III, an oil-soluble industrial dye, is used as a model chemical to demonstrate the ability of these microcapsules to load oil-soluble substances. For microcapsules loaded with Sudan III (Figure 8.2d), their inner oil cores exhibit red color, which is different from the color of the pure oil cores shown in Figure 8.2c. The Sudan III molecules dissolved in the inner oil core is protected by the hydrogel shell and could not permeate through the hydrophilic shell at temperatures below the VPTT. Since the PNIPAM shell can protect the inner oil core against various instability processes such as aggregation and coalescence, the oil-core/hydrogel-shell microcapsules are better microcarriers than emulsions, which are usually employed as delivery systems for oil-soluble drugs but thermodynamically unstable intrinsically. Based on the excellent control of microfluidics on microdroplets, both the double emulsions and microcapsules show very narrow size distributions, with the value of coefficient of variation (CV) less than 3%, exhibiting highly monodisperse core–shell structures.

Incorporation of Fe_3O_4 nanoparticles into the PNIPAM shell enables magnetic-guided targeting delivery of the microcapsules. As demonstrated in Figure 8.3a, first, the microcapsules are randomly dispersed in deionized water at 20 °C. When a magnet is placed under the Petri dish, the microcapsules can be attracted together at site A (Figure 8.3b). After that, the aggregated microcapsules are moved quickly following the arrows from site A to site B under the magnetic guide and finally trapped at the targeted site B (Figure 8.3b–f). Such a magnetic-responsive property can make the microcapsules achieve site- and/or route-specific targeting drug delivery.

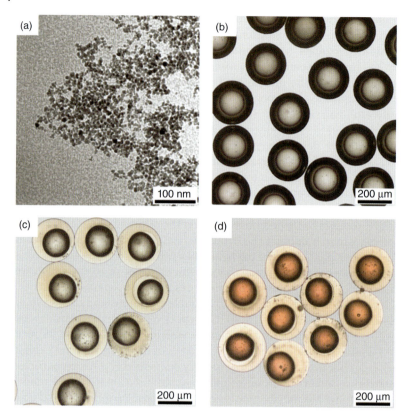

Figure 8.2 TEM image of Fe_3O_4 nanoparticles stabilized by tetramethylammonium hydroxide (a) and optical micrographs of monodisperse O/W/O double emulsions (b), microcapsules with pure oil core (c) and with Sudan III-loaded oil core (d) in water at 20 °C. (Wang et al. 2009 [11]. Reproduced with permission of John Wiley & Sons.)

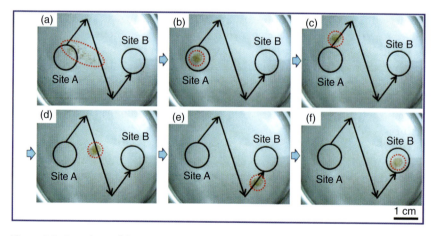

Figure 8.3 Snapshots of the magnetic-guided targeting delivery of core–shell microcapsules loaded with Sudan III in water. (Wang et al. 2009 [11]. Reproduced with permission of John Wiley & Sons.)

Once the microcapsules are delivered to the desired site, burst release of the loaded substances can be triggered by local heating to the temperature above the VPTT. Figure 8.4a illustrates the burst release of the microcapsules when temperature is increased from 20 to 60 °C. With the increasing temperature, the thermo-responsive PNIPAM shell of the microcapsule shrinks dramatically. Since the inner oil core is incompressible but the internal pressure in oil core keeps increasing due to the shell shrinkage, the PNIPAM shell finally ruptures because of the limited mechanical strength. With the shrinkage and rupture of PNIPAM shell, the inner oil core gets squeezed out of the microcapsule within a very short time and spreads fast into the surrounding environment (Figure 8.4a3,a4). As a result, the burst release from the microcapsule is complete, leaving just a hollow cavity without any leftover inside the microcapsule (Figure 8.4b). With such burst-release behaviors, high local drug concentration can be rapidly obtained. To evaluate the release rate, the burst-release behavior of the inner oil core during the first 3.2 s after the PNIPAM shell ruptures is investigated. As shown in Figure 8.4c, the radius of the circular edge of released oil phase increases by ~250 μm within 3.2 s. Such a spread speed is much faster than that in diffusion-driven release systems. Such a quick release and spread rate may make the microcapsules to be of specific interest and significance especially in cases where released substances need to cross some media with high viscosity or low permeability.

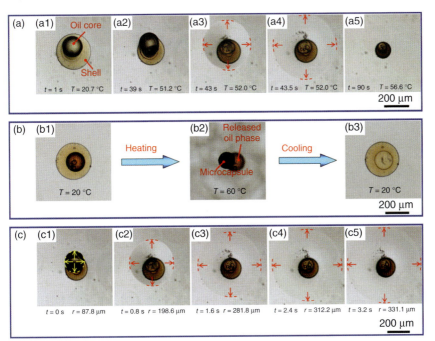

Figure 8.4 Optical microscope snapshots of the burst-release process (a) and complete release performance (b) of the core–shell microcapsules. (c) Burst-release process of inner oil core from the microcapsule (shown in (a)) during the first 3.2 s of release. Time-dependent increase of the radius (r) of the circle edge of released oil phase is used to evaluate the release rate. (Wang et al. 2009 [11]. Reproduced with permission of John Wiley & Sons.)

8.3.2 Fabrication of Core–Shell Microcapsules for Thermo-Triggered Burst Release of Nanoparticles

Nowadays, nanoparticles have attracted considerable interests for drug delivery. In some cases, encapsulation of nanoparticles within microcapsules can provide additional functions for drug delivery: the nanoparticles-in-microcapsule system can provide combined styles for drug release such as sequential release and reduce unwanted initial burst release [14]. They can also protect the nanoparticles against physiological fluids to reduce degradation/aggregation and macrophage clearance [15, 16]. Similar encapsulation and delivery systems can also be found in some plants in nature. For example, the ripe and turgid fruit of squirting cucumber (Figure 8.5a) squirts a stream of mucilaginous liquid containing its seeds into air for wide dispersal, upon its own ripeness or being disturbed by external forces (Figure 8.5b). Inspired by the squirting cucumber, a novel type of core–shell microcapsules for burst release of nanoparticles is developed. The proposed microcapsule is composed of a PNIPAM hydrogel shell, and an oil core with innermost water drops containing nanoparticles (Figure 8.5c). Upon heating, the nanoparticles in the innermost water drops, shielded by both the oil core and the thermo-responsive shell, can be squirted out with a high momentum (Figure 8.5d), just like the seed-ejecting of ripe squirting cucumber.

The fabrication process for such microcapsules is similar to that introduced in Section 8.3.1, except no magnetic nanoparticles are used, and homogenizer-produced W/O emulsions are used instead of inner oil core for generating the emulsion templates.

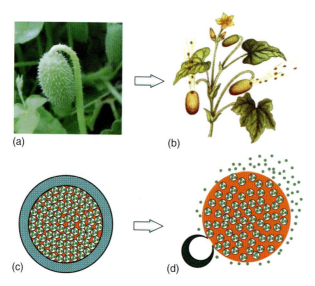

Figure 8.5 (a) Squirting cucumber. (b) Schematic illustration of squirting cucumbers ejecting seeds together with a stream of mucilaginous liquid. (c, d) Core–shell microcapsule with PNIPAM shell and oil core containing innermost water drops with nanoparticles (c) for burst release upon increasing the temperature above the VPTT (d). (Liu *et al*. 2010 [10]. Reproduced with permission of The Royal Society of Chemistry.)

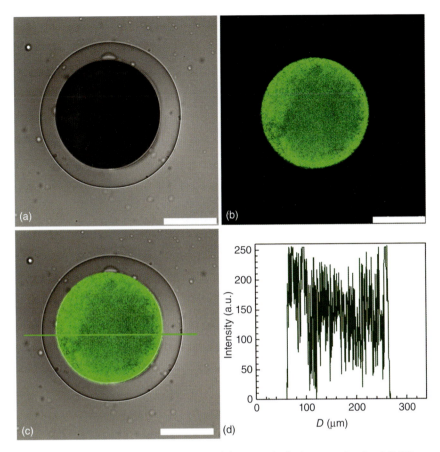

Figure 8.6 Morphological characterization of the core–shell microcapsules. (a–c) CLSM images of core–shell microcapsule at room temperature, observed by the transmission channel (a), green channel (b), and overlay of green channel and transmission channel (c). Scale bars are 100 μm. (d) The fluorescence intensity profile corresponding to (c). (Liu et al. 2010 [10]. Reproduced with permission of The Royal Society of Chemistry.)

As shown in Figure 8.6a, the PNIPAM shell is transparent at room temperature, while the inner core is dark. Meanwhile, the CLSM (confocal laser scanning microscope) fluorescent images show that no leakage of nanoparticles from the core–shell microcapsule is observed (Figure 8.6b,c). This result is also further confirmed by the fluorescence intensity profile, in which the intensity inside the microcapsule is quite high (from 50 to 260), while the intensity outside is nearly zero (Figure 8.6d). To study the burst release of nanoparticles from the microcapsules upon heating, a glass slide with a drop of microcapsule suspension is placed on a thermostatic stage under a microscope. On increasing the temperature from 20 to 50 °C, the PNIPAM shell shrinks rapidly and finally ruptures and releases the contained oil core, together with the innermost water drops and the encapsulated nanoparticles into the surrounding water (Figure 8.7). During the burst-release process, the large bright area in Figure 8.7 indicates the considerable wide distribution of the released fluorescent nanoparticles.

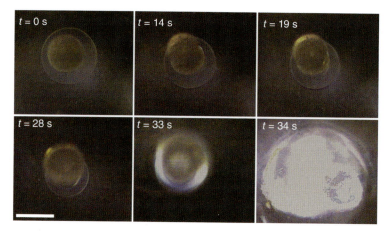

Figure 8.7 Dark-field microscope snapshots showing the burst release of fluorescent nanoparticles from core–shell microcapsules by increasing the temperature from 20 to 50 °C. Scale bars are 200 μm. (Liu et al. 2010 [10]. Reproduced with permission of The Royal Society of Chemistry.)

8.3.3 Fabrication of Core–Shell Microcapsules for Direction-Specific Thermo-Responsive Burst Release

The core–shell microcapsules for direction-specific thermo-responsive burst release are composed of a PNIPAM shell and an eccentric magnetic microparticle as well as an eccentric oil core. The eccentric oil core allows encapsulation of oil-soluble substances, while the eccentric magnetic microparticle enables magnetic-guided remote manipulation of the microcapsules for site-specific targeting delivery and direction-specific controlled release. The PNIPAM shell can protect the encapsulated oil-soluble substances and achieve burst release via a thermo-trigger.

The microcapsules are fabricated with microfluidic-prepared quadruple-component (oil 1 + oil 2)-in-water-in-oil ((O1 + O2)/W/O) double emulsions as templates (Figure 8.8). Typically, soybean oil containing 2% (w/v) PGPR 90, 0.25% (w/v) BDK, and 0.5% (w/v) LR300 is used as the inner oil phase 1 (O1). Ferrofluid (1 ml) dispersed in mixture of 2 ml cyclohexane and 2 ml butyl acetate is used as the oil phase 2 (O2). NIPAM (1 mol l^{-1}), MBA (0.05 mol l^{-1}), 1 wt% Pluronic F127, and 0.5 wt% 2,2′-azobis(2-amidi-nopropanedihydrochloride) are dissolved in deionized water as the middle aqueous phases (W1 and W2). Ten percent (w/v) PGPR 90 and 0.25% (w/v) BDK dispersed in soybean oil are used as the outer oil phase (O). All solutions are supplied to the device as shown in Figure 8.8a for generating the (O1 + O2)/W/O double emulsions. The generated double emulsions are collected in a collection solution of soybean oil containing 10% (w/v) PGPR 90 and 0.25% (w/v) BDK (Figure 8.8b). Then, the double emulsions are kept for ~2 h to allow evaporation of ethyl acetate and cyclohexane in the magnetic oil drops (O2) for constructing the solid magnetic microparticle (Figure 8.8c). The middle aqueous phase is converted into hydrogel shell by UV polymerization for 10 min in an ice-water bath. The

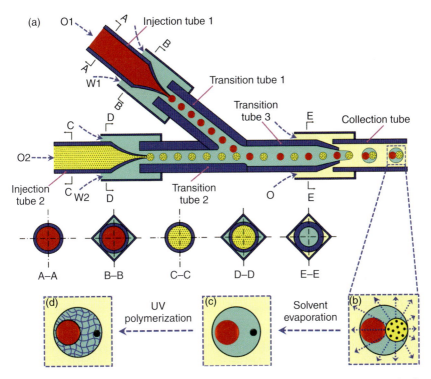

Figure 8.8 Schematic illustration of the fabrication of core–shell PNIPAM microcapsules from (O1 + O2)/W/O double emulsions. (a) Microfluidic device for generating the double emulsions. Illustrations A–A, B–B, C–C, D–D, and E–E are cross-section images of the microfluidic device in relevant positions. (b, c) Solvent evaporation process for constructing the eccentric solid magnetic microparticle. (d) UV-initiated polymerization of the emulsion template for fabricating the core–shell microcapsules. (Liu et al. 2014 [17]. Reproduced with permission of The Royal Society of Chemistry.)

solid magnetic microparticle and soybean oil core loading with LR300 are encapsulated inside the resultant core–shell microcapsules after polymerization (Figure 8.8d).

Figure 8.9a shows (O1 + O2)/W/O quadruple-component double emulsions collected in a vessel at ~11 min after generation. The emulsions are highly monodisperse, and each contains one red oil drop and one black magnetic drop. The generated emulsions are kept for ~2 h to allow complete solvent evaporation of the butyl acetate and cyclohexane in the magnetic droplet, leading to a decreased size and an increased density of the magnetic drop. Then, after the solvent evaporation for preparing the inner magnetic microparticle and UV-initiated polymerization for fabricating the PNIPAM shell, core–shell PNIPAM microcapsules with an eccentric magnetic microparticle and an eccentric oil core can be obtained (Figure 8.9b,c). The CLSM image of the resultant core–shell microcapsules in deionized water at 25 °C (Figure 8.9d) shows no leakage of oil-soluble substances from the microcapsule.

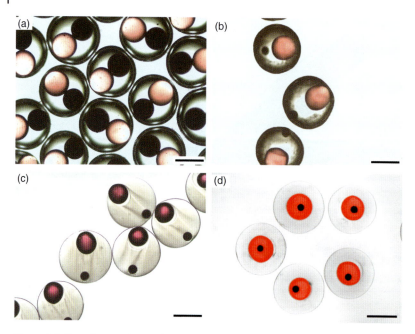

Figure 8.9 Template synthesis of core–shell microcapsules from (O1 + O2)/W/O double emulsions. (a–c) Optical micrographs of the (O1 + O2)/W/O double emulsions at ~11 min after generation (a), and the resultant core–shell microcapsules in soybean oil (b) and in deionized water (c). (d) CLSM image of the resultant core–shell microcapsules in deionized water. Scale bars are 200 μm. (Liu *et al.* 2014 [17]. Reproduced with permission of The Royal Society of Chemistry.)

The direction-specific burst release of oil-soluble substances from the core–shell microcapsules is shown in Figure 8.10. Fluorochrome LR300 is used as the oil-soluble model substance to demonstrate the site-specific targeted delivery and direction-specific burst-release property in a "T"-shaped microchannel geometry. First, due to the eccentric magnetic core, magnetic-guided targeting transport of the microcapsule from the top of the microchannel to the left place can be achieved (Figure 8.10a1–a3,b1–b3). Then, the microcapsule can be manipulated and rotated on microscale using an external magnet to control the release direction of the oil core to the "funnel"-shaped left microchannel (Figure 8.10a3,a4,b3,b4). Next, by increasing the temperature above the VPTT, the inner oil core together with the loaded oil-soluble Fluorochrome LR300 can be burst-released into the "funnel"-shaped left microchannel to achieve direction-specific burst release (Figure 8.10a4,a5,b4–b7). Finally, the microcapsules can then be removed from the targeted site by using an external magnet (Figure 8.10a5,a6,b8,b9). Thus, the microcapsules with oil-soluble substances loaded in the oil core can be site- and/or route-specifically delivered to a desired site under magnetic guide, and the loaded oil-soluble substances can then be burst-released in a controllable specific direction with dual magnetic and thermal manipulation.

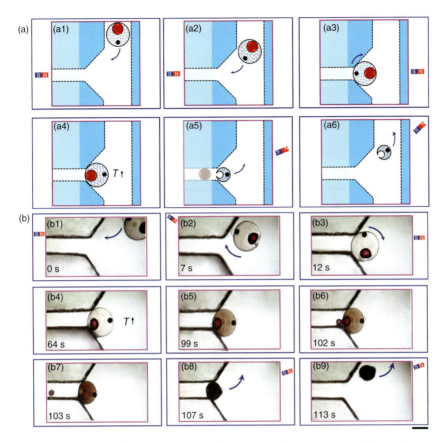

Figure 8.10 Core–shell microcapsules for direction-specific burst release of oil-soluble substances. (a) Scheme illustration of the experimental setup. (b) Optical micrographs showing the magnetic-guided movement (b1, b2) and rotation (b3, b4) of the microcapsules, the burst release of oil-soluble substances from the microcapsule by heating from 20 to 60 °C (b4–b7), and the magnetic removal of microcapsule (b8, b9). Scale bar is 200 μm. (Liu *et al.* 2014 [17]. Reproduced with permission of The Royal Society of Chemistry.)

8.4 Smart Core–Shell Microcapsules for Alcohol-Responsive Burst Release

The core–shell microcapsules for alcohol-responsive burst release are composed of a PNIPAM shell and an oil core, as shown in Figure 8.11a,b. It has been reported that, in water-rich solvents, water molecules can form cage-like structures around the hydrophobic groups of PNIPAM polymer chains and form hydrogen bonds with the amide groups [18]. So the PNIPAM shell remains swollen to well-protect the inner oil core. Upon addition of alcohol, water and alcohol molecules form complexes due to their solvent–solvent interactions, thus resulting in decreased number of water molecules surrounding PNIPAM polymer chains. In a certain alcohol concentration range ($C_{c1} < C < C_{c2}$, as shown in Figure 8.11b), all the water molecules previously solvating PNIPAM

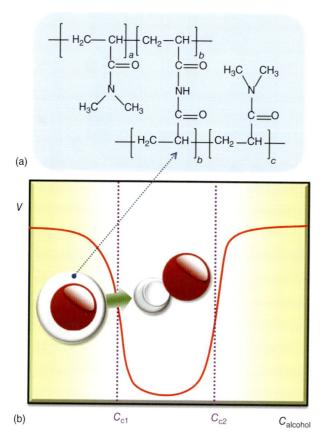

Figure 8.11 Core–shell microcapsules for alcohol-responsive burst release. (a) Chemical structure of PNIPAM shell. (b) Schematic illustration of the concept of alcohol-responsive burst release. (Liu et al. 2012 [7]. Reproduced with permission of American Chemical Society.)

polymer chains are bonded with alcohol molecules. This results in a reduced solvency of PNIPAM with water, and shrinking of the PNIPAM shell due to the polymer–polymer interactions [18–24]. Such a shell shrinkage can then lead to burst release of the inner oil core upon an alcohol trigger.

Figure 8.12 shows the time-dependent deswelling ratio of core–shell microcapsules in alcohol solutions. To exactly measure the real size change of the PNIPAM shell, the oil cores are removed with isopropyl alcohol. As shown in Figure 8.12, generally, higher alcohol concentration leads to faster shrinking of the PNIPAM shell. It is worth noting that the deswelling rate is crucial for ejection of the inner oil core. During deswelling process, the PNIPAM hydrogel shell can form a dense skin layer that is hard to be ruptured [25–27]. Thus, if the microcapsules cannot achieve fast shrinking to generate enough mechanical force to rupture the shell before the skin layer formation, the PNIPAM shell will fail to break. For example, although the microcapsule size can decrease to nearly their minimum sizes in 20% methanol solution at 25 °C, it is found that the microcapsules fail to eject the oil cores at such conditions due to their

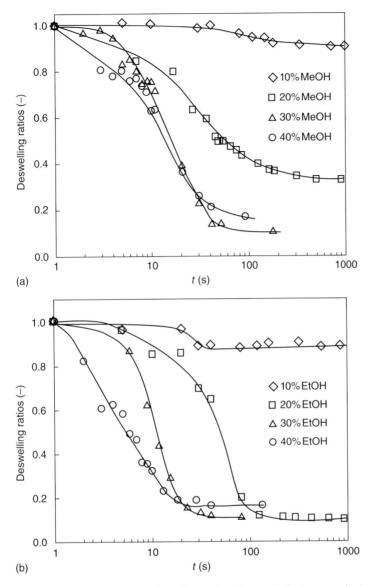

Figure 8.12 Time-dependent deswelling ratios of core–shell microcapsules in aqueous methanol solution (a) and ethanol solution (b) at 25 °C. (Liu et al. 2012 [7]. Reproduced with permission of American Chemical Society.)

slow shrinking process (Figure 8.12a). In contrast, the microcapsules, which can shrink to their minimum sizes within ∼40 s upon 30% or 40% alcohol concentration at 25 °C, can successfully achieve burst release of their inner oil core. Such burst-release processes are demonstrated in Figure 8.13. At 25 °C, initially, the core–shell microcapsules are completely swollen in deionized water. Then, deionized water surrounding the microcapsules is removed, and alcohol solution with a certain concentration at the same temperature is added to the

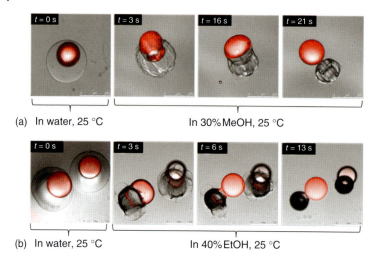

Figure 8.13 Snapshots of the burst-release process of oil cores from core–shell microcapsules upon addition of 30% methanol (a) and 40% ethanol (b) at 25 °C. Scale bars are 250 μm. (Liu et al. 2012 [7]. Reproduced with permission of American Chemical Society.)

microcapsules. Upon alcohol addition, the PNIPAM shell shrinks and squeezes the inner oil core out for burst release.

8.5 Smart Core–Shell Microcapsules for K$^+$-Responsive Burst Release

The concept of the K$^+$-recognition core–shell microcapsules for K$^+$-responsive burst release is schematically illustrated in Figure 8.14. The microcapsule possesses an oil-core/hydrogel-shell structure, with the shell composed of benzo-15-crown-5-acrylamide (B15C5Am) units as K$^+$ sensors, NIPAM units as actuators, and acrylamide (AAm) units as hydrophilicity adjustors. The 15-crown-5 moiety can selectively recognize K$^+$ via formation of stable 2:1 "sandwich" complexes [28–36]. The formation of "sandwich" complexes in the shell can shift the VPTT of PNIPAM to a lower value [34]. Therefore, at a temperature between the two VPTTs, the microcapsule shell shrinks isothermally from a swollen state to a shrunken state as a result of K$^+$ recognition [35], for burst release of the inner oil core. Moreover, since the VPTT of PNIPAM-based copolymer can be increased by introducing hydrophilic monomers, incorporation of AAm into the PNIPAM-based shell enables adjustment of the operation temperature for K$^+$ recognition from 32 °C to body temperature.

K$^+$, Na$^+$, and Ca^{2+} are chosen as the test ions to study their effects on the VPTT shift, because these ions are important for signal transduction in biological systems. Meanwhile, to minimize the salting-out effect [37], nitrate is chosen as a model salt for all ions. As shown in Figure 8.15a, PNIPAM microcapsules show similar phase transition behavior in 0.1 M K$^+$ solution as that in deionized water. By contrast, for the microcapsules with shell composed

8.5 Smart Core–Shell Microcapsules for K⁺-Responsive Burst Release

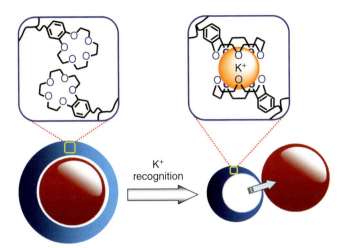

Figure 8.14 Schematic illustration of the concept of K⁺-recognition core–shell microcapsules for burst release. The K⁺-triggered volume shrinking of the microcapsule based on the formation of a 2:1 "sandwich" complex between 15-crown-5 receptors and K⁺ can lead to burst release of the encapsulated oil core. (Liu et al. 2011 [8]. Reproduced with permission of The Royal Society of Chemistry.)

of poly(*N*-isopropylacrylamide-*co*-benzo-15-crown-5-acrylamide) (PNB), they exhibit nearly the same phase transition behavior in response to Na^+ and Ca^{2+} as that in water, but different phase transition behaviors in response to K^+, showing significant negative shift of VPTT (Figure 8.15a,b). The K^+-triggered negative shift of VPTT becomes more remarkable with the increasing B15C5Am amount (Figure 8.15a) or the K^+ concentration (Figure 8.15c). If a proper operation temperature is chosen (e.g., 27 °C), the K^+-triggered isothermal volume shrinking of PNB microcapsules in response to 0.1 M K^+ is as large as 70–75% when the molar ratio of B15C5Am to NIPAM is 10%. The VPTT of PNB with 10% B15C5Am in deionized water is 33 °C (Figure 8.15a), while that of PNB with 10% B15C5Am and 10% AAm increases to 46 °C due to the introduction of AAm units (Figure 8.15d). In response to 0.1 M K^+, the VPTT of poly(NIPAM-*co*-AAm-*co*-B15C5Am) (PNAB) shifts negatively to 35 °C (Figure 8.15d). This means that the isothermal K^+ recognition of PNAB microcapsules can be operated at body temperature (37 °C). Thus, in response to 0.1 M K^+, PNAB microcapsules can exhibit K^+-triggered isothermal volume shrinking as large as 80–85% at 37 °C.

Figure 8.16 shows the K^+-triggered burst-release behavior of the core–shell microcapsules. Notice that the shell thicknesses of the core–shell microcapsules are not so uniform, due to the lighter density of the oil phase as compared to that of the aqueous phase in the O/W/O emulsions. Such non-uniform shell thickness benefits the burst-release mechanism, because the thinnest point in the shell is usually the point where shell rupture occurs [10]. As shown in Figure 8.16, the PNAB microcapsules remain stable in deionized water at 37 °C. When the environmental water is replaced by 0.2 M K^+ solution with a temperature of 37 °C, the PNAB microcapsules also exhibit very fast shrinkage and finally squeeze the inner oil core out for K^+-triggered burst release.

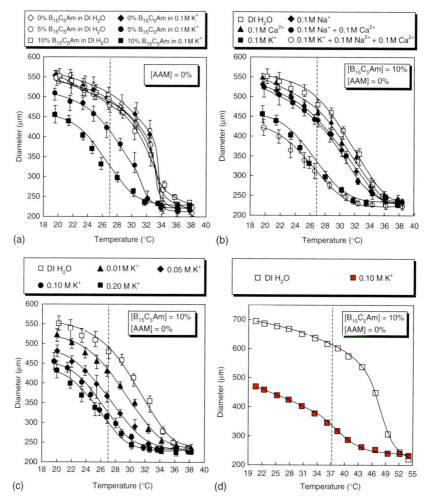

Figure 8.15 Volume-phase transition behaviors of core–shell PNIPAM, poly(NIPAM-co-B15C5Am) and poly(NIPAM-co-AAm-co-B15C5Am) microcapsules. (a) Effect of B15C5Am content in the microcapsule shell on the volume-phase transition behaviors. (b) Effect of cation species on the volume-phase transition behaviors. (c) Effect of K^+ concentration on the volume-phase transition behaviors. (d) Volume-phase transition behaviors of poly(NIPAM-co-AAm-co-B15C5Am) microcapsules in response to 0.1 M K^+. (Liu et al. 2011 [8]. Reproduced with permission of The Royal Society of Chemistry.)

8.6 Smart Core–Shell Microcapsules for pH-Responsive Burst Release

8.6.1 Concept of the Core–Shell Microcapsules for pH-Responsive Burst Release

The concept of the pH-responsive core–shell microcapsule for acid-triggered burst release is schematically illustrated in Figure 8.17. The core–shell microcapsule is composed of a terephthalaldehyde-cross-linked chitosan hydrogel shell

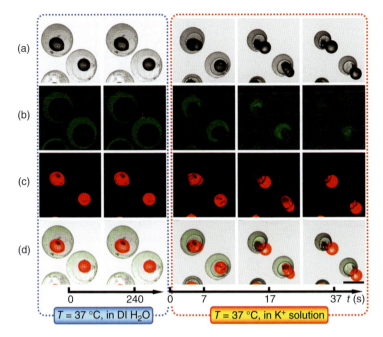

Figure 8.16 CLSM images of K$^+$-triggered burst-release behaviors of poly(NIPAM-co-AAm-co-B15C5Am) microcapsules in response to 0.2 M K$^+$ at 37 °C. The molar ratios of B15C5Am to NIPAM and AAm to NIPAM are both 10%. (a) Transmission channel images. (b) Green channel images. (c) Red channel images. (d) Overlay images. The scale bar is 500 μm. (Liu et al. 2011 [8]. Reproduced with permission of The Royal Society of Chemistry.)

and an inner oil core. It has been reported that chitosan can be cross-linked by terephthalaldehyde via formation of a Schiff base between the amino group of chitosan and the aldehyde group of dialdehyde compounds in neutral medium (Figure 8.17a). The stability of such a cross-linking largely depends on the environmental pH [38, 39]. In neutral medium, the microcapsules maintain structural integrity, which protect the substances loaded in the inner oil core from being released before reaching the site with low pH (e.g., stomach). In acidic environment with low pH, the amino groups of chitosan are protonated and positively charged; thus the intramolecular electrostatic repulsion and enhanced hydrophilicity make chitosan hydrogel swell dramatically. With more and more amino groups protonated, the Schiff base bonding becomes unstable [38] and leads to final decomposition of the cross-linked chitosan hydrogel (Figure 8.17a). Therefore, based on this reversible cross-linking between chitosan and terephthalaldehyde, acid-triggered burst release can be achieved for such pH-responsive core–shell microcapsules, as shown in Figure 8.17b. Moreover, it can be expected that, faster protonation of the amino groups can be achieved with lower pH value, thus faster decomposition as well as burst-release process can be obtained.

8.6.2 Fabrication of the Core–Shell Chitosan Microcapsules

The strategy for microcapsule preparation is to use monodisperse O/W/O double emulsions from microfluidics as templates for microcapsule fabrication

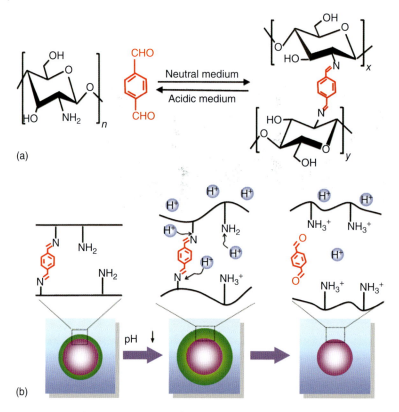

Figure 8.17 Schematic illustration of the cross-linking reaction between chitosan and terephthalaldehyde in neutral medium for shell fabrication (a) and the acid-triggered burst-release behavior of the core–shell chitosan microcapsule (b). (Liu et al. 2011 [9]. Reproduced with permission of The Royal Society of Chemistry.)

via interfacial cross-linking. Typically, 2.0 wt% water-soluble chitosan for shell fabrication, 1.5 wt% Pluronic F127 for emulsion stabilization, and 2.0 wt% hydroxyethylcellulose for viscosity adjustment are dissolved in water as middle fluid. The pH value of the middle fluid is adjusted to 6.7 by addition of 1.0 mol l^{-1} NaOH. A mixture of soybean oil and benzyl benzoate (1 : 2, v/v) containing 2.0 wt% cross-linker terephthalaldehyde is used as inner fluid, while soybean oil containing 8.0 wt% PGPR is used as outer fluid. After injection of the three fluids into the microfluidic device (Figure 2.5a) for emulsion generation, the obtained O/W/O double emulsions were collected in a container and left to stand overnight for complete cross-linking of the chitosan in the water phase. The cross-linking reaction occurs at the inner O/W interface of the double emulsions as soon as the inner oil fluid contacts the middle water fluid in the microfluidic device. The terephthalaldehyde in the inner oil fluid diffuses into the middle water fluid to cross-link the chitosan for microcapsule fabrication. Both the microfluidic emulsion generation process and subsequent interfacial cross-linking reaction are all performed at 20 °C. The resultant core–shell chitosan microcapsules are washed using a mixture of ethyl acetate and isopropanol with volume ratio of 1 : 5 to remove the inner and outer oil phases, and finally dispersed into water. A mixture of butyl acetate and benzyl benzoate with volume ration of 2 : 1, containing 2.0 wt% terephthalaldehyde and 1 mg ml^{-1} LR300 is used as the inner fluid for preparing microcapsules containing LR300. The resultant microcapsules are washed with 1.0 wt% OP-10 aqueous solution to remove the outer soybean oil but keep the inner oil core inside the microcapsules. Such one-pot method has competitive advantages for fabrication of chitosan-based core–shell microcapsules with more controllable structure and simpler procedure.

8.6.3 Core–Shell Chitosan Microcapsules for pH-Responsive Burst Release

To estimate the capability of pH-responsive core–shell microcapsules for acid-triggered burst release, the decomposition of the microcapsule shell in the pH range 1.5–4.7 is systematically investigated. The core–shell chitosan microcapsules are first immersed in deionized water, followed by changing the environmental water by quickly adding HCl or phosphate buffer solution with different pH values. The acid-triggered decomposition behaviors of the microcapsule shell are studied under CLSM microscope. By adding buffer solution with pH 3.1, the microcapsules swell first and then start to collapse and finally decompose completely within 127 s (Figure 8.18). Figure 8.19a shows the acid-triggered decomposition process of the chitosan microcapsules in buffer solutions with different pH values. All the microcapsules first swell and then gradually collapse

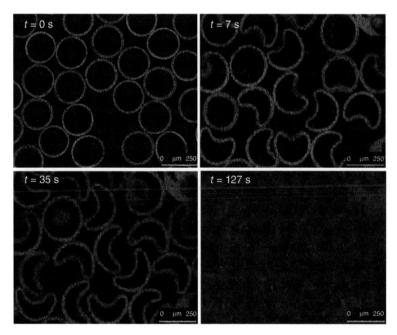

Figure 8.18 CLSM microscope snapshots of the acid-triggered decomposition process of core–shell chitosan microcapsules. pH 3.1 buffer solution is added at $t = 0$ s. Scale bars are 250 μm. (Liu et al. 2011 [9]. Reproduced with permission of The Royal Society of Chemistry.)

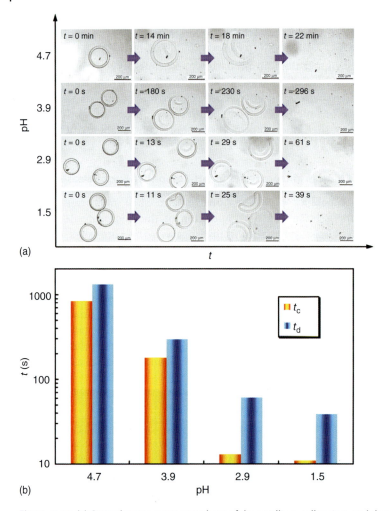

Figure 8.19 (a) Optical microscope snapshots of the swelling, collapsing, and decomposing of the chitosan microcapsules in various pH solutions at 37 °C. Scale bars are 200 μm. (b) Time period from addition of buffer solution to the moment that the microcapsules start to collapse (t_c) and are decomposed completely (t_d) at 37 °C in various pH solutions. (Liu et al. 2011 [9]. Reproduced with permission of The Royal Society of Chemistry.)

and finally decompose. Moreover, as shown in Figure 8.19b, lower environmental pH can lead to faster acid-triggered swelling as well as the shell decomposition. With environmental pH 4.7, it takes 22 min for the complete decomposition of the microcapsules, whereas when the environmental pH value decreases to 1.5, the microcapsules decompose rapidly in 39 s. This pH-dependent decomposition can be utilized to develop smart gastric delivery systems for releasing antacid agents at a rate depending on the pH value of the gastric juice.

To demonstrate the feasibility of the microcapsules to encapsulate oil-soluble substances for pH-responsive burst release, LR300, a red fluorescent dye, is encapsulated in the cross-linked chitosan microcapsules as a model oil-soluble

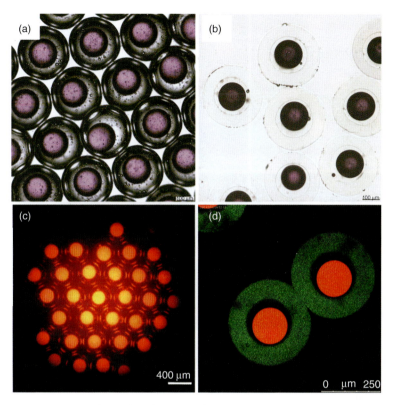

Figure 8.20 Optical micrographs (a, b) and CLSM images (c, d) of the O/W/O double emulsions containing LR300 (a, c) and the resultant chitosan microcapsules washed by OP-10 (b, d). Scale bars are 100 μm in (a, b), 400 μm in (c), and 250 μm in (d). (Liu et al. 2011 [9]. Reproduced with permission of The Royal Society of Chemistry.)

substance. Figure 8.20a,b shows the optical micrographs of the O/W/O double emulsions containing LR300 in the inner oil core (Figure 8.20a) and the resultant LR300-loaded chitosan microcapsules (Figure 8.20b). The average size of the resultant chitosan microcapsules (∼341 μm) is larger than that of their double emulsion templates (∼305 μm). The increased size of the microcapsule results from the washing process with the detergent OP-10, the hydrophilicity of which can make the chitosan hydrogel shell swell. Moreover, it has been reported that the chitosan cross-linked with glutaraldehyde or terephthalaldehyde can display autofluorescence properties due to the formation of Schiff's bases [40, 41]. As shown in Figure 8.20c, for the O/W/O double emulsions without the cross-linking reaction, their middle water phase surrounding the LR300-contained oil core is dark, indicating that there is no cross-linking of the chitosan. In contrast, the shell of the resultant microcapsules exhibits green fluorescence, which confirms the cross-linking between terephthalaldehyde and chitosan for fabricating the hydrogel shell (Figure 8.20d).

Figure 8.21 shows the acid-triggered burst-release behaviors of core–shell chitosan microcapsules loaded with LR300. When the LR300-loaded chitosan

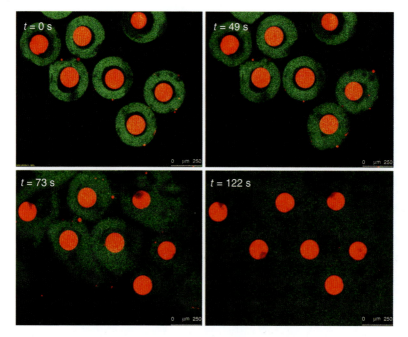

Figure 8.21 CLSM microscope snapshots of the acid-triggered burst-release process of core–shell chitosan microcapsules. pH 3.1 buffer solution is added at $t = 0$ s. Scale bars are 250 μm. (Liu et al. 2011 [9]. Reproduced with permission of The Royal Society of Chemistry.)

microcapsules are subjected to buffer solution with pH 3.1, the microcapsules decompose rapidly and leave the LR300-loaded inner oil cores exposed to the environmental solution in 122 s. Because the microcapsule shell is already in a considerable swollen state due to the use of OP-10 before changing the pH, the microcapsule shell does not swell much during the whole decomposition process. According to the data in Figure 8.19, such burst-release behaviors are controllably responding to the environmental pH values. The core–shell chitosan microcapsules are promising for achieving acid-triggered stomach-targeted delivery with prompt onset and complete release characteristics.

8.7 Summary

In this chapter, flexible microfluidic strategies for controllable fabrication of monodisperse core–shell microcapsules with an oil core for controllable encapsulation and a stimuli-responsive shell for controlled release are introduced. The resultant core–shell microcapsules, with a shell consisting of thermo-responsive, alcohol-responsive, ion-responsive, and pH-responsive hydrogels, allow controllable encapsulation of oil-soluble components and nanoparticles for controlled release under triggers including temperature, alcohol, K^+, and pH. Based on the high controllability of microfluidics over the composition and structures of double emulsions, core–shell microparticles with controllable core compartment and versatile functional shells can be developed. These smart core–shell microparticles are promising as microcarriers for protection of actives and controlled release of drugs.

References

1 De Geest, B.G., De Koker, S., Immesoete, K., Demeester, J., De Smedt, S.C., and Hennink, W.E. (2008) Self-exploding beads releasing microcarriers. *Adv. Mater.*, **20**, 3687–3691.
2 Chu, L.Y., Yamaguchi, T., and Nakao, S. (2002) A molecular-recognition microcapsule for environmental stimuli-responsive controlled release. *Adv. Mater.*, **14**, 386–389.
3 De Rose, R., Zelikin, A.N., Johnston, A.P.R., Sexton, A., Chong, S.-F., Cortez, C., Mulholland, W., Caruso, F., and Kent, S.J. (2008) Binding, internalization, and antigen presentation of vaccine-loaded nanoengineered capsules in blood. *Adv. Mater.*, **20**, 4698–4703.
4 Yu, A.M., Wang, Y.J., Barlow, E., and Caruso, F. (2005) Mesoporous silica particles as templates for preparing enzyme-loaded biocompatible microcapsules. *Adv. Mater.*, **17**, 1737–1741.
5 Kreft, O., Skirtach, A.G., Sukhorukov, G.B., and Moehwald, H. (2007) Remote control of bioreactions in multicompartment capsules. *Adv. Mater.*, **19**, 3142–3145.
6 Kreft, O., Prevot, M., Moehwald, H., and Sukhorukov, G.B. (2007) Shell-in-shell microcapsules: a novel tool for integrated, spatially confined enzymatic reactions. *Angew. Chem. Int. Ed.*, **46**, 5605–5608.
7 Liu, L., Song, X.-L., Ju, X.-J., Xie, R., Liu, Z., and Chu, L.-Y. (2012) Conversion of alcoholic concentration variations into mechanical force via core-shell capsules. *J. Phys. Chem. B*, **116**, 974–979.
8 Liu, Z., Liu, L., Ju, X.-J., Xie, R., Zhang, B., and Chu, L.-Y. (2011) K^+-recognition capsules with squirting release mechanisms. *Chem. Commun.*, **47**, 12283–12285.
9 Liu, L., Yang, J.P., Ju, X.J., Xie, R., Liu, Y.M., Wang, W., Zhang, J.J., Niu, C.H., and Chu, L.Y. (2011) Monodisperse core-shell chitosan microcapsules for pH-responsive burst release of hydrophobic drugs. *Soft Matter*, **7**, 4821–4827.
10 Liu, L., Wang, W., Ju, X.-J., Xie, R., and Chu, L.-Y. (2010) Smart thermo-triggered squirting capsules for nanoparticle delivery. *Soft Matter*, **6**, 3759–3763.
11 Wang, W., Liu, L., Ju, X.-J., Zerrouki, D., Xie, R., Yang, L., and Chu, L.-Y. (2009) A novel thermo-induced self-bursting microcapsule with magnetic-targeting property. *ChemPhysChem*, **10**, 2405–2409.
12 Chu, L.-Y., Utada, A.S., Shah, R.K., Kim, J.-W., and Weitz, D.A. (2007) Controllable monodisperse multiple emulsions. *Angew. Chem. Int. Ed.*, **46**, 8970–8974.
13 Wang, W., Zhang, M.-J., and Chu, L.-Y. (2014) Functional polymeric microparticles engineered from controllable microfluidic emulsions. *Acc. Chem. Res.*, **47**, 373–384.

14 Hasan, A.S., Socha, M., Lamprecht, A., El Ghazouani, F., Sapin, A., Hoffman, A., Maincent, P., and Ubrich, N. (2007) Effect of the microencapsulation of nanoparticles on the reduction of burst release. *Int. J. Pharm.*, **344**, 53–61.

15 Wanakule, P., Liu, G.W., Fleury, A.T., and Roy, K. (2012) Nano-inside-micro: disease-responsive microgels with encapsulated nanoparticles for intracellular drug delivery to the deep lung. *J. Controlled Release*, **162**, 429–437.

16 Bhavsar, M.D. and Amiji, M.M. (2007) Gastrointestinal distribution and in vivo gene transfection studies with nanoparticles-in-microsphere oral system (NiMOS). *J. Controlled Release*, **119**, 339–348.

17 Liu, Y.-M., Wu, W., Ju, X.-J., Wang, W., Xie, R., Mou, C.-L., Zheng, W.-C., Liu, Z., and Chu, L.-Y. (2014) Smart microcapsules for direction-specific burst release of hydrophobic drugs. *RSC Adv.*, **4**, 46568–46575.

18 Crowther, H.M. and Vincent, B. (1998) Swelling behavior of poly N-isopropylacrylamide microgel particles in alcoholic solutions. *Colloid Polym. Sci.*, **276**, 46–51.

19 Acharya, A., Goswami, A., Pujari, P.K., Sabharwal, S., and Manohar, S.B. (2002) Positron annihilation studies of poly(N-isopropyl acrylamide) gel in mixed solvents. *J. Polym. Sci., Part A: Polym. Chem.*, **40**, 1028–1036.

20 Chee, C.K., Hunt, B.J., Rimmer, S., Soutar, I., and Swanson, L. (2011) Time-resolved fluorescence anisotropy studies of the cononsolvency of poly(N-isopropyl acrylamide) in mixtures of methanol and water. *Soft Matter*, **7**, 1176–1184.

21 Schild, H.G., Muthukumar, M., and Tirrell, D.A. (1991) Cononsolvency in mixed aqueous-solutions of poly(N-isopropyl acrylamide). *Macromolecules*, **24**, 948–952.

22 Patil, P.N., Kathi, S., Dutta, D., and Pujari, P.K. (2010) Understanding the swelling of poly(N-isopropyl acrylamide) gels through the study of free volume hole size distributions using positron annihilation spectroscopy. *Polym. Bull.*, **65**, 577–587.

23 Zhu, P.W. and Napper, D.H. (1999) Light scattering studies of poly(N-isopropylacrylamide) microgel particles in mixed water-acetic acid solvents. *Macromol. Chem. Phys.*, **200**, 1950–1955.

24 Lele, A.K., Karode, S.K., Badiger, M.V., and Mashelkar, R.A. (1997) Prediction of re-entrant swelling behavior of poly(N-isopropyl acrylamide) gel in a mixture of ethanol-water using lattice fluid hydrogen bond theory. *J. Chem. Phys.*, **107**, 2142–2148.

25 Beines, P.W., Klosterkamp, I., Menges, B., Jonas, U., and Knoll, W. (2007) Responsive thin hydrogel layers from photo-cross-linkable poly(N-isopropylacrylamide) terpolymers. *Langmuir*, **23**, 2231–2238.

26 Oni, Y., Theriault, C., Hoek, A.V., and Soboyejo, W.O. (2011) Effects of temperature on diffusion from PNIPA-based gels in a BioMEMS device for localized chemotherapy and hyperthermia. *Mater. Sci. Eng., C*, **31**, 67–76.

27 Shiotani, A., Mori, T., Niidome, T., Niidome, Y., and Katayama, Y. (2007) Stable incorporation of gold nanorods into N-isopropylacrylamide hydrogels and their rapid shrinkage induced by near-infrared laser irradiation. *Langmuir*, **23**, 4012–4018.

References

28 Chen, C.Y., Cheng, C.T., Lai, C.W., Wu, P.W., Wu, K.C., Chou, P.T., Chou, Y.H., and Chiu, H.T. (2006) Potassium ion recognition by 15-crown-5 functionalized CdSe/ZnS quantum dots in H_2O. *Chem. Commun.*, 263–265.

29 Lin, S.Y., Liu, S.W., Lin, C.M., and Chen, C.H. (2002) Recognition of potassium ion in water by 15-crown-5 functionalized gold nanoparticles. *Anal. Chem.*, **74**, 330–335.

30 Patel, G., Kumar, A., Pal, U., and Menon, S. (2009) Potassium ion recognition by facile dithiocarbamate assembly of benzo-15-crown-5-gold nanoparticles. *Chem. Commun.*, 1849–1851.

31 Xia, W.S., Schmehl, R.H., and Li, C.J. (1999) A highly selective fluorescent chemosensor for K^+ from a bis-15-crown-5 derivative. *J. Am. Chem. Soc.*, **121**, 5599–5600.

32 Yamauchi, A., Hayashita, T., Nishizawa, S., Watanabe, M., and Teramae, N. (1999) Benzo-15-crown-5 fluoroionophore/gamma-cyclodextrin complex with remarkably high potassium ion sensitivity and selectivity in water. *J. Am. Chem. Soc.*, **121**, 2319–2320.

33 Kim, J., McQuade, D.T., McHugh, S.K., and Swager, T.M. (2000) Ion-specific aggregation in conjugated polymers: highly sensitive and selective fluorescent ion chemosensors. *Angew. Chem. Int. Ed.*, **39**, 3868–3872.

34 Mi, P., Chu, L.-Y., Ju, X.-J., and Niu, C.H. (2008) A smart polymer with ion-induced negative shift of the lower critical solution temperature for phase transition. *Macromol. Rapid Commun.*, **29**, 27–32.

35 Mi, P., Ju, X.-J., Xie, R., Wu, H.-G., Ma, J., and Chu, L.-Y. (2010) A novel stimuli-responsive hydrogel for K+-induced controlled-release. *Polymer*, **51**, 1648–1653.

36 Yamauchi, A., Hayashita, T., Kato, A., Nishizawa, S., Watanabe, M., and Teramae, N. (2000) Selective potassium ion recognition by benzo-15-crown-5 fluoroionophore/γ-cyclodextrin complex sensors in water. *Anal. Chem.*, **72**, 5841–5846.

37 Inomata, H., Goto, S., Otake, K., and Saito, S. (1992) Effect of additives on phase-transition of *N*-isopropylacrylamide gels. *Langmuir*, **8**, 687–690.

38 Li, M., Cheng, S., and Yan, H. (2007) Preparation of crosslinked chitosan/poly(vinyl alcohol) blend beads with high mechanical strength. *Green Chem.*, **9**, 894–898.

39 Hejazi, R. and Amiji, M. (2004) Stomach-specific anti-H-pylori therapy. Part III: effect of chitosan microspheres crosslinking on the gastric residence and local tetracycline concentrations in fasted gerbils. *Int. J. Pharm.*, **272**, 99–108.

40 Wei, W., Wang, L.-Y., Yuan, L., Wei, Q., Yang, X.-D., Su, Z.-G., and Ma, G.-H. (2007) Preparation and application of novel microspheres possessing autofluorescent properties. *Adv. Funct. Mater.*, **17**, 3153–3158.

41 Wei, W., Yuan, L., Hu, G., Wang, L.-Y., Wu, H., Hu, X., Su, Z.-G., and Ma, G.-H. (2008) Monodisperse chitosan microspheres with interesting structures for protein drug delivery. *Adv. Mater.*, **20**, 2292–2296.

9

Microfluidic Fabrication of Monodisperse Hole–Shell Microparticles

9.1 Introduction

Polymeric microparticles with core–shell structures show great potential as encapsulation microcarriers for controlled load/release of substances [1–3], protection of actives [4], and confined microreaction of biochemical reactants [5]. Core–shell structures with solid shells enable effective encapsulation; however, transport of the encapsulants through the shell is usually difficult to achieve. Incorporation of holes in the shell can provide more versatility, because the hole structures can facilitate the mass transport of substances through the shell based on the size or functional selectivity of the holes. This can produce microparticles with porous shells for a wide range of applications such as controlled capture of particles [1, 6], controlled release of substances [7, 8], and removal of pollutants [9]. Especially, for microparticles with pores in the shell, a single, defined hole in the shell can provide a very versatile structure for selectively capturing particles for classification and separation, or capturing cells for confined culture. Moreover, such a hole–shell structure also facilitates the employment of the symmetry-breaking principle to create asymmetric structures for directional moving and actuating.

Typically, hole–shell microparticles in colloidal scale can be fabricated by using particle-template or emulsion-template methods, such as buckling of silicon droplets during polymerization [10], self-assembly of phase-separated polymers [11], selective polymerization of phase-separated droplets [12], diffusion-induced escape of monomers [13] or solvents [14–16] from microparticles during fabrication, and freeze-drying of polymeric particles that are swollen by solvent [1, 17]. These colloidal-scale hole–shell microparticles can provide excellent performance when microparticles with sizes less than a few microns are required. By contrast, larger microparticles can provide additional versatility when the size requirements are not limited to very small particles. These microparticles usually have sizes of tens of microns or larger and can be typically fabricated by using emulsion droplets as templates. With microfluidic techniques to produce the emulsion templates [18–20], microparticles with improved control over their size monodispersity and structures can be achieved. The microparticle structure largely depends on the configurations between the core droplet and the shell droplet in the double emulsion templates. For example, with shell droplet completely wetted on the core droplet, aqueous-biphasic droplets and water-in-oil-in-water (W/O/W) double emulsions can, respectively, produce hole–shell microparticles with bowl shape [6] and fishbowl shape [21]. Full versatility of the hole–shell microparticles requires precise and independent control of the shape and size of both the single hole and the hollow core structure and the functionality of the surface of the hollow core. This requires precise control of the configurations and interfacial properties of the double emulsion templates.

The authors' group developed new microfluidic strategies for fabrication of hole–shell microparticles from double emulsions. Such microparticles can be produced by employing the density mismatch between inner and outer droplets in double emulsions or dewetting behaviors of outer droplet on inner droplet for creating evolved double emulsions as templates. In this chapter, the microfluidic fabrication of hole–shell microparticles from these two strategies and their performances in motion generation, sensing and actuating, and controlled capturing are introduced.

9.2 Microfluidic Strategy for Fabrication of Monodisperse Hole–Shell Microparticles

Generally, hole–shell microparticles can be fabricated by using double emulsions with eccentric core–shell structures. Such eccentric core–shell structures can be created by employing the density mismatch between inner and outer droplets in double emulsions, or dewetting behaviors of the outer droplet on the inner droplet. For the strategy that is based on the density mismatch, the inner droplet with smaller density can float up in the outer droplet to create double emulsions with eccentric core–shell structures. Template synthesis of microparticles from the evolved double emulsions, followed by release of the inner droplet to create a hole in the shell, can produce microparticles with hole–shell structures. Alternatively, for the strategy that is based on the dewetting behaviors, double emulsions with more controllable eccentric core–shell structures can be created by adjusting the interfacial energy to make the outer droplet partially dewet on the inner droplet. Such emulsion templates allow template synthesis of microparticles with versatile and controllable hole–shell structures. Detailed fabrication strategies are introduced in the following sections.

9.3 Hole–Shell Microparticles for Thermo-Driven Crawling Movement

9.3.1 Concept of the Hole–Shell Microparticles for Thermo-Driven Crawling Movement

The novel hole–shell microparticle for thermo-driven crawling movement, which is also called the thermo-driven soft microcrawlers, is designed with a

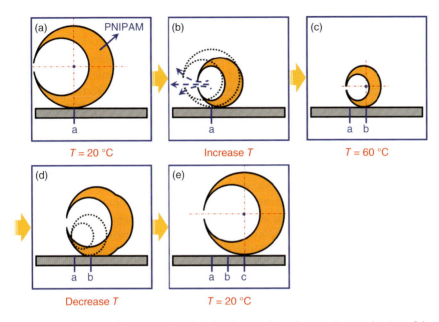

Figure 9.1 Schematic illustration showing the thermo-driven locomotion mechanism of the microcrawler via asymmetry swelling/shrinking volume transitions of the PNIPAM hydrogel body. (Wang et al. 2013 [22]. Reproduced with permission of Institute of Physics.)

thermo-responsive hydrogel body and a bell-like inner cavity (Figure 9.1a). To construct the water-containing hydrogel body of the microcrawler, poly(N-isopropylacrylamide) (PNIPAM) hydrogel, which enables switching between swollen and shrunken states when changing temperature across its volume-phase transition temperature (VPTT) (~32 °C), is used as the thermo-responsive soft material. Double emulsions from microfluidics are used as templates to fabricate the jellyfish-like structure, with an eccentric cavity inside the microcrawler. The thermo-responsive volume-phase transitions of PNIPAM hydrogel allow actuation of microcrawler using a wireless and remotely controllable mechanism. As shown in Figure 9.1a–e, the asymmetric structure of the PNIPAM microcrawler can undergo asymmetric swelling/shrinking volume changes in response to temperature changes to generate locomotion. First, the PNIPAM hydrogel body of the microcrawler is swollen, with water held in the inner cavity (Figure 9.1a). Upon increasing the temperature above the VPTT, the PNIPAM body shrinks and expels the water out. So, the water in the inner cavity is then quickly pushed out through the hole. This can produce a water jet to propel the microcrawler forward (Figure 9.1a–c). Then, on decreasing the temperature below the VPTT again, the shrunken microcrawler swells, with the bell swelling faster than the head; this can generate force onto the ground to make the microcrawler crawl forward (Figure 9.1c–e). The dramatic volume changes of thermo-responsive PNIPAM hydrogel is crucial for generating momentum to push the microcrawler forward, which provides a novel model to transform thermal stimuli into directional movement.

9.3.2 Fabrication of Hole–Shell Microparticles for Thermo-Driven Crawling Movement

Monodisperse O/W/O double emulsions generated from microfluidic device (Figure 9.2a) are used as templates for synthesizing microcrawlers. Typically, 4 ml deionized water with 0.04 g Pluronic F127, 0.452 g N-isopropylacrylamide (NIPAM), 0.0308 g N,N'-methylene-bis-acrylamide (MBA), and 0.024 g ammonium persulfate is used as the middle fluid. Soybean oil with 5% (w/v) polyglycerol polyricinoleate (PGPR 90) and 0.1% (w/v) Sudan Red, and soybean oil with 5% (w/v) PGPR 90 are, respectively, used as the inner and outer fluids. Monodisperse O/W/O double emulsions generated from the microfluidic device are collected in a collection solution, which is soybean oil with 5% (w/v) PGPR 90 and 0.2% (w/v) 2,2-dimethoxy-2-phenylacetophenone (BDK) as the photoinitiator. Then, the collected double emulsions are kept for 30 min (Figure 9.2b) to make the oil core float up in the double emulsions, due to the smaller density of oil core as compared with that of the middle aqueous layer (Figure 9.2c). These emulsions with eccentric oil core are converted into hydrogel microcrawlers by UV treatment for 15 min at a temperature below the VPTT (Figure 9.2d). Under

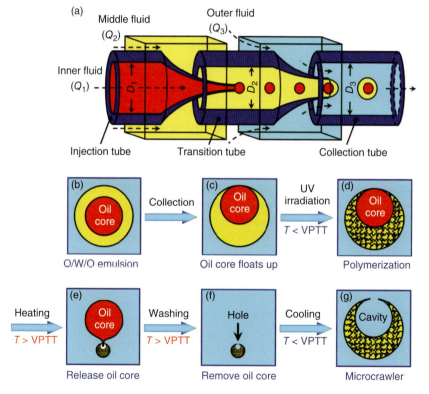

Figure 9.2 Schematic illustration showing the fabrication process of microcrawlers from double emulsions. Microfluidic device (a) for generating O/W/O double emulsions as templates for synthesis of microcrawlers (b–g) (Wang *et al.* 2013 [22]. Reproduced with permission of Institute of Physics.)

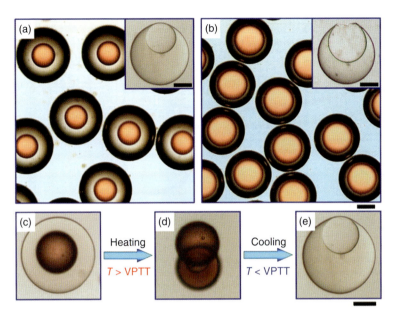

Figure 9.3 Template synthesis of microcrawlers. (a, b) Optical images of double emulsions containing oil cores with different sizes and the resultant microcrawlers (insets). (c–e) Formation of acaleph-like microcrawler via thermo-triggered release of the inner oil core. Scale bars are 100 μm. (Wang et al. 2013 [22]. Reproduced with permission of Institute of Physics.)

UV light, the activated photoinitiator BDK diffuses to the interface between the outer oil phase and middle aqueous phase, where it triggered the polymerization of NIPAM in the middle aqueous phase to build PNIPAM hydrogel. Then, the inner oil core is squeezed out by local heating to the temperature above the VPTT. This can create a ruptured hole in the PNIPAM hydrogel shell. After that, microcrawlers are obtained by washing off the oil phase with ethanol and water (Figure 9.2e–g). Since the size of the double emulsions and the inner oil cores can be accurately controlled with microfluidics (Figure 9.3a,b), excellent manipulation on the size of the resultant microcrawlers and the inner cavity can be achieved, as shown in the insets of Figure 9.3a,b. Figure 9.3c–e shows the process for creating the inner cavity upon heating. Upon increasing the temperature above the VPTT, the thermo-responsive PNIPAM shell shrinks dramatically and finally squeezes the eccentric oil core out (Figure 9.3c,d), which creates an eccentric cavity inside the resultant microcrawler and a hole at the thinnest site of the shell (Figure 9.3e).

9.3.3 Effect of Inner Cavity on the Thermo-Responsive Volume-Phase Transition Behaviors of Hole–Shell Microparticles

The eccentric cavity in the microcrawlers, which utilizes the symmetry-breaking principle, plays a key role in the thermo-driven crawling movement. To investigate the effect of eccentric cavity on the thermo-responsive volume changes for crawling movement, the dynamics of shrinking and swelling of microcrawlers in aqueous solution of 0.05 wt% sodium dodecyl sulfate (SDS) are studied by quickly

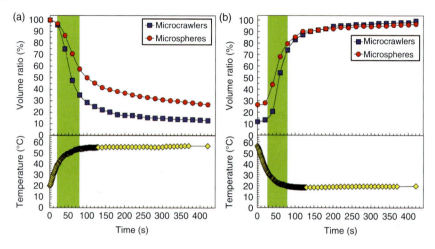

Figure 9.4 Dynamic shrinking (a) and shrinking (b) behaviors of the microcrawlers and voidless PNIPAM microspheres in response to temperature changes in 0.05 wt% SDS solution. (Wang et al. 2013 [22]. Reproduced with permission of Institute of Physics.)

heating from 20 to 60 °C and then subsequently cooling back to 20 °C (Figure 9.4). SDS is used to minimize the adhesion effect of glass container on the motion behavior of PNIPAM microcrawlers, which allows the microcrawlers to move smoothly in the glass holder [23]. PNIPAM microspheres without cavity but with the same diameter as that of the microcrawler are used as the control group.

As shown in Figure 9.4a, the microcrawlers with eccentric cavity shrink faster than the voidless PNIPAM microspheres upon quickly heating from 20 to 60 °C. When both curves reach a plateau at $t = 400$ s, the volume ratio of the microcrawlers is smaller (∼13%) than that of the microspheres (∼26%), also indicating a faster shrinking process of microcrawler than microsphere. Similar behaviors can also be observed for their swelling process upon cooling back to 20 °C, where the microcrawlers swell faster than the microspheres (Figure 9.4b). The different shrinking and swelling kinetics between microcrawlers and microspheres are caused by the presence/absence of the eccentric cavity. Generally, the thermo-response rate of PNIPAM hydrogel is governed by diffusion-limited transport of water in and out of the cross-linked PNIPAM networks. Upon heating, the PNIPAM hydrogel shrinks and expels the inner water out of the networks. As compared with the voidless PNIPAM microspheres, PNIPAM microcrawlers with eccentric cavity provide larger specific surface area, including the inner surface of cavity and the outer surface of microcrawler, for expelling water. The cavity offers much less resistance to transport water as compared with the cross-linked PNIPAM networks of microcrawler, while the hole structure in the cavity shell provides a convenient exit for water inside the cavity to escape. Similarly, upon cooling, the microcrawlers provide larger specific surface area and less resistant channel for intake of water.

Meanwhile, during the shrinking and swelling process, the volume change rates of the microcrawler are proportional to the heating and cooling rates, as shown in Figure 9.4. With rapid increase/decrease of temperature at $t = 20$–80 s, the microcrawlers show dramatic volume shrinking/swelling changes. The sharp volume change near the VPTT is crucial for generating momentum to drive the microcrawler forward; such a volume change can be further improved by faster heating and cooling processes [24].

9.3.4 Hole–Shell Microparticles for Thermo-Driven Crawling Movement

The asymmetry cavity structure and faster response rate of the microcrawlers allow the PNIPAM hydrogel around the microcrawler bell serving as chemical muscles for thermo-driven locomotion. To study the thermo-driven locomotion of microcrawlers, the microcrawler is horizontally placed in a glass holder (Figure 9.1b) containing 0.05 wt% SDS solution, which is sealed with cover glass to prevent water evaporation. The distance between the geometric centers of the sample before and after each temperature change circulation is measured to determine the moving distance. As a control group, the thermo-responsive behaviors of the voidless PNIPAM microspheres with $V_{cavity}:V_{microcrawler} = 0\%$ are first investigated by quickly heating from 20 to 60 °C and then subsequently cooling back to 20 °C (Figure 9.5). However, the symmetric shrinking/swelling process of the homogeneous microspheres makes the composition of horizontal forces equal to zero and results in no locomotion.

By contrast, microcrawlers with eccentric cavity the asymmetric shrinking/swelling process can achieve thermo-driven movement behaviors. Figure 9.6 shows the thermo-driven locomotion behaviors of microcrawlers with eccentric cavities of different sizes during temperature change circulations. Cyclic shrinking/swelling volume changes are conducted by repeated heating/cooling to achieve thermo-driven step-by-step motion. As shown in Figure 9.6a, the microcrawler with cavity of $V_{cavity}:V_{microcrawler} = 9.7\,vol\%$ moves with distances of 6.82 and 8.16 µm for the first and second temperature circulation, respectively. As shown in Figure 9.6b, the microcrawler with larger cavity of $V_{cavity}:V_{microcrawler} = 30.7\,vol\%$ moves farther with a distance of 13.08 µm for the first temperature circulation and 36.37 µm for the second. The moving tracks of microcrawlers, illustrated by labeling their geometric centers at different stages, show that the microcrawlers move in an opposite direction to the open hole, as the arrows shown in Figure 9.6. Thus, repeated shrinking/swelling

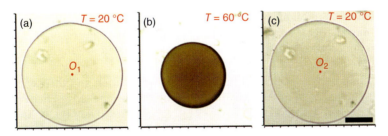

Figure 9.5 Thermo-responsive behaviors of PNIPAM microspheres with $V_{cavity}:V_{microcrawler} = 0\%$. O_i presents the coordinate point of the geometric center of the microsphere. The temperature is quickly changed between 20 and 60 °C. Scale bar is 100 µm. (Wang et al. 2013 [22]. Reproduced with permission of Institute of Physics.)

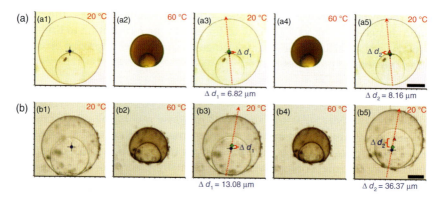

Figure 9.6 Thermo-driven locomotion of microcrawlers with cavities of different sizes. (a) $V_{cavity}:V_{microcrawler} = 9.7$ vol% and (b) $V_{cavity}:V_{microcrawler} = 30.7$ vol%. The symbols ○, ◇, and △ present the geometric centers of the spherical microcrawlers at different stages. The arrows indicate the moving direction of the thermo-driven movement, which is just in the opposite direction to the open hole. $\triangle d_i$ is the moving distance of the microcrawler after the ith ($i = 1,2$) shrinking/swelling circulation. Scale bars are 100 μm. (Wang et al. 2013 [22]. Reproduced with permission of Institute of Physics.)

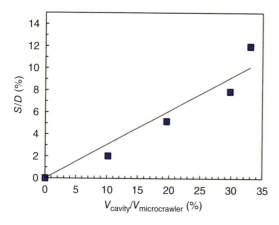

Figure 9.7 Effect of $V_{cavity}/V_{microcrawler}$ on the scaled moving distance (S/D) of microcrawlers. S is the moving distance of microcrawlers per circulation, and D is the outer diameter of microcrawlers. (Wang et al. 2013 [22]. Reproduced with permission of Institute of Physics.)

volume changes of the microcrawlers successfully transform thermal stimuli into directional mechanical movement. Although each movement is still tiny as compared to the microcrawler size (e.g., 36.37 μm vs 360 μm in Figure 9.6b5), the microcrawler can achieve a long distance walk under repeated heating/cooling triggers.

The relationship between the moving distance per circulation and the cavity size is further studied by systematically investigating the thermo-driven behaviors of microcrawlers with different $V_{cavity}/V_{microcrawler}$ ratios (Figure 9.7). The results show that the moving distance per circulation increases with increasing $V_{cavity}/V_{microcrawler}$. This indicates that the moving speed of the microcrawler depends on both the sizes of microcrawler and inner cavity. This concept of thermo-driven microcrawlers offers a novel model for designing biomimetic soft microrobots with symmetry-breaking principle.

9.4 Hole–Shell Microparticles for Pb^{2+} Sensing and Actuating

9.4.1 Fabrication of Hole–Shell Microparticles for Pb^{2+} Sensing and Actuating

For template synthesis of hole–shell microparticles for Pb^{2+} sensing and actuating, typically, soybean oil containing 2% (w/v) PGPR 90, 1% (w/v) BDK, and 0.05% (w/v) Sudan III is used as the inner oil phase 1 (O$_1$); isopentyl acetate with 5% (w/v) polystyrene (PS) and 20% (v/v) ferrofluid is used as the inner oil phase 2 (O$_2$); and 1 mol l^{-1} NIPAM, 0.1 mol l^{-1} benzo-18-crown-6-acrylamide (B18C6Am), 0.02 mol l^{-1} MBA, 1 wt% Pluronic F127, 0.5 wt% 2,2'-azobis(2-amidi-nopropane dihydrochloride) (V50), and 6 wt% glycerin are dissolved in DI water as the middle aqueous phase (W). Soybean oil containing 10% (w/v) PGPR 90 and 1% (w/v) BDK is used as the outer oil phase (O). Monodisperse (O$_1$ + O$_2$)/W/O double emulsions, each containing a red oil droplet and a black ferrofluid droplet, are generated from the microfluidic device as templates (Figures 9.8a and 9.9a). The ferrofluid droplet contains oleic-acid-modified magnetic nanoparticles (OA-MNPs) with mean diameter of ~12 nm, as shown in the TEM image of Figure 9.9b. The generated double emulsions are then collected in a vessel containing soybean oil with 10% (w/v) PGPR 90 and 1% (w/v) BDK (Figure 9.8b) and then kept for 40 min to allow solvent evaporation of isopentyl acetate and n-hexane in the ferrofluid droplet (O$_2$) for constructing the solid magnetic PS core (Figure 9.8c). This leads to shrinkage of the ferrofluid droplet and precipitation of the contained PS polymers and OA-MNPs for forming solid magnetic PS core in the double emulsions (Figure 9.9c). Then, the emulsion templates are treated with UV light for 15 min at a temperature lower than the VPTT to convert the middle aqueous phase into poly(PNIPAM-co-B18C6Am) (PNB) hydrogel for microparticle construction. The soybean oil droplet and magnetic core are encapsulated inside the microparticle after polymerization (Figure 9.8d). Then, by heating to 65 °C, the inner soybean oil droplet is squeezed out by the shrinking PNIPAM hydrogel shell, resulting in a cavity in the microparticle and a cracked hole in the shell (Figure 9.8e). Finally, PNB microparticles for Pb^{2+} sensing and actuating (Figure 9.8f,g), which is also called microactuators, each with a cavity and an eccentric magnetic core can be obtained after washing with DI water. The inset of an optical micrograph in Figure 9.9c clearly shows the eccentric magnetic core and hollow cavity structure of PNB microactuator.

9.4.2 Magnetic-Guided Targeting Behavior of Poly(NIPAM-co-B18C6Am) Hole–Shell Microparticles

The eccentric OA-MNPs-containing PS core enables wireless manipulation of the microactuators with an external magnetic field for both translational and rotational motions. The magnetic-guided movement of the microactuator is performed in a glass holder (size: ø 6 mm) with a T-shaped microchannel containing aqueous solution with 0.4 wt% bovine serum albumin (BSA) at 41.2 °C

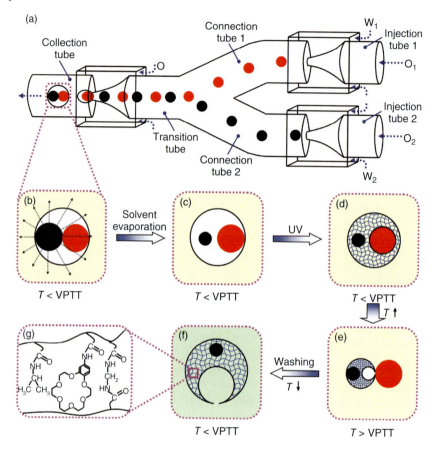

Figure 9.8 Schematic illustration of the process for fabricating PNB microactuators. (a) Microfluidic device for generating ($O_1 + O_2$)/W/O double emulsions. (b, c) Solvent evaporation process for constructing the magnetic solid PS core. (d–g) UV-initiated polymerization of emulsion templates (d) followed by releasing the soybean oil core by heating (e) for fabricating PNB microactuators (f) with cross-linked hydrogel shell (g). (Liu et al. 2013 [25]. Reproduced with the permission of American Chemical Society.)

(Figure 9.10). The BSA is used to prevent the adhesion of the microactuator to the glass surface [26, 27]. A magnet is manually placed near the glass holder and moved to guide the microactuator movement (Figure 9.10a). With external magnetic field for guiding the movement, the PNB microactuator placed at the top of the microchannel can move through every corner of the T-shaped microchannel via both rotational and directed movements (Figure 9.10b–i). The magnetic-responsive property allows the microactuators to achieve site- and/or route-specific targeting delivery.

9.4.3 Effects of Pb^{2+} on the Thermo-Responsive Volume Change Behaviors of Poly(NIPAM-co-B18C6Am) Hole–Shell Microparticles

The effect of Pb^{2+} on the thermo-responsive volume change behaviors of the PNB microactuators is investigated by using similar microactuators with PNIPAM

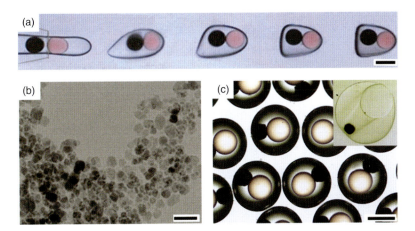

Figure 9.9 Template synthesis of PNB microactuators from quadruple-component $(O_1+O_2)/W/O$ double emulsions. (a) High-speed snapshots showing the generation of emulsion templates in microfluidic device. (b) TEM image of oleic-acid-modified magnetic nanoparticles. (c) Optical micrograph of the double emulsions at 425 s after formation and the resultant PNB microactuator (inset) in DI water. Scale bars are 200 μm in (a) and (c), and 30 nm in (b). (Liu et al. 2013 [25]. Reproduced with the permission of American Chemical Society.)

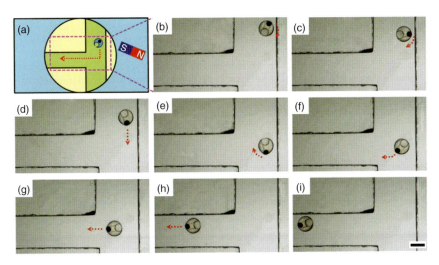

Figure 9.10 Magnetic-guided movement of the PNB microactuator. (a) Schematic illustration of the experimental setup. (b–i) Optical micrographs showing the movement of a PNB microactuator under magnetic guide in aqueous solution with 0.4 wt% BSA at 41.2 °C. Scale bar is 200 μm. (Liu et al. 2013 [25]. Reproduced with the permission of American Chemical Society.)

hydrogel body as the control group. Figure 9.11 shows the thermo-responsive volume changes of PNB and PNIPAM microactuators in DI water and $0.02\,\text{mol}\,\text{l}^{-1}$ Pb^{2+} aqueous solution. V_T and V_0 are, respectively, the microactuator volumes at temperature T and at the initial temperature (~19 °C). VR_1 and VR_2 are the volume ratios (V_T/V_0) for the microactuators and

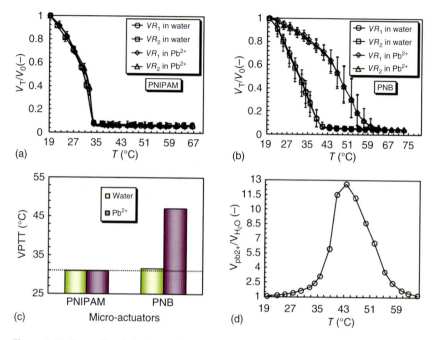

Figure 9.11 Deswelling behaviors of PNIPAM (a) and PNB (b) microactuators. V_T and V_0 are, respectively, the equilibrium volumes at temperature T and initial temperature (19 °C), and VR_1 and VR_2 represent the volume ratio (V_T/V_0) for the microactuators and the inner cavity, respectively. (c) The VPTT of PNIPAM and PNB microactuators in DI water and Pb^{2+} solution. (d) Temperature-dependent volume change ratio of PNB microactuators in Pb^{2+} solution to that in DI water. The Pb^{2+} concentration is 0.02 mol l^{-1}. (Liu et al. 2013 [25]. Reproduced with the permission of American Chemical Society.)

the inner cavity, respectively. With increasing temperature, the volumes of all microactuators decrease, indicating excellent thermo-responsive properties. The temperature at which the V_T/V_0 value decreases to half of the total change value is determined as the VPTT of the microactuator. The PNIPAM microactuators in both solutions show similar volume change behaviors with VPTT ≈ 31 °C (Figure 9.11a,c). By contrast, as shown in Figure 9.11b,c, the PNB microactuators show a thermo-responsive volume change behavior with a higher VPTT in 0.02 mol l^{-1} Pb^{2+} solution (VPTT$_2$ ≈ 47 °C) as compared to that in DI water (VPTT$_1$ ≈ 31.5 °C), due to the formation of complexes between crown ether and Pb^{2+} in the hydrogel networks. The B18C6Am moieties can selectively capture Pb^{2+} into their cavities through supramolecular "host–guest" complexation. This can cause charged polymer chains, thus producing repulsion forces among each other and also leading to osmotic pressure to resist the shrinkage of PNB networks during heating. As a result, an increased VPTT is achieved due to the more swollen networks of PNB microactuators in 0.02 mol l^{-1} Pb^{2+} solution. Therefore, to realize isothermal Pb^{2+}-induced volume swelling of the PNB microactuators, the optimal operating temperature should be located between the VPTT$_1$ (31.5 °C) and VPTT$_2$ (47 °C).

9.4 Hole–Shell Microparticles for Pb^{2+} Sensing and Actuating

A parameter defined as $V_{Pb^{2+}}/V_{H_2O}$ is used to determine the optimal operation temperature for PNB microactuators. $V_{Pb^{2+}}$ and V_{H_2O} are, respectively, the average volumes of PNB microactuators in 0.02 mol l^{-1} Pb^{2+} solution and DI water at each temperature. As shown in Figure 9.11d, with increasing temperature, the value of $V_{Pb^{2+}}/V_{H_2O}$ first increases and then decreases, with the maximum value obtained at 43.6 °C. This indicates that the optimal operation temperature for PNB microactuators to achieve the best Pb^{2+}-responsive property is ~43.6 °C.

9.4.4 Effects of Hollow Cavity on the Time-Dependent Volume Change Behaviors of Poly(NIPAM-co-B18C6Am) Hole–Shell Microparticles

The effect of hollow cavity on the time-dependent volume change behavior of the PNB microactuator is investigated by employing voidless PNB microsphere with same diameter as that of the microactuator as the control group. As shown in Figure 9.12, both the PNB microactuator and microsphere are shrunken in DI water at 40.4 °C ($t = 0$ s). After adding 0.02 mol l^{-1} Pb^{2+}, PNB

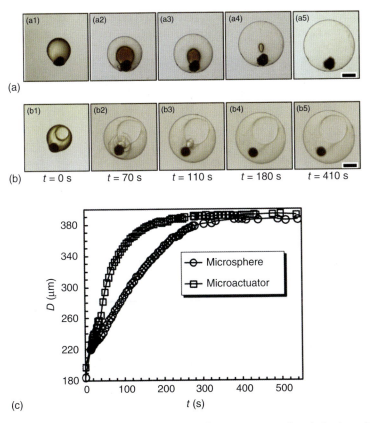

Figure 9.12 Optical micrographs of the Pb^{2+}-responsive swelling behaviors of voidless PNB microsphere (a) and PNB microactuator (b) at 40.4 °C. t represents the time right after adding 0.02 mol l^{-1} Pb^{2+}. (c) Dynamic diameter changes of voidless PNB microsphere and microactuator upon adding 0.02 mol l^{-1} Pb^{2+} at 40.4 °C. Scale bars are 100 μm. (Liu et al. 2013 [25]. Reproduced with the permission of American Chemical Society.)

microactuator (Figure 9.12b) shows a much faster swelling process than the voidless microsphere (Figure 9.12a). The time-dependent changes of diameters of the microactuator and microsphere are plotted in Figure 9.12c. After adding 0.02 mol l^{-1} Pb^{2+}, the diameter of PNB microactuator increases faster than that of the microsphere for the first ~220 s and then reaches the same value at $t = $ ~400 s. It is reported that the diffusion-limited transport of water molecules in and out of the cross-linked networks determines the response rate of a cross-linked microgel [24]. Therefore, as compared with the voidless PNB microsphere, the hollow cavity together with the hole in the shell can provide the microactuator with larger surface area and much less resistance for transporting Pb^{2+} ions, thus resulting in a faster Pb^{2+}-response rate.

9.4.5 Micromanipulation of Poly(NIPAM-co-B18C6Am) Hole–Shell Microparticles for Preventing Pb^{2+} Leakage from Microcapillary

The use of PNB microactuators for micromanipulation in microenvironment is demonstrated by using them to prevent Pb^{2+} leaking from microcapillary with Pb^{2+}-contaminated solution flowing through (Figure 9.13). To prevent the adhesion of microactuators on the glass surface, 0.1 wt% BSA is added

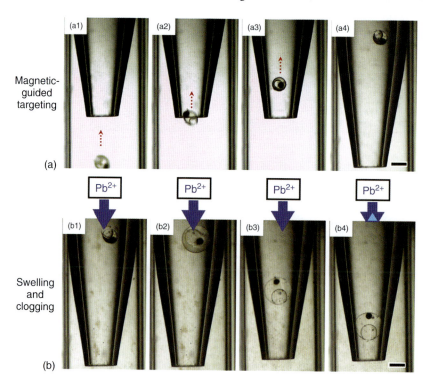

Figure 9.13 Micromanipulation of PNB microactuator for preventing leakage of Pb^{2+} from microcapillary. (a) Magnetic-guided transport of the microactuator into the microcapillary. (b) Clogging of the microcapillary by Pb^{2+}-responsive swelling of the microactuator upon addition of Pb^{2+}-contaminated flow. Scale bars are 200 μm. (Liu et al. 2013 [25]. Reproduced with the permission of American Chemical Society.)

into the aqueous solution to achieve a smooth magnetic-guided movement. Based on the visible precipitation that formed when BSA contacts with Pb^{2+}, the BSA can also be used as an indicator to show the inflow of Pb^{2+} solution (Figure 9.13b). As shown in Figure 9.13a, under the guide of an external magnet, the initially shrunken PNB microactuator can be moved into the tapered microcapillary. When Pb^{2+} solution flows through the microcapillary, the microactuator isothermally swells in response to Pb^{2+} (Figure 9.13b1,b2). Then, the microactuator is moved to the microcapillary tip, with an inner diameter of 260 μm, and then clogs the microcapillary with its swollen volume to prevent the Pb^{2+}-contaminated solution flowing out (Figure 9.13b3,b4). The results show the excellent maneuverability and fast Pb^{2+} response of PNB microactuator for micromanipulation in Pb^{2+}-contaminated microenvironment.

9.5 Hole–Shell Microparticles for Controlled Capture and Confined Microreaction

9.5.1 Microfluidic Fabrication of Hole–Shell Microparticles

For fabrication of hole–shell microparticles for controlled capture, W/O/W double emulsions generated from glass-capillary microfluidic device (Figure 9.14a) are used as initial templates. The microfluidic device is assembled basically according to the literature [18], except an additional square capillary used as the observation tube is connected to the collection tube for observing the generated W/O/W emulsions. The excellent controllability of the fabrication strategy is demonstrated by starting with preparation of hole–shell microparticles from ordinary core–shell W/O/W emulsions.

Typically, for the production of core–shell W/O/W emulsions, aqueous solution containing 0.01 g ml^{-1} poly(N-isopropylacrylamide-co-methyl methacrylate-co-allylamine) nanogels labeled with fluorescein isothiocyanate (FITC-PNIPAM nanogel) is used as the inner fluid (IF). The FITC-PNIPAM nanogel is used for independently modifying the core surface of the resultant hole–shell microparticles. Photocurable ethoxylated trimethylolpropane triacrylate (ETPTA) monomer with 1% (v/v) 2-hydroxy-2-methyl-1-phenyl-1-propanone (HMPP) is used as the middle fluid (MF). Aqueous solution with 1% (w/v) Pluronic F127 and 5% (w/v) glycerol is used as the outer fluid (OF). As shown in Figure 9.14b, IF is first emulsified into droplets in the transition tube, with FITC-PNIPAM nanogels absorbed at the droplet surface for stabilization. Then, these water drops of IF are encapsulated in the oil shell of MF in the collection tube, resulting in monodisperse core–shell W/O/W emulsions, as shown in Figure 9.14c,d. After being collected in vessels, the inner droplet of the W/O/W emulsions is drawn to top of the oil shell due to the density mismatch between IF and MF (Figures 9.14e1 and 9.15b1). Then, the eccentric core–shell W/O/W emulsions are treated with UV irradiation for 5 min. This can produce fishbowl-shaped hole–shell microparticles with large hollow core and small single hole in the shell (Figures 9.14e4 and 9.16a). The formation of the single hole in the shell is due to the volume contraction and fast polymerization

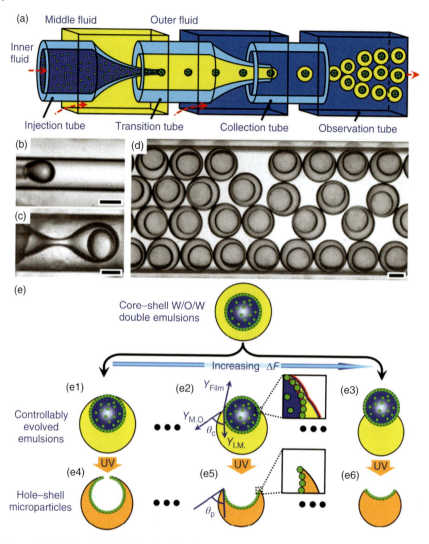

Figure 9.14 Microfluidic fabrication of hole–shell microparticles from controllably evolved W/O/W emulsions. (a) Microfluidic device for generation of monodisperse W/O/W emulsions with inner droplet containing nanoparticles as interface for stabilization. (b–d) Optical micrographs showing the generation of inner droplets (b) and core–shell W/O/W emulsions (c, d) in microfluidic device. (e) Schematic illustration showing the template synthesis of hole–shell microparticles from controllably evolved W/O/W emulsions. Functional nanoparticles are dispersed in the inner droplet to modify the core surface of the resultant hole–shell microparticles. Scale bars are 100 μm. (Wang et al. 2013 [28]. Reproduced with permission of John Wiley & sons.)

(within ~1 s) [29] of the ETPTA oil shell under UV irradiation. The volume of oil shell, which decreases to ~59.1 vol% after polymerization, can squeeze the inner droplet out of the oil shell, while the fast *in situ* polymerization fixes this morphology and produces the hole–shell microparticles with single hole.

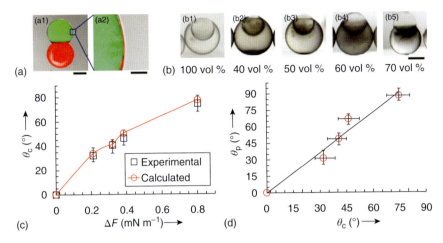

Figure 9.15 Controllable evolution of the W/O/W emulsions via adjustment of adhesion energy. (a) CLSM images of evolved W/O/W emulsions, with inner droplet containing FITC-PNIPAM nanogels and oil shell containing 0.01% (w/v) fluorescent dye LR300. Scale bars are 100 μm in (a1) and 20 μm in (a2). (b) Optical micrographs of side views of evolved W/O/W emulsions with oil phase containing different fractions of ETPTA. Scale bar is 100 μm. (c) Effect of adhesion energy ΔF on the contact angle θ_c of the W/O/W emulsions. The calculated values of θ_c are obtained from the force balance of $\gamma_{I.M.}$, $\gamma_{M.O.}$ and γ_{Film} with an equation: $\theta_c = \arccos[(\gamma_{Film}^2 - \gamma_{I.M.}^2 - \gamma_{M.O.}^2)/2\gamma_{I.M.}\gamma_{M.O.}]$. (d) Relationship between the measured value of the shape angle θ_p of microparticles and the measured value of contact angle θ_c of the W/O/W emulsion templates. (Wang et al. 2013 [28]. Reproduced with permission of John Wiley & sons.)

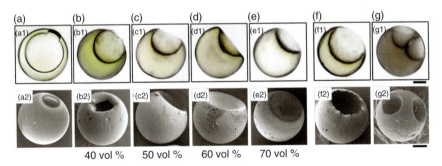

Figure 9.16 Hole–shell microparticles synthesized from controllably evolved W/O/W emulsion templates. (a–e) Optical (a1–e1) and SEM (a2–e2) images of hole–shell microparticles with fishbowl shape (a), bowl shape (b, c) and truncated-sphere shape (d, e). (f) Hole–shell microparticles with large core and large hole, which are fabricated by increasing the volume of inner droplet in W/O/W emulsions. (g) Microparticles with dual hole–shell structures which are fabricated by increasing the number of inner droplet in W/O/W emulsions. Scale bars are 50 μm. (Wang et al. 2013 [28]. Reproduced with permission of John Wiley & sons.)

9.5.2 Precise Control over the Hole–Shell Structure of the Microparticles

By inducing the adhesion between the aqueous inner droplet and outer aqueous phase, the W/O/W emulsions can controllably evolve from core–shell configuration to acorn-shaped configuration. For generation of W/O/W emulsions with evolved acorn shape, ETPTA and organic solvent benzyl benzoate (BB) containing 5% (w/v) surfactant PGPR 90 and 1% (v/v) HMPP is used as the middle fluid (MF). The volume ratio of ETPTA and BB at 40 : 60, 50 : 50, 60 : 40, and 70 : 30 are used in the MF to produce W/O/W emulsions with different evolved acorn shapes. Since PGPR 90 is poorly soluble in ETPTA, the solvent quality is reduced and leads to adhesion of the aqueous inner droplet with the outer aqueous phase, as shown in Figure 9.15a. The thin oil film that separates the inner droplet and outer aqueous phase (Figure 9.15a2) is associated with an adhesion energy [30, 31], which can be expressed as $\Delta F = \gamma_{I.M.} + \gamma_{M.O.} - \gamma_{Film}$, where $\gamma_{I.M.}$ and $\gamma_{M.O.}$ are, respectively, the interfacial tensions between inner phase and middle phase, and between middle phase and outer phase, and γ_{Film} is the tension of the thin oil film. The contact angle θ_c between the adhesive inner droplet and outer aqueous phase (Figure 9.14e2) directly reflect the value of ΔF [32]; such a contact angle θ_c also determines the shape angle θ_p of the resultant hole–shell microparticles (Figure 9.14e5). Therefore, by accurate manipulation of ΔF, controllable evolution of the W/O/W emulsions into different acorn shapes can be achieved for template synthesis of hole–shell microparticles with advanced structures (Figure 9.14e). Furthermore, the core surface of the resultant hole–shell microparticles can be independently modified by introducing functional nanoparticles into the inner droplet and anchoring them at the W/O interface via UV polymerization.

Precise control of the morphologies of the evolved W/O/W emulsions is investigated by studying the contact angle θ_c as a function of ΔF (Figure 9.15b,c). For calculation of ΔF, $\gamma_{I.M.}$ and $\gamma_{M.O.}$ are measured by using pendent droplet method, while γ_{Film} is obtained from adhesion experiments between droplets of IF and OF in MF phase, according to force balance [30, 31]. By fine adjustment of ETPTA fraction in the MF phase, ΔF can be flexibly tuned for precise manipulation of the contact angle θ_c. As shown in Figure 9.15b1, for core/shell W/O/W emulsions with MF phase containing 100 vol% ETPTA as the shell, $\theta_c = 0°$ is observed, indicating $\Delta F = 0$, theoretically. As shown in Figure 9.15b2–b5, for acorn-shaped W/O/W emulsions with oil shell containing both ETPTA and BB, the value of their contact angle θ_c increases with increasing the ΔF. The experimental value of the contact angle θ_c and their calculated values are in good agreement (Figure 9.15c). With precise manipulation on the θ_c value, these controllably evolved W/O/W emulsions allow template synthesis of hole–shell microparticles with well-controlled shape angle θ_p for structure control (Figure 9.15d). Thus, hole–shell microparticles, with controllable structures ranging from bowl shape (Figure 9.16b,c), to truncated-sphere shape (Figure 9.16d,e) can be produced. The structures of the hole–shell microparticle and the single hole in the shell show good monodispersity. For example, the CV values of the sizes of hole–shell microparticles and their holes shown in Figure 9.16c are 1.35% and 2.26%, respectively, indicating highly monodisperse structures. Moreover, besides fishbowl-shaped hole–shell microparticles with large core and small hole, by tuning flow rates to change the inner droplet volume, hole–shell microparticles with large core and large hole can also be produced (Figure 9.16f). Meanwhile, by changing the inner droplet number

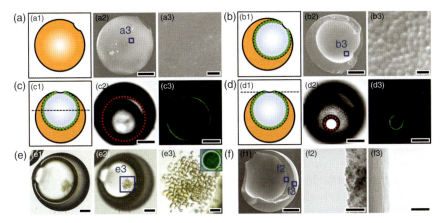

Figure 9.17 Hole–shell microparticles with flexibly functionalized core surface. (a, b) Schematic illustrations (a1, b1) and SEM images of the smooth outer surface (a2, a3) and rough core surface modified with FITC-PNIPAM nanogels (b2, b3). (c, d) Schematic illustrations and CLSM images that focus on the equator cross-section of the hollow core (c) and the hole mouth (d) of the hole–shell microparticle. The red circles in (c2) and (d2) indicate the location of FITC-PNIPAM nanogels shown in the CLSM images (c3, d3). (e) Optical micrographs of hole–shell microparticles captured with algae cells. (f) SEM (f1) and TEM (f2, f3) images showing the core surface of hole–shell microparticles modified with gold nanoparticles (f2) (f1) and their unmodified outer surface (f3). Scale bars are 500 nm in (a3) and (b3), 10 μm in (e3), 50 nm in (f2) and (f3), and 50 μm for the rest. (Wang et al. 2013 [28]. Reproduced with permission of John Wiley & sons.)

via flow rate adjustment, microparticles with dual hole–shell structures can be fabricated (Figure 9.16g).

9.5.3 Precise Control over the Functionality of Hollow Core Surface

Flexibility of the fabrication strategy for independently functionalizing the core surface of hole–shell microparticles is first demonstrated by coating the core surface with FITC-PNIPAM nanogels. As shown in the SEM images of Figure 9.17a,b, a smooth outer surface (Figure 9.17a) and a rough core surface coated with densely packed nanogels (Figure 9.17b) can be observed for the hole–shell microparticle. The CLSM (confocal laser scanning microscope) images in Figure 9.17c,d also confirm the locations of FITC-PNIPAM nanogels on the core surface. Based on the thermo-responsive property of PNIPAM for volume-phase transitions between a hydrophilic swollen state and a hydrophobic shrunken state, these anchored FITC-PNIPAM nanogels allow control of the wettability of the modified core surface by changing the temperature. Since cell behaviors can be influenced by surface wettability [33], the hole–shell microparticles contain core surface with such a controllable wettability could be potential for cell capture and confined culture. Figure 9.17e demonstrates the cell capture by using the hole–shell microparticles to capture algae cells. Based on the versatility of the modification strategy, the hole–shell microparticles can be flexibly modified with functional microparticles, such as gold nanoparticles with excellent catalytic property [34], as shown in Figure 9.17f, for modifying their core surface.

9.5.4 Hole–Shell Microparticles for Controlled Capture and Confined Microreaction

The diverse uses of hole–shell microparticles are further demonstrated by using the hole–shell microparticles for selective capture of microspheres, and confined synthesis of functional materials. Under ultrasonic treatment, microspheres with diameter fitting the size of the single hole and hollow core can be trapped into the hole–shell microparticles based on "lock–key" size match for 1 : 1 capture, as shown in Figure 9.18a. Moreover, under ultrasonic vibration, the size-screening effect of the single hole allows selective loading of microspheres (red color) with diameters smaller than the size of the single hole from larger ones in water for size classification (Figure 9.18b). The hole–shell microparticles can be further incorporated with magnetic nanoparticles in their shell for magnetic-guided collection of the captured microspheres for easy separation. Furthermore,

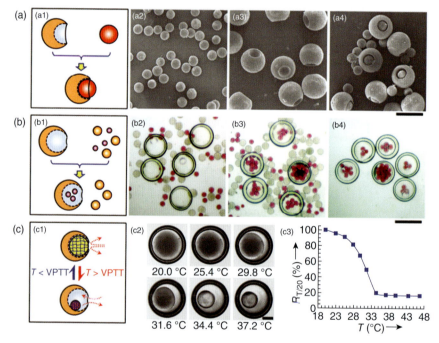

Figure 9.18 Hole–shell microparticles for controlled capture (a), size classification (b) and confined microreaction (c). (a) SEM images showing bowl-shaped microparticles for capturing microspheres based on "lock–key" size match. (b) Optical micrographs showing fishbowl-shaped microparticles for selectively capturing smaller microspheres from larger microspheres for size classification. (c) Confined synthesis of thermo-responsive PNIPAM microgel in fishbowl-shaped microparticle for controlling the "open–close" of the single hole. Optical micrographs (c2) and temperature-dependent volume change curve (c3) showing the thermo-responsive volume change property of the inside PNIPAM microgel. $R_{T/20}$ is defined as the ratio of microgel volume at T (°C) to that at 20 °C. Scale bars are 200 μm in (a) and (b), and 50 μm in (c). (Wang et al. 2013 [28]. Reproduced with permission of John Wiley & sons.)

the hole–shell microparticles can be employed as microreactors for confined synthesis of thermo-responsive PNIPAM microgels inside the core. Such a PNIPAM microgel inside the hole–shell microparticle can serve as a smart gate for controlling the "open–close" of the single hole (Figure 9.18c), thus providing a simple way to control the mass transfer into and out of the hollow core.

9.6 Summary

In this chapter, microfluidic strategies for fabrication of controllable hole–shell microparticles from evolved double emulsions with eccentric core–shell structures are introduced. Such eccentric core–shell structures can be created by employing the density mismatch between inner and outer droplets in double emulsions or partial dewetting of the outer droplet on the inner droplet. The resultant hole–shell microparticles with thermo-responsive shell can be used as soft robotics for motion generation and as actuators for sensing and actuating. The resultant hole–shell microparticles with polyETPTA shell and well-defined hollow core and single hole can be used as microcontainers for controlled capturing and confined microreaction. The strategies introduced in the chapter provide new approach for controllable fabrication of hole–shell microparticles for developing novel soft robotics, sensors, actuators, microcarriers, and microreactors.

References

1 Im, S.H., Jeong, U.Y., and Xia, Y.N. (2005) Polymer hollow particles with controllable holes in their surfaces. *Nat. Mater.*, **4**, 671–675.
2 Zhu, Y.F., Shi, J.L., Shen, W.H., Dong, X.P., Feng, J.W., Ruan, M.L., and Li, Y.S. (2005) Stimuli-responsive controlled drug release from a hollow mesoporous silica sphere/polyelectrolyte multilayer core-shell structure. *Angew. Chem. Int. Ed.*, **44**, 5083–5087.
3 Amstad, E., Kim, S.-H., and Weitz, D.A. (2012) Photo- and thermoresponsive polymersomes for triggered release. *Angew. Chem. Int. Ed.*, **51**, 12499–12503.
4 De Rose, R., Zelikin, A.N., Johnston, A.P.R., Sexton, A., Chong, S.-F., Cortez, C., Mulholland, W., Caruso, F., and Kent, S.J. (2008) Binding, internalization, and antigen presentation of vaccine-loaded nanoengineered capsules in blood. *Adv. Mater.*, **20**, 4698–4703.
5 Ota, S., Yoshizawa, S., and Takeuchi, S. (2009) Microfluidic formation of monodisperse, cell-sized, and unilamellar vesicles. *Angew. Chem. Int. Ed.*, **48**, 6533–6537.
6 Ma, S., Thiele, J., Liu, X., Bai, Y., Abell, C., and Huck, W.T.S. (2012) Fabrication of microgel particles with complex shape via selective polymerization of aqueous two-phase systems. *Small*, **8**, 2356–2360.
7 Chu, L.Y., Yamaguchi, T., and Nakao, S. (2002) A molecular-recognition microcapsule for environmental stimuli-responsive controlled release. *Adv. Mater.*, **14**, 386–389.

8 Dinsmore, A.D., Hsu, M.F., Nikolaides, M.G., Marquez, M., Bausch, A.R., and Weitz, D.A. (2002) Colloidosomes: selectively permeable capsules composed of colloidal particles. *Science*, **298**, 1006–1009.

9 Guan, G., Zhang, Z., Wang, Z., Liu, B., Gao, D., and Xie, C. (2007) Single-hole hollow polymer microspheres toward specific high-capacity uptake of target species. *Adv. Mater.*, **19**, 2370–2374.

10 Sacanna, S., Irvine, W.T.M., Chaikin, P.M., and Pine, D.J. (2010) Lock and key colloids. *Nature*, **464**, 575–578.

11 Minami, H., Kobayashi, H., and Okubo, M. (2005) Preparation of hollow polymer particles with a single hole in the shell by SaPSeP. *Langmuir*, **21**, 5655–5658.

12 Ma, G.H., Su, Z.G., Omi, S., Sundberg, D., and Stubbs, J. (2003) Microencapsulation of oil with poly(styrene-N,N-dimethylaminoethyl methacrylate) by SPG emulsification technique: effects of conversion and composition of oil phase. *J. Colloid Interface Sci.*, **266**, 282–294.

13 Han, J., Song, G., and Guo, R. (2006) A facile solution route for polymeric hollow spheres with controllable size. *Adv. Mater.*, **18**, 3140–3144.

14 Li, M. and Xue, J. (2011) Facile route to synthesize polyurethane hollow microspheres with size-tunable single holes. *Langmuir*, **27**, 3229–3232.

15 Lim, Y.T., Kim, J.K., Noh, Y.-W., Cho, M.Y., and Chung, B.H. (2009) Multifunctional silica nanocapsule with a single surface hole. *Small*, **5**, 324–328.

16 Yow, H.N. and Routh, A.F. (2008) Colloidal buckets formed via internal phase separation. *Soft Matter*, **4**, 2080–2085.

17 Yin, W. and Yates, M.Z. (2008) Effect of interfacial free energy on the formation of polymer microcapsules by emulsification/freeze-drying. *Langmuir*, **24**, 701–708.

18 Chu, L.-Y., Utada, A.S., Shah, R.K., Kim, J.-W., and Weitz, D.A. (2007) Controllable monodisperse multiple emulsions. *Angew. Chem. Int. Ed.*, **46**, 8970–8974.

19 Wang, W., Xie, R., Ju, X.-J., Luo, T., Liu, L., Weitz, D.A., and Chu, L.-Y. (2011) Controllable microfluidic production of multicomponent multiple emulsions. *Lab Chip*, **11**, 1587–1592.

20 Utada, A.S., Lorenceau, E., Link, D.R., Kaplan, P.D., Stone, H.A., and Weitz, D.A. (2005) Monodisperse double emulsions generated from a microcapillary device. *Science*, **308**, 537–541.

21 Wang, J., Hu, Y., Deng, R., Xu, W., Liu, S., Liang, R., Nie, Z., and Zhu, J. (2012) Construction of multifunctional photonic crystal microcapsules with tunable shell structures by combining microfluidic and controlled photopolymerization. *Lab Chip*, **12**, 2795–2798.

22 Wang, W., Yao, C., Zhang, M.-J., Ju, X.-J., Xie, R., and Chu, L.-Y. (2013) Thermo-driven microcrawlers fabricated via a microfluidic approach. *J. Phys. D: Appl. Phys.*, **46**, 114007.

23 Zhou, M.-Y., Xie, R., Yu, Y.-L., Chen, G., Ju, X.-J., Yang, L., Liang, B., and Chu, L.-Y. (2009) Effects of surface wettability and roughness of microchannel on flow behaviors of thermo-responsive microspheres therein during the phase transition. *J. Colloid Interface Sci.*, **336**, 162–170.

24 Chu, L.-Y., Kim, J.-W., Shah, R.K., and Weitz, D.A. (2007) Monodisperse thermoresponsive microgels with tunable volume-phase transition kinetics. *Adv. Funct. Mater.*, **17**, 3499–3504.

25 Liu, Y.-M., Wang, W., Zheng, W.-C., Ju, X.-J., Xie, R., Zerrouki, D., Deng, N.-N., and Chu, L.-Y. (2013) Hydrogel-based microactuators with remote-controlled locomotion and fast Pb^{2+}-response for micromanipulation. *ACS Appl. Mater. Interfaces*, **5**, 7219–7226.

26 Delamarche, E., Bernard, A., Schmid, H., Bietsch, A., Michel, B., and Biebuyck, H. (1998) Microfluidic networks for chemical patterning of substrate: design and application to bioassays. *J. Am. Chem. Soc.*, **120**, 500–508.

27 Wyss, H.M., Franke, T., Mele, E., and Weitz, D.A. (2010) Capillary micromechanics: measuring the elasticity of microscopic soft objects. *Soft Matter*, **6**, 4550–4555.

28 Wang, W., Zhang, M.-J., Xie, R., Ju, X.-J., Yang, C., Mou, C.-L., Weitz, D.A., and Chu, L.-Y. (2013) Hole-shell microparticles from controllably evolved double emulsions. *Angew. Chem. Int. Ed.*, **52**, 8084–8087.

29 Kim, S.-H., Jeon, S.-J., and Yang, S.-M. (2008) Optofluidic encapsulation of crystalline colloidal arrays into spherical membrane. *J. Am. Chem. Soc.*, **130**, 6040–6046.

30 Poulin, P. and Bibette, J. (1998) Adhesion of water droplets in organic solvent. *Langmuir*, **14**, 6341–6343.

31 Aronson, M.P. and Princen, H.M. (1980) Contact angles associated with thin liquid-films in emulsions. *Nature*, **286**, 370–372.

32 Bremond, N. and Bibette, J. (2012) Exploring emulsion science with microfluidics. *Soft Matter*, **8**, 10549–10559.

33 Sun, T. and Qing, G. (2011) Biomimetic smart interface materials for biological applications. *Adv. Mater.*, **23**, H57–H77.

34 Daniel, M.C. and Astruc, D. (2004) Gold nanoparticles: assembly, supramolecular chemistry, quantum-size-related properties, and applications toward biology, catalysis, and nanotechnology. *Chem. Rev.*, **104**, 293–346.

10

Microfluidic Fabrication of Controllable Multicompartmental Microparticles

10.1 Introduction

Microparticles with compartments are widely used as encapsulation systems for protection of active species [1–3], controlled release of various substances [4–8], and confined microreaction of chemicals [9–11]. Multicompartmental microparticles that enable the isolated co-encapsulation and on-demand release of different components are important as delivery carriers for diverse incompatible actives and/or as microreactor vessels for the reaction of diverse active reactants [12, 13]. However, in the area of fabricating microcapsules for the encapsulation of different components, the major challenge is creation of separate multiple compartments into a single capsule for isolated co-encapsulation and triggered release, especially with precise control over the structure of multicompartments and the encapsulation level and release behavior of each component.

Typically, several strategies are employed to create multiple compartments into a single microparticle for separate encapsulation of diverse components [14–17]. Polyelectrolyte microparticles [16, 17] and alginate microparticles [14], each with compartment-in-compartment structures, are developed by stepwise encapsulation of one capsule in another capsule for remote control of microbioreactions and immunoprotection, respectively. However, troublesome multistep process is required for controllable encapsulation of different components in separate compartments of these microparticles. For simple encapsulation of different components in one step, multicompartmental titania microparticles are developed by compound-fluidic electrospray technique [15]. However, for the abovementioned microparticles, accurate control over the structure uniformity of the microparticles and the encapsulation characteristic of each component still remain difficult to achieve. Moreover, using these techniques, it is difficult to engineer microparticles with more hierarchical multicompartment structures.

With great control of droplets [18–21], emulsions generated from microfluidic techniques offer excellent templates for the synthesis of multicompartmental microparticles [22–26]. Based on microfluidic technique, the authors' group developed microparticles with versatile and controllable multicompartment structures from multiple emulsions. In this chapter, the microfluidic fabrication of these multicompartmental microparticles and their performances for co-encapsulation and controlled release are introduced.

Microfluidics for Advanced Functional Polymeric Materials, First Edition. Liang-Yin Chu and Wei Wang.
© 2017 Wiley-VCH Verlag GmbH & Co. KGaA. Published 2017 by Wiley-VCH Verlag GmbH & Co. KGaA.

10.2 Microfluidic Strategy for the Fabrication of Controllable Multicompartmental Microparticles

Microfluidic strategy for fabrication of multicompartmental microparticles generally employs the phase-separated droplet-in-droplet structures as templates to engineer the inner compartments. For example, multicomponent double emulsions, which enable controllable co-encapsulation of different inner droplets [27], can be used as templates for the creation of microparticles with separate multicompartments at the same inner level for isolated co-encapsulation of diverse incompatible components. By converting the outer layer of the double emulsions into solid or hydrogel shell, simple one-step encapsulation of different components into one single microparticle can be realized. Meanwhile, the inner structure of the microparticles and the encapsulation characteristics of different components can be precisely manipulated based on the excellent control of microfluidics over the number, ratio, and size of different inner droplets [27]. Moreover, higher order multiple emulsions such as triple emulsions and quadruple emulsions can provide more hierarchical phase-separated structures with multiply nested droplets as templates for fabricating more complex multicompartmental microparticles. After converting the outer layer of triple emulsions into solid or hydrogel shell, their second inner droplets and innermost droplets can be used as separate compartments for encapsulating both oil- and water-soluble components. Further improvement of the multicompartment structures can be achieved by using quadruple emulsions as templates. These microfluidic strategies provide new opportunities to design multicompartmental microparticles for controllable and isolated co-encapsulation of diverse components and on-demand release of these components with different triggering mechanisms. In the following sections, multicompartmental microparticles, with multi-core/shell structures engineered from multicomponent double emulsions and with Trojan-horse-like structures engineered from triple emulsions, are introduced in detail.

10.3 Multi-core/Shell Microparticles for Co-encapsulation and Synergistic Release

10.3.1 Microfluidic Fabrication of Multi-core/Shell Microparticles

For the synthesis of the multicompartmental microparticles with multi-core/shell structures, monodisperse O/W/O double emulsions containing two types of inner droplets with distinct contents are generated in a two-branch microfluidic device as templates. Typically, soybean oil with 5% (w/v) polyglycerol polyricinoleate (PGPR 90) is used as the oil phase. Aqueous solution with 1% (w/v) Pluronic F127 and 5% (w/v) glycerol is used as the aqueous phase. As shown in Figure 10.1, inner fluids $F_{i\text{-}1}$ and $F_{i\text{-}2}$ are the oil phases with and without Sudan III, respectively; middle fluid $F_{m\text{-}1}$, $F_{m\text{-}2}$, $F_{m\text{-}3}$, and $F_{m\text{-}4}$ are aqueous phases containing 11.3% (w/v) N-isopropylacrylamide (NIPAM) monomer, 0.77% (w/v) N,N'-methylenebisacrylamide (MBA) cross-linker,

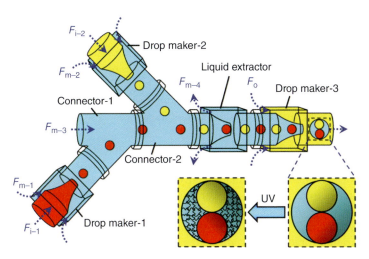

Figure 10.1 Microfluidic production of controlled quadruple-component O/W/O double emulsions for synthesizing multicompartmental microparticles. (Wang et al. 2012 [28]. Reproduced with permission of De-Gruyter.)

and 0.6% (w/v) ammonium persulfate (APS) initiator; and outer fluid F_o is the oil phase. To generate the double emulsion templates, the aqueous phase and oil phase fluids are injected into the microfluidic device through syringe pumps. First, F_{i-1} and F_{i-2} are separately emulsified into oil droplets, respectively, by F_{m-1} and F_{m-2} in drop maker-1 and drop maker-2. After being collected by the connectors, these oil droplets flow into the main microchannel. Then, the middle fluids together with the oil droplets are further emulsified by F_o in drop maker-3. This allows isolated co-encapsulation of the different oil droplets into the double emulsions. The F_{m-3} injected into connector-1 and the F_{m-4} drained out of the liquid extractor are, respectively, designed to add and remove liquid from the main stream to tune the volume fraction of disperse phase in continuous phase. The generated double emulsions are collected in a collection solution, which is oil phase with 1% (w/v) photoinitiator 2,2-dimethoxy-2-phenylacetophenone (BDK) as photoinitiator. To synthesize the multi-core/shell microparticles, the NIPAM-containing middle aqueous layer of the O/W/O double emulsions is converted into thermo-responsive hydrogel shell by UV polymerization and their different inner oil droplets become separate compartments for storing diverse components (Figure 10.1). This enables one-step encapsulation of different components into one single microparticle; meanwhile, the encapsulation of each component in the microparticle can be individually and precisely optimized by controlling the emulsion structures.

10.3.2 Multi-core/Shell Microparticles for Controllable Co-encapsulation

With excellent control of the structure of multicomponent multiple emulsions, microfluidic technique allows precise control of the co-encapsulation characteristics of different encapsulated component. Since the inner oil droplets are individually generated in different drop-making units, the encapsulation of the component in each inner oil droplet can be individually optimized by changing the composition of each inner fluid. Moreover, the formation rate and size of the inner oil droplets can be adjusted by changing the flow rates of the dispersed and continuous phase, and the microchannel dimension. Thus, by accurate control of the number and ratio of the inner oil droplets, further optimization of the encapsulation of each component in the double emulsions, as well as in the resultant microparticles, can be realized.

First, the precise control of the number and ratio of inner oil droplets with microfluidics is shown in Figure 10.2 to demonstrate the excellent manipulation of the multicomponent encapsulation in double emulsions. Because the inner red and transparent droplets are periodically formed in different drop makers, these two types of inner droplets are regularly and alternately aligned in the main microchannel. Thus, the formation rates and the numbers of inner red and transparent droplets in the droplet array can be individually tuned by changing the flow rates [27]. After further emulsification, monodisperse quadruple-component double emulsions containing one red droplet and one transparent droplet (1R1T) (Figure 10.2a), one red droplet and two transparent droplets (1R2T) (Figure 10.2b), two red droplets and one transparent droplet (2R1T) (Figure 10.2c), and two red droplets and two transparent droplets (2R2T)

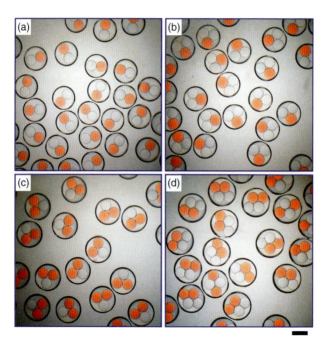

Figure 10.2 Optical micrographs of monodisperse quadruple-component O/W/O double emulsions. Each of the emulsions contains one red droplet and one transparent droplet (1R1T) in (a), one red droplet and two transparent droplets (1R2T) in (b), two red droplets and one transparent droplet (2R1T) in (c), and two red droplets and two transparent droplets (2R2T) in (d). Scale bar is 200 μm. (Wang et al. 2012 [28]. Reproduced with permission of De-Gruyter.)

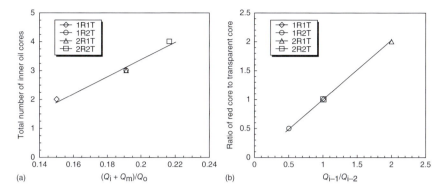

Figure 10.3 Effect of relative flow rates on the number (a) and ratio (b) of the co-encapsulated different oil cores. Both the values of Q_{i-1}/Q_{m-1} and Q_{i-2}/Q_{m-2} for drop maker-1 and drop maker-2 are fixed at 1/3. (Wang et al. 2012 [28]. Reproduced with permission of De-Gruyter.)

(Figure 10.2d) can be controllably formed. Figure 10.3a,b shows the effect of relative flow rates on the number and ratio of the inner oil droplets. The value of Q_i/Q_m is kept at 1/3 for drop maker-1 and drop maker-2 to ensure the generation of oil droplets with the same size. With increasing $(Q_i + Q_m)/Q_o$, the total number of the inner oil droplets in the double emulsions increases (Figure 10.3a), while with increasing Q_{i-1}/Q_{i-2}, the ratio of inner red droplets to transparent droplets increases (Figure 10.3b). These controllable multicomponent double emulsions provide excellent templates for the synthesis of multi-core/shell microparticles.

Based on the precise control of the emulsion structures, the internal structure of the multicompartments and the encapsulation characteristic of each component in the microparticles can be precisely engineered. As illustrated in Figure 10.4, the structures of the resultant multi-core/shell microparticles are similar to those of their templates shown in Figure 10.2. The optical

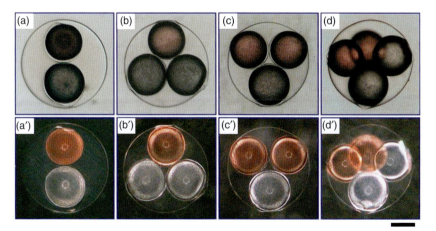

Figure 10.4 Bright-field (a–d) and dark-field (a′–d′) optical micrographs of multi-core/shell microparticles with different oil droplets for isolated co-encapsulation of diverse lipophilic components. Each of the microparticles is with 1R1T structure in (a, a′), 1R2T structure in (b, b′), 2R1T structure in (c, c′), and 2R2T structure in (d, d′). Scale bar is 100 μm. (Wang et al. 2012 [28]. Reproduced with permission of De-Gruyter.)

micrographs of the microparticles in bright-field (Figure 10.4a–d) and dark-field (Figure 10.4a′–d′) clearly show their multi-core/shell structures, which contain different inner oil droplets with well-controlled number and ratio. At temperatures below the volume-phase transition temperature (VPTT) of PNIPAM (poly(N-isopropylacrylamide)), the PNIPAM hydrogel shell, which is in a swollen and hydrophilic state, can provide protection to their inner oil droplets against coalescence. The individual oil droplets in the microparticles work as separate compartments for encapsulation of different components, which ensures the proper isolation of the encapsulated components from each other. The multi-core/shell microparticles provide excellent systems for controllable co-encapsulation of diverse incompatible components.

10.3.3 Multi-core/Shell Microparticles for Synergistic Release

Based on the thermo-responsive swelling/shrinking volume transitions of PNIPAM network upon heating, the PNIPAM shell can serve as an excellent thermo-triggered actuator for controlled release of the encapsulated diverse components. As illustrated in Figure 10.5a,b, upon increasing the environmental

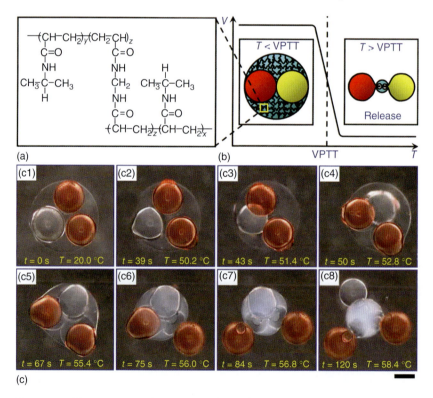

Figure 10.5 Thermo-triggered synergistic release of co-encapsulated different components from the multi-core/shell microparticle. (a) Chemical structure of PNIPAM shell. (b) Illustration showing the thermo-triggered synergistic release upon heating. (c) Dark-field optical snapshots showing the thermo-triggered release process of the co-encapsulated different components. Scale bar is 100 μm. (Wang et al. 2012 [28]. Reproduced with permission of De-Gruyter.)

temperature above the VPTT, the initially swollen PNIPAM shell can shrink and squeeze out the inner oil droplets for controlled release. Figure 10.5c shows the thermo-triggered controlled release of the co-encapsulated diverse components from a multi-core/shell microparticle with 2R1T structure. Upon increasing the temperature from 20 to 60 °C, the thermo-responsive PNIPAM shell shrinks dramatically and keeps extruding the inner oil droplets. This can produce increasing pressures on both the oil droplets and the PNIPAM shell. Then, because of the limited mechanical strength, the PNIPAM shell finally ruptures, but still keeps shrinking due to the increasing temperature. Thus, the inner oil droplets are squeezed out of the microparticle through the ruptures on the PNIPAM shell, leading to synergistic release of the encapsulated diverse components. These multi-core/shell microparticles for the isolated co-encapsulation and controlled synergistic release of diverse components show great potential as microcarriers and microreactors for incompatible actives and reactants.

10.4 Trojan-Horse-Like Microparticles for Co-delivery and Programmed Release

10.4.1 Fabrication of Trojan-Horse-Like Microparticles from Triple Emulsions

Triple emulsions provide multilayered structures with multiple nested droplets as compartments for template synthesis of Trojan-horse-like microparticles. The outer layer of the triple emulsions can become the microparticle shell, while the innermost droplets, protected by the next inner droplets and the shell, can become separated compartments for the encapsulation of different components. To demonstrate this potential, triple emulsions are used as templates to fabricate a Trojan-horse-like microparticle made with a thermo-sensitive PNIPAM shell for controlled release of actives. To fabricate such Trojan-horse-like microparticles from triple emulsions, typically, as shown in Figure 10.6, poly(dimethylsiloxane) oil (PDMS) (viscosity: 100 cSt) containing 2 wt% Dow Corning 749 is used as the outermost fluid; an aqueous solution containing 10 wt% glycerol, 2% (w/v) poly(vinyl alcohol) (PVA), 11.3% (w/v) NIPAM, 0.8% (w/v) sodium acrylate, 0.77% (w/v) MBA, and 0.6% (w/v) APS is used as the middle fluid (II); PDMS oil (viscosity: 10 cSt) containing 5 wt% Dow Corning 749

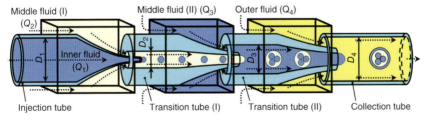

Figure 10.6 Schematic illustration of the microfluidic device for the generation of triple emulsion templates (Chu et al. 2007 [20]. Reproduced with permission of John Wiley & Sons.)

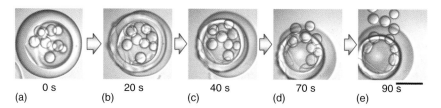

Figure 10.7 Trojan-horse-like microparticles for pulsed release. Scale bar is 200 μm. (Chu et al. 2007 [20]. Reproduced with permission of John Wiley & Sons.)

and 8% (v/v) accelerator N,N,N',N'-tetramethylethylenediamine (TEMED) is used as the middle fluid (I); and an aqueous solution containing 10 wt% glycerol and 2% (w/v) PVA is used as the inner fluid. Since TEMED is both oil- and water soluble, during emulsion generation, the TEMED in middle fluid (I) can diffuse into middle fluid (II) to activate the initiator APS via a redox reaction for polymerizing the monomers. This can produce Trojan-horse-like microparticles, each consisting of a thermo-sensitive PNIPAM shell and an inner oil droplet containing several water droplets inside, as shown in Figure 10.7a.

Upon increasing the environmental temperature from 25 to 50 °C, above the VPTT of PNIPAM, the thermo-sensitive PNIPAM shell rapidly shrinks by expelling water. However, because of the incompressible inner oil, the hydrogel shell finally breaks, providing pulsed release of the innermost water droplets into the environmental medium, as shown in Figure 10.7b–e. Thus, the microparticles can provide a release process with a Trojan-horse-like style, which can protect the innermost water droplets in the hydrogel shell until the thermo-trigger for controlled release. Moreover, besides the innermost water droplets for encapsulating water-soluble actives, the inner oil droplet in the microparticle can also be employed as compartment for encapsulating oil-soluble actives. Thus, co-delivery of both water- and oil-soluble components in one single microparticle can be achieved. These results show the utility of microfluidic techniques for controllable generation of multicompartmental microparticles with multiple internal compartments that remain separate from each other and highlight their potential for creation of highly engineered structures for controlled release.

10.5 Summary

In this chapter, flexible microfluidic strategies for controllable fabrication of multicompartmental microparticles are introduced. Accurate control of the internal compartment structures can be achieved by precisely manipulating the nested droplet-in-droplet structures of the emulsion templates. Double emulsions containing different inner droplets enable template synthesis of microparticles with separate compartments located in the same internal level. Triple emulsions containing a droplet encapsulated with smaller droplets allow fabrication of Trojan-horse-like microparticles with hierarchical inner compartments. Even more hierarchical and complex multicompartmental microparticles can be developed from higher order multiple emulsions. For example, quadruple emulsions with an additional liquid layer allow engineering of a second shell into the multicompartmental microparticles to construct compartment-in-compartment structures for encapsulating distinct components without cross-contamination. These multicompartmental microparticles from microfluidic multiple emulsions can provide separate compartments for controllable and isolated co-encapsulation of diverse incompatible components. Moreover, controlled programmed release of these encapsulated multiple components can be achieved by engineering different release mechanisms into the functional shell materials. These controllable multicompartmental microparticles are promising for applications in fields such as drug delivery, controlled release, and microreaction.

References

1 Choi, S.-W., Zhang, Y., and Xia, Y. (2009) Fabrication of microbeads with a controllable hollow interior and porous wall using a capillary fluidic device. *Adv. Funct. Mater.*, **19**, 2943–2949.
2 De Rose, R., Zelikin, A.N., Johnston, A.P.R., Sexton, A., Chong, S.-F., Cortez, C., Mulholland, W., Caruso, F., and Kent, S.J. (2008) Binding, internalization, and antigen presentation of vaccine-loaded nanoengineered capsules in blood. *Adv. Mater.*, **20**, 4698–4703.
3 Kim, J.W., Jung, M.O., Kim, Y.J., Ryu, J.H., Kim, J., Chang, I.S., Lee, O.S., and Suh, K.D. (2005) Stabilization of enzyme by exclusive volume effect in hydrophobically controlled polymer microcapsules. *Macromol. Rapid Commun.*, **26**, 1258–1261.
4 Liu, L., Wang, W., Ju, X.-J., Xie, R., and Chu, L.-Y. (2010) Smart thermo-triggered squirting capsules for nanoparticle delivery. *Soft Matter*, **6**, 3759–3763.
5 Wang, W., Liu, L., Ju, X.-J., Zerrouki, D., Xie, R., Yang, L., and Chu, L.-Y. (2009) A novel thermo-induced self-bursting microcapsule with magnetic-targeting property. *ChemPhysChem*, **10**, 2405–2409.
6 De Geest, B.G., McShane, M.J., Demeester, J., De Smedt, S.C., and Hennink, W.E. (2008) Microcapsules ejecting nanosized species into the environment. *J. Am. Chem. Soc.*, **130**, 14480–14482.
7 Skirtach, A.G., Javier, A.M., Kreft, O., Koehler, K., Alberola, A.P., Moehwald, H., Parak, W.J., and Sukhorukov, G.B. (2006) Laser-induced release of encapsulated materials inside living cells. *Angew. Chem. Int. Ed.*, **45**, 4612–4617.
8 Chu, L.Y., Yamaguchi, T., and Nakao, S. (2002) A molecular-recognition microcapsule for environmental stimuli-responsive controlled release. *Adv. Mater.*, **14**, 386–389.
9 He, Q., Duan, L., Qi, W., Wang, K., Cui, Y., Yan, X., and Li, J. (2008) Microcapsules containing a biomolecular motor for ATP biosynthesis. *Adv. Mater.*, **20**, 2933–2937.

10 Duan, L., He, Q., Wang, K., Yan, X., Cui, Y., Moewald, H., and Li, J. (2007) Adenosine triphosphate biosynthesis catalyzed by F0F1 ATP synthase assembled in polymer microcapsules. *Angew. Chem. Int. Ed.*, **46**, 6996–7000.

11 Choi, W.S., Park, J.H., Koo, H.Y., Kim, J.Y., Cho, B.K., and Kim, D.Y. (2005) "Grafting-from" polymerization inside a polyelectrolyte hollow-capsule microreactor. *Angew. Chem. Int. Ed.*, **44**, 1096–1101.

12 Sun, B.J., Shum, H.C., Holtze, C., and Weitz, D.A. (2010) Microfluidic melt emulsification for encapsulation and release of actives. *ACS Appl. Mater. Interfaces*, **2**, 3411–3416.

13 Johnston, A.P.R., Such, G.K., and Caruso, F. (2010) Triggering release of encapsulated cargo. *Angew. Chem. Int. Ed.*, **49**, 2664–2666.

14 Kim, J., Arifin, D.R., Muja, N., Kim, T., Gilad, A.A., Kim, H., Arepally, A., Hyeon, T., and Bulte, J.W.M. (2011) Multifunctional capsule-in-capsules for immunoprotection and trimodal imaging. *Angew. Chem. Int. Ed.*, **50**, 2317–2321.

15 Chen, H., Zhao, Y., Song, Y., and Jiang, L. (2008) One-step multicomponent encapsulation by compound-fluidic electrospray. *J. Am. Chem. Soc.*, **130**, 7800–7801.

16 Kreft, O., Skirtach, A.G., Sukhorukov, G.B., and Moehwald, H. (2007) Remote control of bioreactions in multicompartment capsules. *Adv. Mater.*, **19**, 3142–3145.

17 Kreft, O., Prevot, M., Moehwald, H., and Sukhorukov, G.B. (2007) Shell-in-shell microcapsules: a novel tool for integrated, spatially confined enzymatic reactions. *Angew. Chem. Int. Ed.*, **46**, 5605–5608.

18 Abate, A.R. and Weitz, D.A. (2009) High-order multiple emulsions formed in poly(dimethylsiloxane) microfluidics. *Small*, **5**, 2030–2032.

19 Shah, R.K., Shum, H.C., Rowat, A.C., Lee, D., Agresti, J.J., Utada, A.S., Chu, L.-Y., Kim, J.-W., Fernandez-Nieves, A., Martinez, C.J., and Weitz, D.A. (2008) Designer emulsions using microfluidics. *Mater. Today*, **11**, 18–27.

20 Chu, L.-Y., Utada, A.S., Shah, R.K., Kim, J.-W., and Weitz, D.A. (2007) Controllable monodisperse multiple emulsions. *Angew. Chem. Int. Ed.*, **46**, 8970–8974.

21 Utada, A.S., Lorenceau, E., Link, D.R., Kaplan, P.D., Stone, H.A., and Weitz, D.A. (2005) Monodisperse double emulsions generated from a microcapillary device. *Science*, **308**, 537–541.

22 Kim, S.-H., Hwang, H., Lim, C.H., Shim, J.W., and Yang, S.-M. (2011) Packing of emulsion droplets: structural and functional motifs for multi-cored microcapsules. *Adv. Funct. Mater.*, **21**, 1608–1615.

23 Seiffert, S., Thiele, J., Abate, A.R., and Weitz, D.A. (2010) Smart microgel capsules from macromolecular precursors. *J. Am. Chem. Soc.*, **132**, 6606–6609.

24 Gong, X., Wen, W., and Sheng, P. (2009) Microfluidic fabrication of porous polymer microspheres: dual reactions in single droplets. *Langmuir*, **25**, 7072–7077.

25 Kim, S.-H., Jeon, S.-J., and Yang, S.-M. (2008) Optofluidic encapsulation of crystalline colloidal arrays into spherical membrane. *J. Am. Chem. Soc.*, **130**, 6040–6046.

26 Nie, Z.H., Xu, S.Q., Seo, M., Lewis, P.C., and Kumacheva, E. (2005) Polymer particles with various shapes and morphologies produced in continuous microfluidic reactors. *J. Am. Chem. Soc.*, **127**, 8058–8063.
27 Wang, W., Xie, R., Ju, X.-J., Luo, T., Liu, L., Weitz, D.A., and Chu, L.-Y. (2011) Controllable microfluidic production of multicomponent multiple emulsions. *Lab Chip*, **11**, 1587–1592.
28 Wang, W., Luo, T., Ju, X.-J., Xie, R., Liu, L., and Chu, L.-Y. (2012) Microfluidic preparation of multicompartment microcapsules for isolated co-encapsulation and controlled release of diverse components. *Int. J. Nonlinear Sci. Numer. Simul.*, **13**, 325–332.

11

Microfluidic Fabrication of Functional Microfibers with Controllable Internals

11.1 Introduction

Microfibers show great potential in fabrication of 3D matrix for tissue engineering scaffolds [1–4], implantable therapeutic systems [5], medical patches [6–8], and energy storage systems [9, 10]. Especially, controllable creation of micro-sized compartmental internals within microfibers is crucial for enhancing their functions for many applications. For example, microfibers with tubular microchannels can serve as artificial microvessels for transporting fluid and culturing vascular cells [11–14], and as bio-microreactors for enzyme-immobilized biocatalysis [15]. Microfibers with well-tailored multicompartments can be used to controllably load actives that benefit cell growth or wound healing for developing advanced tissue engineering scaffolds or medical patches and to encapsulate phase change materials for fabricating energy storage systems. Therefore, development of microfibers with well-controlled micro-sized compartmental internals exhibits significant importance for their myriad applications.

Typically, microfibers can be generated by electrospinning [16, 17], wet spinning [18, 19], and microfluidic approaches [11, 20–29]. The electrospinning and wet spinning are well-established techniques that can produce fibers with submicron and micron diameters [30]. However, it is difficult for these techniques to individually and precisely create controllable complex internals within the fibers. Recently developed microfluidic techniques, with excellent manipulation on microflows, show great potential for manufacturing microfibers with controllable structures [4, 11, 20, 31]. Typically, based on the cylinder core flow that is hydrodynamically shaped by continuous phase in microfluidic device, solid microfibers can be produced by using the core flow as templates [11, 22, 25, 32, 33]. Creation of tubular microchannel compartment in microfibers can be achieved by using tubular flows with cylinder-in-cylinder structures [14, 15, 21]. The change in flow rates offers a simple and efficient approach to control the microfiber structures. By using droplet-containing continuous flow in microfluidic channel as templates, more complex microfibers with separate compartments containing oil and particles are developed for multi-encapsulation [31]. However, the requirement of devices with outlet of different dimensions for controlling the microfiber diameter is usually troublesome. Therefore, better control of the internal structures, such as the size and shape of the compartment and the distance between compartments, is still desired.

In this chapter, simple and versatile microfluidic strategies for controllable fabrication of microfibers with controllable tubular, peapod-like and spindle-knot-like internals are introduced. The microfibers with core–sheath structures provide tubular internals for encapsulation of phase change materials for temperature regulation. The microfibers with peapod-like internals provide separate oil "peas" as compartments for controllable loading of actives for synergistic drug delivery. The microfibers with magnetic spindle-knots as internals enable excellent magnetic manipulation of the microfibers for controllably assembling spider-web-like structures for water collection.

11.2 Microfluidic Strategy for Fabrication of Functional Microfibers with Controllable Internals

Fabrication of functional microfibers with controllable internals can be achieved by using coaxial three-phase jets containing core flow and sheath flow from microfluidics as templates (Figure 11.1). By incorporating functional monomers or polymers in the sheath flow of the jet templates, *in situ* gelation or solidification of the sheath flow during jet flowing allows converting the sheath flow into the polymeric wall of the microfiber. Meanwhile, the cylindrical core flow becomes the tubular internal of the microfiber (Figure 11.1a,c). When the cylindrical core flow is emulsified into discrete droplets in the sheath flow, *in situ* gelation or solidification of the sheath flow can lead to microfibers with discrete droplets as separate multicompartments (Figure 11.1b–d). Moreover, when further solidifying the discrete droplets into solid particles and drying their hydrogel shell, microfibers with spindle-knot-like internals can be obtained (Figure 11.1e).

11.3 Core–Sheath Microfibers with Tubular Internals for Encapsulation of Phase Change Materials

In this section, the microfluidic strategy for fabrication of functional microfibers with controllable internals is first demonstrated by fabricating core–sheath poly(vinyl butyral) (PVB) microfibers with tubular internals containing phase change materials for temperature regulation.

11.3.1 Fabrication of Core–Sheath Microfibers with Tubular Internals

Two-stage glass-capillary microfluidic device is used for generation of core–sheath flow jet as templates for fabricating microfibers with tubular internals (Figure 11.2a). Briefly, two cylindrical capillaries, which are, respectively, used as the injection and transition tubes, are coaxially aligned within a square tube that is used as the collection tube. The inner dimension of the collection tube is 1.0 mm. The injection and transition tubes exhibit outer diameters of 960 µm, and inner diameters of 550 µm. Moreover, the injection and transition tubes contain tapered orifices with inner diameters of 80 and 400 µm, respectively. The microfluidic device is vertically placed for continuous

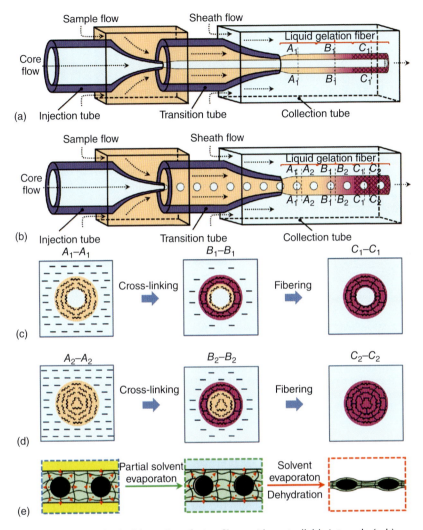

Figure 11.1 Microfluidic fabrication of microfibers with controllable internals. (a, b) Microfluidic devices for fabricating microfibers with tubular (a) and peapod-like (b) internals. (c, d) Different cross-sections of the tubular (c) and peapod-like (c, d) microfibers showing the cross-linking process. (He et al. 2015 [34]. Reproduced with permission of The Royal Society of Chemistry.) (e) Solidification of the inner droplets combined with drying of the microfiber shell for fabricating microfibers with spindle-knot-like internals. (He et al. 2015 [35]. Reproduced with permission of American Chemical Society.)

fabrication of the tubular microfibers (Figure 11.2a). The melting paraffin wax with typical melting point of 27 °C, dimethyl sulfoxide (DMSO) solution containing 14 wt% PVB resin with a molecular weight of 40 000–70 000 Da, and aqueous solution containing 1% (w/v) carboxymethylcellulose sodium (CMC) are used as the inner, middle, and outer fluids, respectively. CMC is used to improve the viscosity of aqueous fluid to ensure the stability of flow. The core–sheath flow jet templates are generated by pumping the three fluids

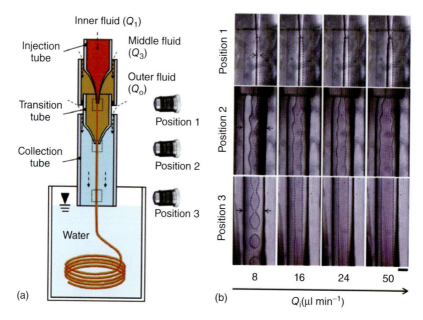

Figure 11.2 Microfluidic generation of core–sheath flow jet templates for synthesizing microfibers with tubular internals. (a) Schematic illustration of the microfluidic device for fabricating microfibers with tubular internals. (b) High-speed snapshots of the core–sheath flow jet, with a core of melting paraffin wax and a sheath of PVB solution, at different inner flow rates. The melting paraffin wax is dyed with LR300 for clear observation. The flow rates of middle and outer fluids are fixed at 100 and 250 μl min^{-1}, respectively. Scale bar is 200 μm. (Wen et al. 2015 [9]. Reproduced with permission of Elsevier.)

into the microfluidic device at appropriate flow rates. In the collection tube, the DMSO in the sheath phase of flow jet is gradually extracted into the outer fluid due to its water-soluble property. Thus, the PVB polymers in the sheath phase are concentrated and solidified to form the wall of tubular microfibers. The generated microfibers are directly collected into water bath for further DMSO extraction. The preparation of the microfibers is carried out in an incubator with constant temperature of 40 °C. Next, the tubular microfibers are immersed in water at 20 °C to crystallize the paraffin wax.

Core–sheath PVB microfibers with different structures of tubular internals are prepared by varying the flow rate (Q_i) of inner fluid, while fixing the flow rates of middle fluid (Q_m) and outer fluid (Q_o) at 100 and 250 ml min^{-1}, respectively. The high-speed snapshots in Figure 11.2b show the Q_i-dependent structure of the flow jets at the outlets of the injection tube (Position 1 in Figure 11.2b), the transition tube (Position 2 in Figure 11.2b), and the collection tube (Position 3 in Figure 11.2b). With increasing Q_i, the diameter of the cylindrical jet of inner fluid increases at Position 1. At Position 2, the flow of inner fluid in the jet template becomes unstable and the flow of middle fluid slightly expands. This is due to the sudden enlargement of microchannel and the low viscosity of paraffin wax (5.6 mPa s at 40 °C). When further flowing downstream at Position 3, the cylindrical jet of inner fluid at Q_i of 8 μl min^{-1} breaks into droplets, while those with Q_i higher than 8 μl min^{-1} become stable. Moreover, with increasing Q_i, the diameter of cylindrical jet of inner fluid increases while the outer diameter of the annular jet of middle fluid remains almost unchanged. This can lead to encapsulation of paraffin wax with higher content in the resultant microfiber. After solvent extraction of DMSO from the PVB-containing middle fluid, the core–sheath PVB microfibers with tubular internals containing high content of paraffin wax are prepared.

11.3.2 Morphological Characterization of the Core–Sheath Microfibers

Figure 11.3a,b shows the morphologies of PVB microfibers containing paraffin wax with (Figure 11.3b) and without (Figure 11.3a) LR300 dye. As compared with the solid PVB microfibers, the core–sheath PVB microfibers show pink color due to the encapsulation of LR300-dyed paraffin wax in the tubular internals. Moreover, the obtained microfibers are of several meters in length, indicating the power of the microfluidic technique for continuous microfiber fabrication. The structures of the core–sheath PVB microfibers are further characterized by scanning electronic microscope (SEM). Before characterization, the samples are mechanically fractured in liquid nitrogen to ensure the integrity of their cross-sectional microstructures, and then soaked in n-octane for 24 h to remove the encapsulated paraffin wax. Figure 11.3c–f shows SEM images showing the cross-sectional structure of the core–sheath PVB microfibers fabricated at different Q_i. All the microfibers exhibit core–sheath structures with hollow tubular internal protected by a PVB shell. As the Q_i increases from 8 to 50 μl min^{-1}, the diameter of the tubular internal increases, while the wall thickness of the microfibers decreases. The microfiber structures are consistent with those of their jet templates as observed at Position 3 in Figure 11.2b. Moreover, the PVB shell of the microfibers shows the typical structures consisting of a dense skin layer and a finger-like porous sublayer due to the solvent extraction. The skin layer is smooth and dense without any visible pores (Figure 11.3g,h), which provides good protection for encapsulation of paraffin wax.

11.3.3 Thermal Property of the Core–Sheath Microfibers

The phase transition temperatures and enthalpies of core–sheath PVB microfibers during the heating/cooling processes are measured by differential scanning calorimeter (DSC) at a nitrogen atmosphere. During the DSC measurement, the heating/cooling rate for increasing the temperature from 0 to 50 °C is set at 5 °C min^{-1}. Microfibers with weight of 5 mg are used as samples for each test. The microfibers after 10 heating/cooling cycles are also measured, which are loaded in a test tube and alternately immersed into water bath at 0 and 50 °C for 10 min for equilibration. The thermal stabilities of core–sheath PVB microfibers are tested by thermogravimetric analyzer at a heating rate of 10 °C min^{-1} from room temperature up to 600 °C under nitrogen atmosphere.

The DSC curves of paraffin wax and core–sheath PVB microfibers containing different contents of paraffin wax are shown in Figure 11.4. The codes PVB/8P, PVB/16P, PVB/24P, and PVB/50P represent the core–sheath PVB microfibers fabricated at Q_i of 8, 16, 24, and 50 μl min^{-1}, respectively. The melting starting

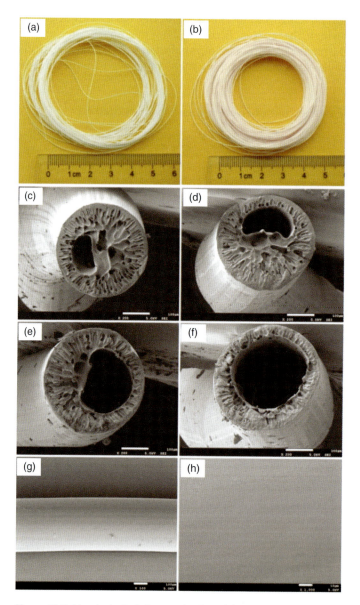

Figure 11.3 Morphological characterization of the core–sheath PVB microfibers. (a, b) Optical photos of solid PVB microfibers (a) and core–sheath PVB microfibers containing paraffin wax dyed with LR300 (b) ($Q_i = 24\,\mu l\,min^{-1}$). (c–h) SEM images of core–sheath PVB microfibers after removing paraffin wax, which are respectively fabricated at Q_i of $8\,\mu l\,min^{-1}$ (c), $16\,\mu l\,min^{-1}$ (d), $24\,\mu l\,min^{-1}$ (e, g, h), and $50\,\mu l\,min^{-1}$ (f). The Q_m and Q_o are fixed at 100 and $250\,\mu l\,min^{-1}$, respectively. (Wen *et al.* 2015 [9]. Reproduced with permission of Elsevier.)

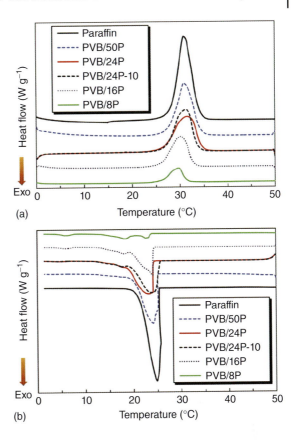

Figure 11.4 DSC curves of paraffin wax and core–sheath PVB microfibers fabricated at different Q_i. (a) Melting peaks, (b) crystallization peaks. (Wen et al. 2015 [9]. Reproduced with permission of Elsevier.)

temperature ($T_{m,o}$) and crystallization starting temperature ($T_{c,o}$) of paraffin wax are 28.21 and 26.57 °C, respectively. For the core–sheath PVB microfibers, their $T_{m,o}$ and $T_{c,o}$ are lower than those of paraffin wax, showing a slight increase with increasing paraffin wax content. The melting peak temperature ($T_{m,p}$) and crystallization peak temperature ($T_{c,p}$) of paraffin wax are 30.89 and 24.83 °C, respectively. Compared with those of paraffin wax, the $T_{m,p}$ and $T_{c,p}$ values of the core–sheath PVB microfibers show no obvious variations except for PVB/8P microfibers. For the PVB/8P microfibers, two exothermic peaks, which are heterogeneous nucleation and homogeneous nucleation during the cooling process, are observed in Figure 11.4b. Correspondingly, the two $T_{c,p}$ values of PVB/8P microfibers are 18.30 and 22.72 °C, respectively. This phenomenon indicates a supercooling behavior of the encapsulated paraffin [36], due to the separate encapsulation of paraffin droplets in PVB/8P microfibers as shown in Figure 11.4b.

However, both endothermic peaks and exothermic peaks of the core–sheath PVB microfibers become larger on increasing the content of paraffin wax. As the Q_i of melting paraffin wax is increased from 8 to 50 μl min^{-1}, the melting enthalpy (ΔH_m) and the crystallization enthalpy (ΔH_c) of core–sheath PVB microfibers becomes ~4 times lager. That is, the ΔH_m increases from 29.43 to 128.2 J g^{-1}, and ΔH_c increases from 20.26 to 124 J g^{-1}. Obviously, the phase change enthalpies of the core–sheath PVB microfibers are strongly dependent on the content of paraffin wax, while their phase change temperatures are not. Based on these DSC data, the encapsulation ratio (R) of paraffin wax in these microfibers, defined as the ratio of the melting enthalpy of the microfibers to that of the paraffin wax, can be calculated by Equation 11.1 [37, 38]:

$$R = \frac{\Delta H_{m,\text{Paraffin/PVB}}}{\Delta H_{m,\text{Paraffin}}} \times 100\% \tag{11.1}$$

where $\Delta H_{m,\text{Paraffin/PVB}}$ and $\Delta H_{m,\text{Paraffin}}$ are denoted as the melting enthalpies of the core–sheath microfibers and paraffin wax, respectively. The calculated R values of PVB/8P, PVB/16P, PVB/24P, and PVB/50P microfibers are 16.03%, 42.76%, 59.86%, and 69.83%, respectively. Since higher R value indicate better phase change performance, the PVB/50P microfibers with R value as high as ~70% show the best thermal property.

Next, the repeatability of the phase change performance of core–sheath PVB microfibers is investigated by comparing the DSC results of freshly prepared PVB/24P microfibers with those after 10 heating/cooling cycles (Figure 11.4). The ΔH_m and ΔH_c of freshly prepared PVB/24P microfibers are 109.9 and 103.3 J g^{-1}, respectively. While for the PVB/24P microfibers after 10 heating/cooling cycles (PVB/24P-10), their ΔH_m and ΔH_c values decrease approximately by only 2% and 0.4%, respectively (Figure 11.4). The results indicate that the core–sheath PVB microfibers show phase change performance with good repeatability. Figure 11.5a shows the TGA (thermal gravimetric analysis) curves of paraffin wax, solid PVB microfibers, and core–sheath PVB microfibers containing paraffin wax fabricated at different Q_i. The paraffin wax is completely decomposed at 220 °C, and the solid PVB microfiber is thermally stable up to 300 °C, both showing a one-stage degradation. For the core–sheath PVB microfibers, two-stage degradation upon heating is observed due to the encapsulation of paraffin wax. Their weight loss at temperatures below 250 °C (i.e., the first stage) is attributable to the decomposition of the encapsulated paraffin wax, while the second drop in the curve above 300 °C (i.e., the second stage) is caused by the decomposition of PVB in the microfibers. Meanwhile, the residual weight of the core–sheath microfibers at 300 °C decreases with the increasing Q_i, indicating the increase of encapsulation content of paraffin wax in the microfibers.

The temperature at which the weight loss of core–sheath PVB microfiber reaches 10% of the original weight ($T_{10\%}$) is used to evaluate the thermal stability. The $T_{10\%}$ values of paraffin wax and solid PVB microfibers are about 136.7 and 370 °C, respectively. With increasing R, the $T_{10\%}$ value of core–sheath PVB microfibers linearly decreases as shown in Figure 11.5b and gradually approaches that of paraffin wax. For the PVB/50P microfibers, their $T_{10\%}$ is 143.1 °C, which is 6.4 °C higher than that of paraffin wax. The results demonstrate that the thermal stability of core–sheath PVB microfibers decreases with increasing R.

11.3.4 Core–Sheath Microfibers for Temperature Regulation

First, the mechanical properties of core–sheath PVB microfibers are measured by electronic universal testing machine incorporated with a sensor with capacity

Figure 11.5 Thermal stability of core–sheath PVB microfibers. (a) TGA curves for paraffin wax, solid PVB microfibers, and core–sheath PVB microfibers containing paraffin wax fabricated at different Q_i. (b) Effect of paraffin wax content on $T_{10\%}$, which is the temperature at which the weight loss reaches 10% of the original weight. (Wen et al. 2015 [9]. Reproduced with permission of Elsevier.)

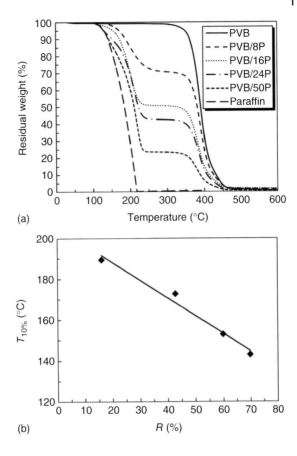

of 10 N. During the test, the sample gauge length between clamps is kept constant at 20 mm, and the strain rate is set at 10 mm min^{-1}. As shown in Figure 11.6, the tensile strength and failure strain of the solid PVB microfibers are 0.01451 N tex^{-1} and 59%, respectively. For the core–sheath PVB microfibers, the encapsulation of paraffin wax improves their failure strain and decreases their tensile strength. With the encapsulation ratio (R) of paraffin wax increasing from 16.03% to 69.83%, the failure strain of core–sheath PVB microfibers increases from 74.39% to 106.00% (Figure 11.6a), while the tensile strength decreases from 0.00682 to 0.00204 N tex^{-1} (Figure 11.6b). For all the samples, the PVB/24P microfibers show relatively high R value, good thermal stability, and moderate mechanical strength and thus are further selected for temperature regulation.

To investigate the thermo-regulating performance of the core–sheath PVB microfibers, time-dependent surface temperatures of model hats covered by the microfibers are monitored during heating/cooling cycles. The PVB/24P microfibers are used to construct the hat by twining on a 3D-printed hat frame (diameter: 2.9 cm, height: 3 cm) made of acrylonitrile butadiene styrene (ABS) copolymers, with solid PVB microfibers as the control group (Figure 11.7a,b). The thickness of the twined microfiber layer on the hat frame is 1.32 mm. For the

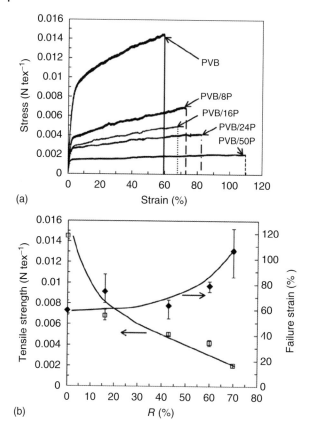

Figure 11.6 Mechanical properties of core–sheath PVB microfibers. (a) Stress–strain curves of solid and core–sheath PVB microfibers. (b) Effect of content of paraffin wax on the mechanical properties of the microfibers. (Wen et al. 2015 [9]. Reproduced with permission of Elsevier.)

heating process, both model hats with initial temperature of 16 °C are placed in an incubator with constant temperature of 40 °C. After equilibrating, the model hats are then cooled down at 16 °C. The surface temperatures of the model hats during the heating and cooling processes are monitored by an infrared camera. To confirm the stability and repeatability of the thermo-regulating performances, the surface temperature of model hats during five heating/cooling cycles are monitored.

As shown in Figure 11.7c,d, during both heating and cooling processes, at $t = 1$ min, two model hats show non-uniform surface temperature distribution. As compared with the hat with solid PVB microfibers, the hat with PVB/24P microfibers shows a lower average surface temperature upon heating and a higher average surface temperature upon cooling (Figure 11.7c1,d1). At $t = 12$ min, the surface temperature distribution of both hats become uniform. Meanwhile, at $t = 12$ min, the average surface temperature of hat with PVB/24P microfibers is 5.1 °C lower during heating process while 4.6 °C higher during the cooling process as compared with that of the hat with solid PVB microfibers (Figure 11.7c2,d2). Finally, at $t = 24$ min, the temperatures of both hats are approximately equal to the environmental temperature (Figure 11.7c3,d3). The average value of the time-dependent surface temperatures in selected regions (dash boxes) of the model hats as shown in Figure 11.7c,d during the

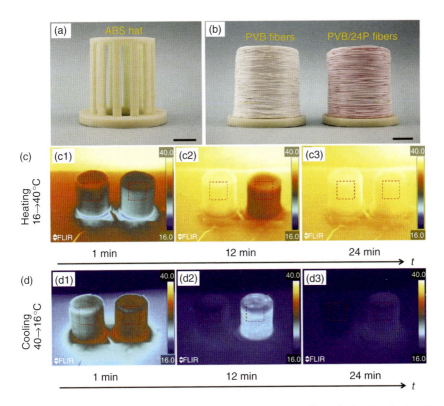

Figure 11.7 Thermo-regulating performance of PVB/24P microfibers during heating/cooling cycles. (a) Optical photos of a 3D-printed ABS hat frame with diameter of 2.9 cm and height of 3 cm. (b) Two model hats based on solid PVB fibers (left) and PVB/24P fibers (right). (c, d) Infrared thermal images showing the surface temperature distribution of the two model hats based on solid PVB fibers (left) and PVB/24P fibers (right) during heating (c) and cooling (d) processes. Scale bars are 1 cm. (Wen et al. 2015 [9]. Reproduced with permission of Elsevier.)

heating/cooling cycles is analyzed from the thermal images and plotted in Figure 11.8. Figure 11.8a shows the thermal-regulating performance of the PVB/24P microfibers during the first heating/cooling cycle. At temperatures below the typical melting point of 27 °C, the surface temperature (T_s) of model hats increases quickly, and the temperature difference is not remarkable. When the temperature is close to 27 °C, the surface temperature of hat with solid PVB microfibers still increases fast, while that of the hat with PVB/24P microfibers rises much slower. Further increasing the surface temperature from 27 to 32 °C takes 3 min for the hat with solid PVB microfibers, but 10 min for the hat with PVB/24P microfibers, indicating the excellent temperature regulation of the PVB/24P microfibers. Similarly, during the cooling process, the T_s values of the hat with PVB/24P microfibers decrease more slowly than that with solid PVB microfibers at temperature below the crystallization starting temperature of paraffin wax (24.13 °C). Such temperature performances result from the absorption or release of a large amount of thermal energy based on the phase change of the encapsulated paraffin wax. Moreover, the time-dependent

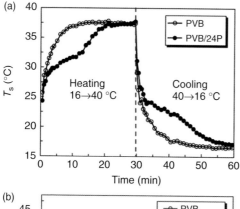
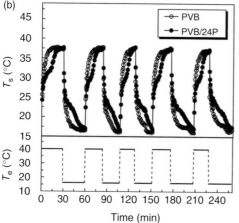

Figure 11.8 Time-dependent surface temperature of model hats during the first heating/cooling cycle (a) and five heating/cooling cycles (b). The surface temperature (T_s) is the average temperature of the selected region (dash boxes) in the model hats as shown in Figure 11.7c,d. T_e denotes the environmental temperature. (Wen et al. 2015 [9]. Reproduced with permission of Elsevier.)

surface temperature of hat with PVB/24P microfibers is repeatable upon repeated heating/cooling cycles, showing an excellent and repeatable temperature regulation performance (Figure 11.8b).

The time-dependent internal temperatures of the model hats under imitated solar irradiation are also monitored by a thermocouple thermometer placed inside the hats. A high-voltage Xenon short arc lamp is used to imitate the solar irradiation (800 W m^{-2}). The environmental temperature is ~25 °C. Figure 11.9a–c shows a doll wearing the model hat with a thermocouple placed inside for monitoring the internal temperature under imitated solar irradiation. Under imitated solar irradiation, the microfibers can absorb imitated solar irradiation and convert it into thermal energy, leading to temperature increase. As shown in Figure 11.9d, similarly, upon such irradiation, the internal temperatures of the two hats show similar increase behaviors at first, but then show different increase behaviors at temperatures higher than ~27 °C. With continuous irradiation, the internal temperature of the hat with solid PVB microfibers increases and reaches equilibrium at 34 °C. For the one with PVB/24P microfibers, the internal temperature increases slowly from ~27 °C and reaches equilibrium at 33 °C under continuous irradiation. At ~27 °C, most of the irradiation-induced thermal energy is absorbed by paraffin wax due to the

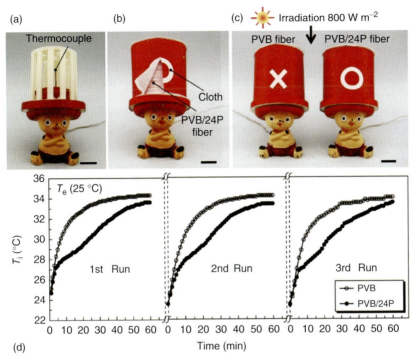

Figure 11.9 Thermo-regulating performance of PVB/24P microfibers under imitated solar irradiation. (a) A thermocouple (±0.1 °C) is placed inside the hat frame at the height of 2.5 cm from the frame bottom. (b) A model hat twined by PVB/24P microfibers and covered by a piece of cloth. (c) Model hats with solid PVB microfibers (left) and PVB/24P microfibers (right). (d) Time-dependent changes of the internal temperature (T_i) inside model hats during the three irradiation processes. Scale bars are 1 cm. (Wen et al. 2015 [9]. Reproduced with permission of Elsevier.)

phase change; thus, their internal temperature increase is mild at ~27 °C [39]. Moreover, for the three irradiation processes shown in Figure 11.9d, the hat with core–sheath PVB microfibers shows excellent thermo-regulation of the internal temperature as compared with the one with solid PVB microfibers. For each irradiation process, the maximal difference between the internal temperatures of two hats occurs during the melting process of paraffin wax, which is 3.6, 3.4, and 3.4 °C, respectively (Figure 11.9d). These results show excellent and repeatable phase change behaviors of core–sheath PVB microfibers for thermo-regulation.

11.4 Peapod-Like Microfibers with Multicompartmental Internals for Synergistic Encapsulation

In this section, the versatility of microfluidic strategy for fabrication of functional microfibers with controllable internals is demonstrated by fabricating peapod-like chitosan microfibers with multicompartment internals for synergistic encapsulation.

11.4.1 Fabrication of Peapod-Like Microfibers with Multicompartmental Internals

The microfluidic device for fabrication of microfibers with peapod-like internals possesses a structure similar to that of the device described in Section 11.3.1 and Figure 11.1, but with different orifice diameters. The tapered orifices of the injection tube and transition tube show inner diameters of 100 and 300 µm, respectively. For fabrication of chitosan microfibers with peapod-like multicompartment internals, aqueous solution containing 4.0 wt% water-soluble chitosan and 2.0 wt% acetic acid is used as the sample flow. Aqueous solution containing 50.0 wt% glutaraldehyde (GA) is used as the sheath flow. Organic solvent benzyl benzoate (BB) is used as the core flow (Figure 11.1b). These three flows are injected into the microfluidic device at appropriate flow rates via syringe pumps to generate stable and controllable jets as templates for microfiber fabrication. First, the core flow is emulsified into monodisperse droplets by the sample flow in the transition tube. Then, the sample flow that carried the BB droplets is hydrodynamically shaped by the sheath flow. This generates a peapod-like jet with cylinder sample flow containing discrete droplets of core flow (Figure 11.1b). Upon contacting, the GA diffuses from the sheath flow into the sample flow and cross-links the chitosan in the sample flow to form chitosan shell encapsulated with liquid droplets. Thus, chitosan microfibers with peapod-like internals are obtained (Figure 11.1c,d). The obtained microfibers are collected in aqueous solution containing 50 wt% GA overnight for complete cross-linking of the chitosan. Moreover, to remove unreacted components such as GA, the resultant microfibers are sufficiently washed and then stored in pure water for further use.

11.4.2 Effects of Flow Rates on the Structures of Peapod-Like Jet Templates and Chitosan Microfibers

To generate jet templates with stable peapod-like structures for fabricating peapod-like chitosan microfibers with multicompartment internals, effects of flow rates on the jet structures are investigated systematically (Figure 11.10); 0.5 wt% Sudan Black and bromoeosin are, respectively, added to the core flow and the sample flow to visualize the jet structure. As shown in Figure 11.10a, when changing flow rates, three representative structures, including stable, unstable, and clogging structures, are observed for the jet templates. Moreover, the ternary phase-like diagram in Figure 11.10b shows the narrow operation region (red square) for obtaining stable peapod-like jets. Under flow conditions shown in this operation region, the core flow can be emulsified into uniform droplets by the sample flow with relatively high Q_{sample}. This can lead to stable sample flow, which carries well-dispersed droplets of the core flow. Then, the droplet-containing sample flow can be further shaped into a stable jet by the sheath flow with relatively high Q_{sheath}, thus producing a stable peapod-like jet template. When increasing Q_{core} and decreasing Q_{sample} (blue circle region), fluctuations of the sample flow occurs. Although oil droplets of the core flow can still be generated, manipulation of the morphology of unstable waving jet by flow rate adjustment remains difficult. Similarly, with flow rates set at Q_{sheath}/Q_T less

11.4 Peapod-Like Microfibers with Multicompartmental Internals

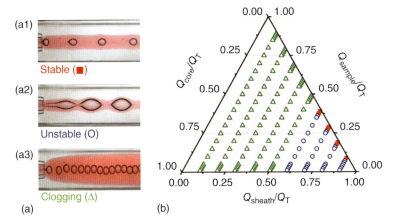

Figure 11.10 Effects of flow rates on the morphology of the peapod-like jet templates. (a) High-speed snapshots showing three typical structures (stable, unstable, and clogging) of peapod-like jet templates. (b) Ternary phase-like diagram showing the flow-rate-dependent structures of the peapod-like jet templates. The total flow rate (Q_T) is the sum of Q_{core}, Q_{sample}, and Q_{sheath}. (He et al. 2015 [34]. Reproduced with permission of The Royal Society of Chemistry.)

than 0.6 (green triangle region), wetting of sample flow onto the microchannel wall is observed. The ternary phase-like diagram provides a valuable guideline for stably generating peapod-like jet templates for microfiber fabrication.

By using the peapod-like jet containing discrete oil droplets as templates, peapod-like chitosan microfibers with multicompartment internals are fabricated. The resultant peapod-like microfibers contain separate oil droplets, each of which can serve as the "pea" compartment for controllable encapsulation. Precise control of the structure of the jet templates as well as the resultant microfibers can be achieved by flow rate adjustment. As shown in Figure 11.11, at a fixed Q_{core} of 20 µl min^{-1}, fine-adjustment of the outer diameter (OD) of the jet, the droplet size, the shell thickness, and the distance between every two droplets can be obtained by changing Q_{sample} and Q_{sheath}. The droplet diameter of the peapod-like microfibers is defined as the average of the vertical diameter and the horizontal diameter of the droplet; the droplet distance is defined as the distance between the centers of two neighboring droplets; the shell thickness is defined as half of the difference between the outer diameter of the most convex interfaces of the peapod-like jet and the vertical diameter of the droplet. Such structure adjustment can provide controllable jet templates for fabricating the peapod-like microfiber. As shown in Figure 11.11a,b, the OD of the jet decreases with increasing Q_{sheath} and increases with increasing Q_{sample} when Q_{core} and Q_{sheath} are fixed. Figure 11.11c shows the effects of Q_{sample} and Q_{sheath} on the droplet distance within the peapod-like chitosan microfiber. The distance between droplets increases with decreasing Q_{sample} and increasing Q_{sheath}. The control of droplet distance defines the number of droplets within the peapod-like chitosan microfiber of per unit length. Meanwhile, as compared to Q_{sheath}, Q_{sample} exhibits more obvious influence on the droplet size, where the droplet diameter decreases with increasing Q_{sample} when Q_{core} and Q_{sheath} are fixed (Figure 11.11d).

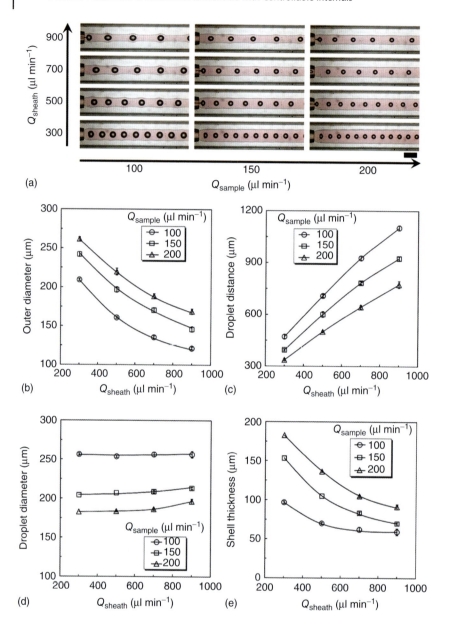

Figure 11.11 Flow-rate-dependent structure of the peapod-like jet templates. (a) High-speed snapshots of peapod-like jet templates generated at different flow rates. Q_{core} is fixed at 20 μl min^{-1}. Scale bar is 500 μm. (b–e) Effects of Q_{sample} and Q_{sheath} on the outer diameter of the peapod-like jet templates (b), the distance between inner droplets (c), the droplet size (d), and the shell thickness of the jet templates (e). (He et al. 2015 [34]. Reproduced with permission of The Royal Society of Chemistry.)

Moreover, the shell thickness of the jet templates can also be adjusted by change of Q_{sheath} and Q_{sample}. As shown in Figure 11.11e, increase in Q_{sample} at fixed Q_{core} and Q_{sheath} leads to increase in the shell thickness, while increase in Q_{sheath} at fixed Q_{core} and Q_{sample} leads to decrease in the shell thickness. All these results show the precise and flexible control of the jet structure for fabricating peapod-like microfibers with multicompartment internals.

The peapod-like chitosan microfibers fabricated by cross-linking of the controllable jet templates can be continuously collected by a spinning procedure (Figure 11.12a). The chitosan microfiber exhibits a uniform peapod-like structure similar to the jet template. Monodisperse oil droplets are uniformly distributed in the chitosan microfiber as separate "pea" compartments (Figure 11.12b). The SEM images of the air-dried chitosan microfiber and its cross-section further confirm the peapod-like structure with uniformly aligned multiple compartments (Figure 11.12c,d). All the results exhibit the power of the microfluidic strategy for preparing chitosan microfibers with controllable multicompartmental internals.

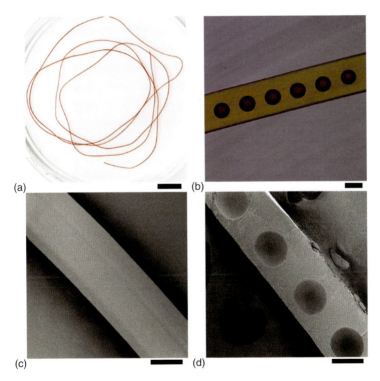

Figure 11.12 Morphological characterization of peapod-like chitosan microfibers with multicompartment internals. (a, b) Digital photo (a) and optical micrograph (b) of peapod-like chitosan microfibers with LR300-dyed oil droplets. (c, d) SEM images of the outer surface (c) and cross-section (d) of the air-dried peapod-like chitosan microfibers. Scale bars are 1 cm in (a), 200 μm in (b), and 100 μm in (c) and (d). (He et al. 2015 [34]. Reproduced with permission of The Royal Society of Chemistry.)

11.4.3 Peapod-Like Chitosan Microfibers with Multicompartment Internals for Synergistic Encapsulation

The hydrophilic chitosan matrix and hydrophobic multicompartments of oil droplets in the peapod-like chitosan microfibers can serve as two separate domains for encapsulation of water-soluble and oil-soluble molecules. First, the ability of the peapod-like chitosan microfibers for encapsulation of oil-soluble molecules is demonstrated by using fluorescent LR300 as the model molecule. The uniformly distributed oil droplets in the peapod-like chitosan microfibers can serve as separate multiple hydrophobic compartments for LR300 encapsulation. Since the cross-linked chitosan matrix can also show green fluorescence, the microfiber morphology can be characterized by confocal laser scanning microscope (CLSM). As shown in Figure 11.13a–c, the oil droplets show red fluorescent color, which confirms the successful encapsulation of LR300 in the cross-linked chitosan microfiber (green color). Compared with the single drug encapsulation, co-delivery of multiple drugs usually enables synergistic efficacy for more efficient treatment of diseases. Such a synergistic encapsulation

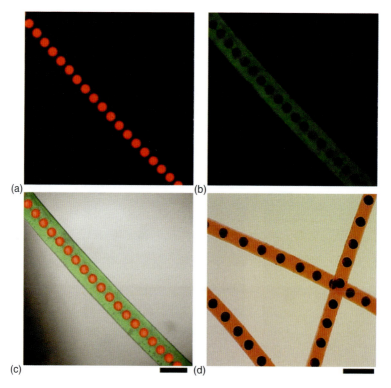

Figure 11.13 Peapod-like chitosan microfibers with multicompartment internals for synergistic encapsulation. (a–c) CLSM images of the peapod-like chitosan microfibers (c) with LR300-dyed oil droplets (red) (a) and fluorescent chitosan matrix (green) (b). (d) Optical micrograph of peapod-like chitosan microfibers containing Sudan Black (black) in the oil cores and bromoeosin (red) in the chitosan matrix. Scale bars are 500 μm. (He *et al.* 2015 [34]. Reproduced with permission of The Royal Society of Chemistry.)

for both oil- and water-soluble drugs is demonstrated by encapsulation of oil-soluble Sudan Black (black color) and water-soluble bromoeosin (red color), respectively, in the inner oil droplets and chitosan matrix (Figure 11.13d). Based on the abovementioned excellent control of the microfiber diameter, the droplet size, and the droplet distance, the encapsulation of each drug in the microfiber can be individually and accurately optimized to ensure optimized efficacy. These peapod-like microfibers are potential for fabricating drug-loaded medical patches for healing wounds. Moreover, because droplets of different compositions can also be introduced with advanced microfluidic devices as described in Chapter 2, more versatile drugs, even incompatible ones, can be loaded in a single microfiber for more diverse synergistic encapsulation.

11.5 Spider-Silk-Like Microfibers with Spindle-Knot Internals for 3D Assembly and Water Collection

In this section, the versatility of microfluidic strategy for fabrication of functional microfibers with controllable internals is demonstrated by fabricating spider-silk-like calcium alginate (Ca-alginate) chitosan microfibers with magnetic spindle-knot internals for 3D assembly and water collection.

11.5.1 Fabrication of Spider-Silk-Like Microfibers with Spindle-Knot Internals

The microfluidic device for fabrication of spider-silk-like microfibers with magnetic spindle-knot internals possesses similar structure as that of the device described in Section 11.3.1 and Figure 11.1, but with different orifice diameters. The tapered orifices of the injection tube and transition tube show inner diameters of 100 and 300 μm, respectively. The inner diameters of the orifice of the injection tube and transition tube are 80 and 200 μm, respectively.

For fabrication of Ca-alginate microfibers with magnetic spindle-knot internals, liquid jet templates with magnetic oil cores are prepared from the microfluidic device. Briefly, an oil mixture of ferrofluid and N-butyl acetate, with volume ratio of 2 : 1, is used as the inner dispersed flow (O_1). An aqueous solution containing 5.0 wt% sodium alginate (Na-alginate) is used as the middle jet flow (W_1). An aqueous solution containing 1.0 wt% hydroxyethyl cellulose (HEC) and 5.0 wt% $CaCl_2$ is used as the outer continuous flow (W_2). Here the N-butyl acetate is used to benefit the easy shear of the inner dispersed flow by the middle jet flow to form monodisperse magnetic oil droplets, since pure ferrofluid is very stable and hard to be emulsified. First, to generate uniform magnetic O_1 droplets, the O_1 is emulsified by the W_1 flow in the transition tube. Then, the W_1 flow containing O_1 droplets is forced into jet that contains regular array of O_1 droplets by W_2 flow. Meanwhile, the Ca^{2+} in W_2 diffuses into the W_1 jet and triggers the cross-linking of alginate in the W_1 jet. Based on the fast cross-linking between alginate and Ca^{2+}, Ca-alginate microfibers with knots of magnetic oil droplets can be continuously fabricated (Figure 11.1b). During the fabrication process, the microfluidic device is vertically fixed to avoid the clogging caused by cross-linked microfibers. The outlet of the collection tube is immersed in an aqueous solution of 5.0 wt% $CaCl_2$ for further cross-linking the collected microfibers. The collected Ca-alginate microfibers with magnetic knots are quickly washed with water to remove excess Ca^{2+} and stored in pure water for further use. To further prepare spider-silk-like Ca-alginate microfibers with magnetic spindle-knots, the newly prepared microfibers are exposed and dried in air for 12 h. During this process, the n-hexane and N-butyl acetate in the magnetic oil droplets can be completely volatilized to produce solid magnetic cores, and the Ca-alginate microfiber shell can be dehydrated to form a spindle shape around the magnetic core for fabrication of the spider-silk-like microfibers with magnetic spindle-knot internals, as typically illustrated in Figure 11.1e.

11.5.2 Morphological Characterization of the Jet Templates and Spider-Silk-Like Microfibers

The formation process of jet templates for fabricating the hydrated Ca-alginate microfibers with magnetic oil cores is monitored by inverted optical microscope equipped with a high-speed digital camera. First, for jet template generation, the O_1 flow is emulsified into monodisperse droplets by W_1 flow in the transition tube of the microfluidic device (Figure 11.14a). To make the edges of the resultant microfibers more visible and clear, 0.1 wt% disperse red is added to the W_1. Then, by shearing of W_2, the W_1 flow that carries the O_1 droplets is forced into a jet (Figure 11.14b). The excellent manipulation of the microflows

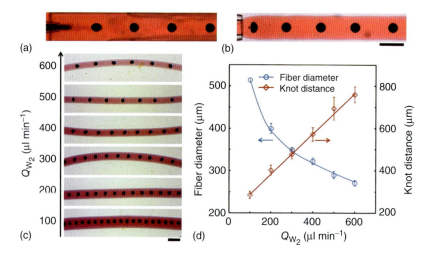

Figure 11.14 Flow-rate-dependent structure of the jet templates and the resultant microfibers. (a, b) High-speed snapshots showing the generation processes of magnetic oil droplets (a) and jet templates (b) in microfluidic device. (c) Optical micrographs of the hydrated microfibers fabricated with different flow rate conditions. (d) Effects of Q_{W_2} on the microfiber diameter (D) (blue line) and knot distance (L) (red line) of the hydrated microfibers. In (c) and (d), Q_{O_1} and Q_{W_1} are fixed at 20 and 200 μl min^{-1}, respectively, and Q_{W_2} is varied from 100 to 600 μl min^{-1}. Scale bars are 500 μm. (He et al. 2015 [35]. Reproduced with permission of American Chemical Society.)

with microfluidics allows precise control of the structures of the jet templates as well as the resultant microfibers. The effects of flow rates on the structures of such peapod-like microfibers in a similar system have been systematically investigated in Section 11.4. In this section, as a typical example, the effect of the outer continuous flow (Q_{W_2}) on the structures of the hydrated microfibers at fixed flow rates of O_1 (Q_{O_1}, 20 µl min^{-1}) and W_1 (Q_{W_1}, 200 µl min^{-1}) is investigated. On increasing Q_{W_2} from 100 to 600 µl min^{-1}, the diameter of microfiber (D) decreases from 513.7 to 270.6 µm (Figure 11.14c,d). Meanwhile, such a Q_{W_2} increase leads to increase of the droplet distance (L), which is defined as the distance between the centers of two neighboring magnetic knots. The control of droplet distance determines the distribution of magnetic knots within the resultant microfibers, which is important for their magnetic patterning and assembling.

By cross-linking of the controllable jet templates, hydrated Ca-alginate microfiber with magnetic oil cores can be continuously produced and collected by a spinning procedure (Figure 11.15a). The hydrated Ca-alginate microfibers show uniform structures similar to that of the jet templates, with monodisperse magnetic oil cores uniformly distributed in the Ca-alginate matrix of the microfiber (Figure 11.15b). After Ca-alginate cross-linking, the oil cores become

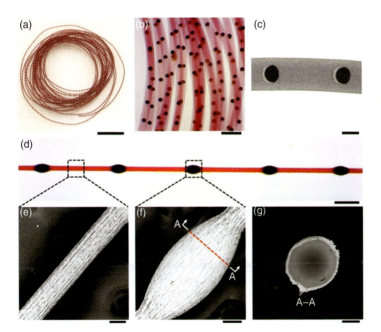

Figure 11.15 Morphological characterization of the hydrated Ca-alginate microfibers and dehydrated spider-silk-like Ca-alginate microfibers with magnetic spindle-knots. (a–c) Digital photo (a), optical micrograph (b), and magnified optical micrograph (c) of the hydrated Ca-alginate microfibers with magnetic oil cores. (d) Optical micrograph of the dehydrated spider-silk-like Ca-alginate microfibers with periodically arrayed magnetic spindle-knots. (e–g) SEM images of the outer surface of the joint part (e), and the outer surface (f) and cross-section (g) of the spindle-knot of the dehydrated spider-silk-like Ca-alginate microfibers. Scale bars are 1 cm in (a), 1 mm in (b), 200 µm in (c), 500 µm in (d), and 50 µm in (e–g). (He et al. 2015 [35]. Reproduced with permission of American Chemical Society.)

smaller than the cavity compartment of the microfiber (Figure 11.15c), due to the gradual evaporation of *n*-hexane and *N*-butyl acetate in the magnetic oil cores. Such an evaporation process lasts for a long time when immersing the microfibers in water. To fabricate dehydrated spider-silk-like microfibers with magnetic spindle-knot internals, the hydrated Ca-alginate microfibers are exposed in air. This significantly accelerates the evaporation process, which is accomplished after 12 h of drying. During the air-drying, microfiber shell shrinks, which deforms the spherical magnetic oil cores into solid magnetic cores with spindle shape. Thus, the resultant dehydrated microfibers exhibit unique spider-silk-like structure, with uniform magnetic spindle-knots periodically distributed along the microfiber (Figure 11.15d). Specifically, during the air-drying process, the hydrated microfibers shrinks from ~361 to ~97 μm in diameter and shrinks by ~14% in length. As a result, the spherical oil cores are deformed and shrink from ~206 to ~168 μm in height and extend from ~206 to ~303 μm in width, leading to the formation of spindle-shaped magnetic cores. Therefore, after air-drying of the solvent of oil cores, spindle-shaped solid magnetic cores can be obtained to produce the spider-silk-like microfibers. As shown in Figure 11.15e–g, the diameters of the joint parts (Figure 11.15e) and spindle-knots (Figure 11.15f) in the dehydrated spider-silk-like Ca-alginate microfibers are 70 and 150 μm, respectively. The Ca-alginate shell of spindle-knot compartment in the dehydrated microfiber is ~10 μm (Figure 11.15g). The dehydrated Ca-alginate microfibers show improved stiffness and strength than the hydrated ones. For mechanical property measurement, hydrated microfibers, with outer diameter of 350 μm, oil core diameter of 200 μm, and knot distance of 550 μm, and their dehydrated counterpart microfibers are used. As measured by electronic universal testing machine, the fracture strain of the hydrated and dehydrated microfibers is ~144.79% and ~25.28%, respectively, while their tensile strength is 0.82 ± 0.03 MPa and 15.39 ± 0.40 MPa, respectively. The Young's modulus of the hydrated and dehydrated CaAlg microfibers is 0.70 ± 0.03 and 533.57 ± 59.90 MPa, respectively. These results indicate that the stiffness and strength of the microfibers are significantly enhanced after dehydration. All the results show the great power of the microfluidic strategy for fabricating spider-silk-like Ca-alginate microfibers with controllable magnetic spindle-knot internals.

11.5.3 Magnetic-Guided Patterning and Assembling of the Ca-Alginate Microfibers

Incorporation of magnetic nanoparticles in the knots can provide the microfibers with magnetic-guided patterning and assembling ability. The magnetic-guided patterning is demonstrated by manually arranging the Ca-alginate microfibers on a magnetic array under water into desired patterns, including letter patterns and multilayered matrix patterns. The magnetic array is constructed by incorporating 6×5 magnetic minipillars (size: ø 1 mm × 3 mm) in polydimethylsiloxane matrix with thickness of 3 mm (Figure 11.16a1). The positions of all magnetic knots in the microfiber are manually adjusted and magnetically attracted on the tip of the magnetic minipillars. The magnetic-guided patterning enables the use of the

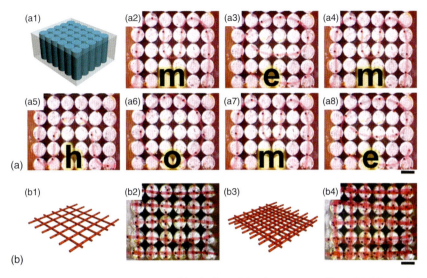

Figure 11.16 Magnetic patterning of the hydrated Ca-alginate microfibers. (a) Schematic illustration (a1) showing the setup of magnetic arrays for microfiber patterning, and optical micrographs (a2–a8) showing different 2D letter patterns ("memhome") arranged from the hydrated Ca-alginate microfibers. (b) Schematic illustrations and optical micrographs of the 3D textile-like patterns constructed from 11 (b1–b2) to 22 (b3–b4) hydrated Ca-alginate microfibers. Scale bars are 1 mm. (He et al. 2015 [35]. Reproduced with permission of American Chemical Society.)

Ca-alginate microfibers for spelling out a word "memhome" (Figure 11.16a2–a8). For the patterning, the Ca-alginate microfibers show enough stiffness against the attractive force between two neighboring magnetic knots. Therefore, bonding of two magnetic knots can be achieved only when the microfiber length between the two magnetic knots is long enough. Moreover, more complex textile-like structures can also be achieved by sequentially patterning multilayer networks of microfibers, each of which consists of 11 (Figure 11.16b1,b2) or 22 (Figure 11.16b3,b4) cross-aligned microfibers. Similarly, based on the magnetic property of the magnetic knots, such magnetic-guided patterning and assembling behaviors can be achieved for the dehydrated CaAlg microfibers.

The magnetic-guided assembly of the Ca-alginate microfibers is performed in a water-containing holder incorporated with different numbers of magnetic minipillars (size: ø 1 mm × 3 mm) for constructing different 3D assembly patterns of microfibers. The 3D magnetic fields are constructed by using magnetic minipillars with numbers $n = 1$, 2, and 4 (Figure 11.17). The hydrated Ca-alginate microfibers are manually adjusted and magnetically attracted on the tip of magnetic minipillars. Figure 11.17a1,b1,c1, respectively, shows the setups of 1, 2, and 4 magnetic minipillars for controllable microfiber assembly. When one magnetic minipillar is used, the microfibers, each with one end attracted on the magnetic minipillar, are assembled into a semi-radial geometry, because their free ends are oriented by a similar semi-radial magnetic field distribution near the magnetic minipillar (Figure 11.17a2,a3). When two magnetic minipillar are used, the microfibers, each with both ends attracted by the magnetic minipillars,

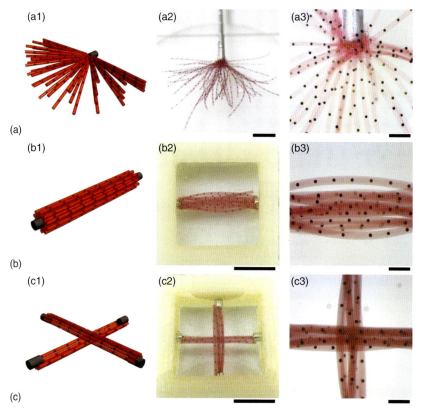

Figure 11.17 Magnetic-guided assembly of the hydrated Ca-alginate microfibers. Schematic illustrations (a1–c1), digital photos (a2–c2), and optical micrographs (a3–c3) of the 3D assembly structures of hydrated Ca-alginate microfibers guided by one (a), two (b), and four (c) magnetic minipillars. Scale bars are 5 mm in (a2–c2) and 1 mm in (a3–c3). (He et al. 2015 [35]. Reproduced with permission of American Chemical Society.)

are parallelly assembled into a columnar shape (Figure 11.17b2,b3), while the microfibers are assembled into a cross-shaped structure for the case involves four magnetic minipillars (Figure 11.17c2,c3). The results show the excellent performance of these Ca-alginate microfibers for controllable patterning and assembling with well-designed magnetic fields.

11.5.4 Water Collection of Dehydrated Ca-Alginate Microfibers with Magnetic Spindle-Knot Internals

The water collection ability of the dehydrated spider-silk-like Ca-alginate microfibers is investigated by using single microfiber and assembled microfibers with a spider-web-like pattern as samples. First, we investigate the water collection ability of a single dehydrated spider-silk-like Ca-alginate microfiber by slightly obliquely placing the microfiber in an upflow of water mist generated from an ultrasonic humidifier (Figure 11.18a). The water collection process is recorded by digital microscope. To quantify the water collection ability, the

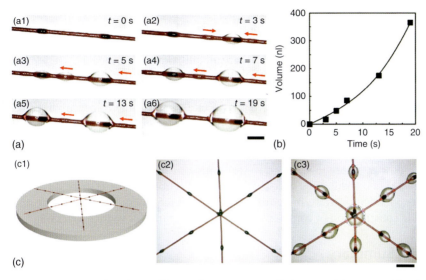

Figure 11.18 Dehydrated microfibers with magnetic spindle-knots and their assembly of spider-web-like structure for water collection. (a) Optical images showing the water collection behaviors on the dehydrated microfiber with magnetic spindle-knots. (b) Time-dependent volume changes of water drop collected on single spindle-knot. (c) Schematic illustration showing the spider-web-like structures assembled from dehydrated microfiber (c1), and optical images showing the spider-web-like structures before (c2) and after (c3) water collection. Scale bars are 500 µm in (a), and 1 mm in (c). (He et al. 2015 [35]. Reproduced with permission of American Chemical Society.)

time-dependent volume of water droplets collected on a single spindle-knot is measured from the snapshots of the recorded movie. For such quantification, both spindle-knots and water droplets are considered as ellipsoids, and the droplet volume is calculated based on its actual height and width.

As shown in Figure 11.18a1, initially, the hydrophilic spider-silk-like Ca-alginate microfiber can adsorb tiny water droplets from the water mist into their cross-linked matrix for slightly swelling. Then, upon continuous water condensation, larger water drops are randomly formed at the spindle-knot part or joint part along the microfiber. Interestingly, although the dehydrated microfiber is placed obliquely for water collection, the water drops formed in the lower joint part can overcome their own gravity and move directionally to the neighboring upper spindle-knot and then coalesce with other water drops at the spindle-knot for water collection (Figure 11.18a2–a6). For horizontally placed dehydrated CaAlg microfibers, similar water collection behaviors from the joint part to the spindle-knot can also be observed. To evaluate their water collection ability, the time-dependent volume changes of water drops collected on a single spindle-knot of the dehydrated microfibers is shown in Figure 11.18b. After collection for $t = 3$ s, water drop with volume of 34.60 nl, which is ∼9 times of that of the spindle-knot (∼3.99 nl), can be collected on the spindle-knot. At $t = 19$ s, a water drops with volume of ∼365 nl can be collected, which is ∼92 times of that of the spindle-knot. These results show the excellent water collection ability of the dehydrated microfibers. Moreover, since the size of the spindle-knot can be easily increased via changing the flow rates during microfiber generation, further enhancement of the water collection ability can be simply achieved by employing larger spindle-knots.

Based on the water collection ability, we further assemble the spider-silk-like microfibers on a circle-shaped magnet to construct an artificial spider web for water collection (Figure 11.18c1). The two ends of each microfiber are fixed on the circle-shaped magnet, and the three magnetized spindle-knots in the central of the microfibers are attracted to each other, resulting in a stable spider-web-like structure (Figure 11.18c1,c2). When placing this spider-web-like structure in the upflow of water mist, large water drops can be collected at each of their spindle-knots, exhibiting excellent water collection ability (Figure 11.18c3). The results demonstrate the potential of these microfibers for controllable and flexible construction of complex assembling structures for water collection.

11.6 Summary

In summary, based on microfluidic technique, simple and flexible approaches have been developed for continuous fabrication of microfibers with controllable tubular, peapod-like and spindle-knot-like internals. The tubular microfibers allow efficient encapsulation of phase change materials with high content for temperature regulation. The peapod-liked microfibers with separate oil cores can serve as multicompartment systems for synergistic encapsulation of multiple drugs. The microfibers with magnetic spindle-knot-like internals can be controllably assembled into spider-web-like structures for excellent water collection. These approaches provide facile and efficient strategy for controllable fabrication of microfibers with complex and well-tailored internals for versatile applications.

References

1 Daniele, M.A., Adams, A.A., Naciri, J., North, S.H., and Ligler, F.S. (2014) Interpenetrating networks based on gelatin methacrylamide and PEG formed using concurrent thiol click chemistries for hydrogel tissue engineering scaffolds. *Biomaterials*, **35**, 1845–1856.

2 Onoe, H., Okitsu, T., Itou, A., Kato-Negishi, M., Gojo, R., Kiriya, D., Sato, K., Miura, S., Iwanaga, S., Kuribayashi-Shigetomi, K., Matsunaga, Y.T., Shimoyama, Y., and Takeuchi, S. (2013) Metre-long cell-laden microfibres exhibit tissue morphologies and functions. *Nat. Mater.*, **12**, 584–590.

3 Kang, E., Choi, Y.Y., Chae, S.K., Moon, J.H., Chang, J.Y., and Lee, S.H. (2012) Microfluidic spinning of flat alginate fibers with grooves for cell-aligning scaffolds. *Adv. Mater.*, **24**, 4271–4277.

4 Kang, E., Jeong, G.S., Choi, Y.Y., Lee, K.H., Khademhosseini, A., and Lee, S.H. (2011) Digitally tunable physicochemical coding of material composition and topography in continuous microfibres. *Nat. Mater.*, **10**, 877–883.

5 Jun, Y., Kim, M.J., Hwang, Y.H., Jeon, E.A., Kang, A.R., Lee, S.H., and Lee, D.Y. (2013) Microfluidics-generated pancreatic islet microfibers for enhanced immunoprotection. *Biomaterials*, **34**, 8122–8130.

6 Supaphol, P., Suwantong, O., Sangsanoh, P., Srinivasan, S., Jayakumar, R., and Nair, S.V. (2012) Electrospinning of biocompatible polymers and their potentials in biomedical applications. *Adv. Polym. Sci.*, **246**, 213–240.

7 Min, S.K., Lee, S.C., Hong, S.D., Chung, C.P., Park, W.H., and Min, B.M. (2010) The effect of a laminin-5-derived peptide coated onto chitin microfibers on re-epithelialization in early-stage wound healing. *Biomaterials*, **31**, 4725–4730.

8 Agarwal, S., Wendorff, J.H., and Greiner, A. (2008) Use of electrospinning technique for biomedical applications. *Polymer*, **49**, 5603–5621.

9 Wen, G.Q., Xie, R., Liang, W.G., He, X.H., Wang, W., Ju, X.J., and Chu, L.Y. (2015) Microfluidic fabrication and thermal characteristics of core-shell phase change microfibers with high paraffin content. *Appl. Therm. Eng.*, **87**, 471–480.

10 Wang, N., Chen, H.Y., Lin, L., Zhao, Y., Cao, X.Y., Song, Y.L., and Jiang, L. (2010) Multicomponent phase change microfibers prepared by temperature control multifluidic electrospinning. *Macromol. Rapid Commun.*, **31**, 1622–1627.

11 Jun, Y., Kang, E., Chae, S., and Lee, S.H. (2014) Microfluidic spinning of micro- and nano-scale fibers for tissue engineering. *Lab Chip*, **14**, 2145–2160.

12 Daniele, M.A., Radom, K., Ligler, F.S., and Adams, A.A. (2014) Microfluidic fabrication of multiaxial microvessels via hydrodynamic shaping. *RSC Adv.*, **4**, 23440–23446.

13 Takei, T., Kishihara, N., Sakai, S., and Kawakami, K. (2010) Novel technique to control inner and outer diameter of calcium-alginate hydrogel hollow microfibers, and immobilization of mammalian cells. *Biochem. Eng. J.*, **49**, 143–147.

14 Lee, K.H., Shin, S.J., Park, Y., and Lee, S.H. (2009) Synthesis of cell-laden alginate hollow fibers using microfluidic chips and microvascularized tissue-engineering applications. *Small*, **5**, 1264–1268.

15 Asthana, A., Lee, K.H., Shin, S.J., Perumal, J., Butler, L., Lee, S.H., and Kim, D.P. (2011) Bromo-oxidation reaction in enzyme-entrapped alginate hollow microfibers. *Biomicrofluidics*, **5**, 24117.

16 Dror, Y., Salalha, W., Avrahami, R., Zussman, E., Yarin, A.L., Dersch, R., Greiner, A., and Wendorff, J.H. (2007) One-step production of polymeric microtubes by co-electrospinning. *Small*, **3**, 1064–1073.

17 Kidoaki, S., Kwon, I.K., and Matsuda, T. (2005) Mesoscopic spatial designs of nano- and microfiber meshes for tissue-engineering matrix and scaffold based on newly devised multilayering and mixing electrospinning techniques. *Biomaterials*, **26**, 37–46.

18 Zhang, J.M., Wang, L.N., Zhu, M.F., Wang, L.Y., Xiao, N.N., and Kong, D.L. (2014) Wet-spun poly(epsilon-caprolactone) microfiber scaffolds for oriented growth and infiltration of smooth muscle cells. *Mater. Lett.*, **132**, 59–62.

19 Grigoryev, A., Sa, V., Gopishetty, V., Tokarev, I., Kornev, K.G., and Minko, S. (2013) Wet-spun stimuli-responsive composite fibers with tunable electrical conductivity. *Adv. Funct. Mater.*, **23**, 5903–5909.

20 Cheng, Y., Zheng, F.Y., Lu, J., Shang, L.R., Xie, Z.Y., Zhao, Y.J., Chen, Y.P., and Gu, Z.Z. (2014) Bioinspired multicompartmental microfibers from microfluidics. *Adv. Mater.*, **26**, 5184–5190.

21 Oh, J., Kim, K., Won, S.W., Cha, C., Gaharwar, A.K., Selimovic, S., Bae, H., Lee, K.H., Lee, D.H., Lee, S.H., and Khademhosseini, A. (2013) Microfluidic fabrication of cell adhesive chitosan microtubes. *Biomed. Microdevices*, **15**, 465–472.

22 Daniele, M.A., North, S.H., Naciri, J., Howell, P.B., Foulger, S.H., Ligler, F.S., and Adams, A.A. (2013) Rapid and continuous hydrodynamically controlled fabrication of biohybrid microfibers. *Adv. Funct. Mater.*, **23**, 698–704.

23 Chae, S.K., Kang, E., Khademhosseini, A., and Lee, S.H. (2013) Micro/nanometer-scale fiber with highly ordered structures by mimicking the spinning process of silkworm. *Adv. Mater.*, **25**, 3071–3078.

24 Choi, C.H., Yi, H., Hwang, S., Weitz, D.A., and Lee, C.S. (2011) Microfluidic fabrication of complex-shaped microfibers by liquid template-aided multiphase microflow. *Lab Chip*, **11**, 1477–1483.

25 Yeh, C.H., Lin, P.W., and Lin, Y.C. (2010) Chitosan microfiber fabrication using a microfluidic chip and its application to cell cultures. *Microfluid. Nanofluid.*, **8**, 115–121.

26 Marimuthu, M., Kim, S., and An, J. (2010) Amphiphilic triblock copolymer and a microfluidic device for porous microfiber fabrication. *Soft Matter*, **6**, 2200–2207.

27 Su, J., Zheng, Y.Z., and Wu, H.K. (2009) Generation of alginate microfibers with a roller-assisted microfluidic system. *Lab Chip*, **9**, 996–1001.

28 Lan, W., Li, S., Lu, Y., Xu, J., and Luo, G. (2009) Controllable preparation of microscale tubes with multiphase co-laminar flow in a double co-axial microdevice. *Lab Chip*, **9**, 3282–3288.

29 Jeong, W., Kim, J., Kim, S., Lee, S., Mensing, G., and Beebe, D.J. (2004) Hydrodynamic microfabrication via "on the fly" photopolymerization of microscale fibers and tubes. *Lab Chip*, **4**, 576–580.

30 Tamayol, A., Akbari, M., Annabi, N., Paul, A., Khademhosseini, A., and Juncker, D. (2013) Fiber-based tissue engineering: progress, challenges, and opportunities. *Biotechnol. Adv.*, **31**, 669–687.

31 Yu, Y., Wen, H., Ma, J.Y., Lykkemark, S., Xu, H., and Qin, J.H. (2014) Flexible fabrication of biomimetic bamboo-like hybrid microfibers. *Adv. Mater.*, **26**, 2494–2499.

32 Lee, B.R., Lee, K.H., Kang, E., Kim, D.S., and Lee, S.H. (2011) Microfluidic wet spinning of chitosan-alginate microfibers and encapsulation of HepG2 cells in fibers. *Biomicrofluidics*, **5**, 022208.

33 Lee, K.H., Shin, S.J., Kim, C.B., Kim, J.K., Cho, Y.W., Chung, B.G., and Lee, S.H. (2010) Microfluidic synthesis of pure chitosan microfibers for bio-artificial liver chip. *Lab Chip*, **10**, 1328–1334.

34 He, X.H., Wang, W., Deng, K., Xie, R., Ju, X.J., Liu, Z., and Chu, L.Y. (2015) Microfluidic fabrication of chitosan microfibers with controllable internals from tubular to peapod-like structures. *RSC Adv.*, **5**, 928–936.

35 He, X.-H., Wang, W., Liu, Y.-M., Jiang, M.-Y., Wu, F., Deng, K., Liu, Z., Ju, X.-J., Xie, R., and Chu, L.-Y. (2015) Microfluidic fabrication of bio-inspired microfibers with controllable magnetic spindle-knots for 3D assembly and water collection. *ACS Appl. Mater. Interfaces*, **7**, 17471–17481.

36 Montenegro, R., Antonietti, M., Mastai, Y., and Landfester, K. (2003) Crystallization in miniemulsion droplets. *J. Phys. Chem. B*, **107**, 5088–5094.

37 Yu, S., Wang, X., and Wu, D. (2014) Microencapsulation of n-octadecane phase change material with calcium carbonate shell for enhancement of thermal conductivity and serving durability: synthesis, microstructure, and performance evaluation. *Appl. Energy*, **114**, 632–643.

38 Alkan, C., Sari, A., Karaipekli, A., and Uzun, O. (2009) Preparation, characterization, and thermal properties of microencapsulated phase change material for thermal energy storage. *Sol. Energy Mater. Sol. Cells*, **93**, 143–147.

39 Wang, Y., Tang, B., and Zhang, S. (2013) Single-walled carbon nanotube/phase change material composites: sunlight-driven, reversible, form-stable phase transitions for solar thermal energy storage. *Adv. Funct. Mater.*, **23**, 4354–4360.

12

Microfluidic Fabrication of Membrane-in-a-Chip with Self-Regulated Permeability

12.1 Introduction

Integration of membranes in microchips has shown great potential for biomedical fields such as analyses [1, 2], separations [3–5], microreactions [6, 7], and cell-based studies [8, 9], due to their portability and less consumption of sample and energy [10]. Especially, microchips containing smart membranes that can change their permeability in response to external chemical/physical stimuli are important and necessary for their use as sensors, separators, and controlled release systems. For example, centimeter-scaled chip that contains two polydimethylsiloxane (PDMS) grids separated by a glucose-responsive bioinorganic gel membrane has been developed for glucose-responsive insulin-delivery in diabetic rats [11]. Such smart-membrane-in-chip is of great significance for broadening the applications of membrane-integrated microchips in biomedical fields.

Generally, membranes in microchips can be fabricated by assembling separately prepared membranes into the microchannels [12, 13]; however, this approach usually suffers from limited sealing performance and troublesome fabrication process. Using permeable materials such as PDMS [14] and calcium alginate hydrogel [15] allows one-step fabrication of both the membrane and the microchip, but only limited materials can be applied to this method. *In situ* formation of membranes in microchannels based on interfacial reactions between parallel laminar flows provides a facile strategy for one-step fabrication of membranes in microchips [16]. Various membranes including nylon membranes [17], polyacrylamide membranes [18], calcium alginate hydrogel [9], and chitosan membranes [19] have been developed within microchips by such an approach for component separation, cell culture, and generation of chemical gradients. As compared to the abovementioned two strategies, this *in situ* membrane formation process provides more versatility for incorporating membranes with flexible geometries in complex microchannels by manipulating the interfaces between parallel laminar flows [7, 19].

Recently, the authors' group developed a simple and versatile microfluidic approach for *in situ* fabrication of nanogel-containing smart membranes in the microchannel of microchips [20]. In this chapter, the microfluidic preparation and responsive performance of the smart-membrane-in-chip are introduced.

12.2 Microfluidic Strategy for Fabrication of Membrane-in-a-Chip

A facile strategy is developed for one-step fabrication of smart membranes in microchips via *in situ* formation of membranes containing responsive nanogels as smart valves in microchannels based on interfacial reactions between parallel laminar flows (Figure 12.1). This fabrication approach is demonstrated by using biocompatible chitosan as substrate material to construct the membrane [20]. Meanwhile, poly(*N*-isopropylacrylamide) (PNIPAM) nanogels, which enable reversible swelling/shrinking volume transitions in response to changes in

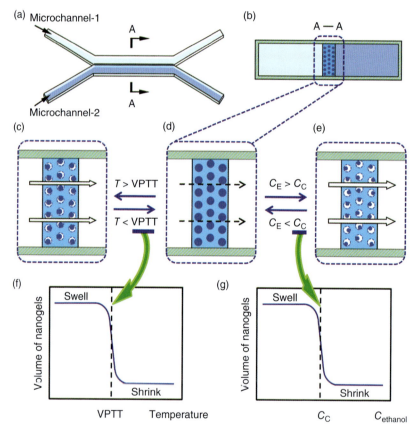

Figure 12.1 Schematic illustration of microfluidic fabrication of nanogel-containing smart membrane in the microchannel of microchip. (a) Fabrication of nanogel-containing membrane in X-shaped microchannel at the laminar interface of fluids from microchannel-1 and microchannel-2. (b) Magnified image showing the cross-sectional view of microchannel and membrane-in-chip. (c–e) Self-regulation of the membrane permeability by changing temperature (c, d) and ethanol concentration (d, e). (f, g) Schematic illustration showing the swelling and shrinking of PNIPAM nanogels when temperature changes across the volume-phase transition temperature (VPTT) (f) and ethanol concentration (C_E) changes across the critical ethanol concentration (C_C) (g). (Sun *et al.* 2014 [20]. Reproduced with permission of The Royal Society of Chemistry.)

temperature and ethanol concentration, are used as smart nanovalves for controlling membrane permeability. To fabricate such a membrane, aqueous phase containing chitosan and PNIPAM nanogels and oil phase containing terephthalaldehyde flow in parallel in the microchannel of a microchip. Cross-linking between the chitosan and terephthalaldehyde at the O/W interface traps the nanogels inside the cross-linked membrane matrix and *in situ* produces a nanogel-containing chitosan membrane within the microchannel. The permeability of the membrane could be self-regulated in response to changes in temperature and ethanol concentration. When the temperature increases from 25 to 40 °C, the embedded PNIPAM nanogels shrink and create voids in the chitosan membrane, resulting in more FITC (fluorescein isothiocyanate) molecules permeating through the membrane. Similarly, the shrinking of the PNIPAM nanogels upon increasing the ethanol concentration from 0 to 30 wt% also leads to increased FITC flux across the membrane (Figure 12.1). Such nanogel-containing smart membranes in microchannels are promising for developing microscale detectors, sensors, separators, and controlled release systems.

12.3 Temperature- and Ethanol-Responsive Smart Membrane in Microchip for Detection

12.3.1 Fabrication of Nanogel-Containing Smart Membrane in Microchip

A microchip that is used as a platform for membrane integration (Figure 12.1a) is fabricated by assembling coverslips (thickness: ~150 μm) on a glass slide [21]. Before assembling, the glass slide is treated with hydrofluoric acid (4 vol%) for hydrophilic modification. Then, half of the glass slide is immersed in a solution of nano superhydrophobic coating for hydrophobic modification. Coverslips are placed on the modified glass slide with half hydrophilic surface and half hydrophobic surface for microchip assembly. The X-shaped gap between the coverslips creates microchannels consisting of a hydrophobic microchannel-1 and a hydrophilic microchannel-2 (Figure 12.1a).

PNIPAM nanogels that are used as nanovalves are synthesized by precipitation polymerization [20]. The SEM image of PNIPAM nanogels shows that they are uniform and spherical with a diameter of ~350 nm (Figure 12.2a). The confocal laser scanning microscope (CLSM) image of PNIPAM nanogels exhibits the successful incorporation of fluorescence dye (Polyfluor 570) in the nanogels (Figure 12.2b). The PNIPAM networks inside the nanogel enable reversible swelling/shrinking volume transitions in response to changes in temperature and ethanol concentration. As shown in Figure 12.2c, upon increasing the temperature from 25 to 40 °C, the average diameter of the nanogels decreases from ~760 to ~450 nm. Such volume change can also be induced by varying the ethanol concentration, and the ethanol-responsive volume change behaviors of PNIPAM nanogels show that the critical ethanol concentration value (C_C) is found to be about 8 wt%. At a fixed temperature (25 °C), when the ethanol concentration is increased from 0 to 30 wt% that above the C_C, a significant shrinkage of the PNIPAM nanogels can also be achieved (Figure 12.2c).

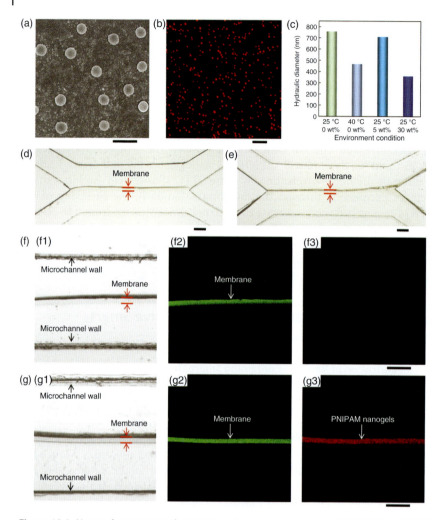

Figure 12.2 Nanogel-containing chitosan membrane in microchip. (a, b) SEM (a) and CLSM (b) images showing the fluorescent PNIPAM nanogels dyed with Polyfluor 570. Scale bars are 1 μm in (a) and 5 μm in (b). (c) The hydraulic diameters of PNIPAM nanogels at different temperatures and ethanol concentrations. (d–g) Optical (d, e) and CLSM images (f, g) of views from the top of the blank chitosan membrane (d, f) and nanogel-containing chitosan membrane (e, g) in microchannels. Scale bars are 250 μm. (Sun et al. 2014 [20]. Reproduced with permission of The Royal Society of Chemistry.)

Such stimuli-responsive volume transitions make the nanogels excellent nanovalves for controlling the membrane permeability.

Typically, for *in situ* fabrication of membranes in the X-shaped microchannels, aqueous phases containing chitosan (0.04 g ml^{-1}) and different contents (Φ) of PNIPAM nanogels (Φ = 0, 10, 20, 40 wt%), with pH ~6.3 adjusted by NaOH (1 M), are flowed in microchannel-2. Meanwhile, mixed oil phase of benzyl benzoate (50 vol%) and soybean oil (50 vol%), which contains terephthalaldehyde (0.02 g ml^{-1}), is co-flowed in microchannel-1. The aqueous phase and oil phase are, respectively, injected by syringe pump at 5 and $1\,\mathrm{ml\,h^{-1}}$ to reach a stable interface between these two laminar flows at 25 °C. After the stable interface is formed, the syringe pumps are shut down to stop the flow of the aqueous and oil phases, which allows interfacial cross-linking between chitosan and terephthalaldehyde for 70 s. Then, the aqueous phase is flowed again, but with a decreased flow rate ($0.5\,\mathrm{ml\,h^{-1}}$) for further reaction for 15 min. After the reaction, the aqueous phase is replaced by deionized water ($1\,\mathrm{ml\,h^{-1}}$), and oil phase is changed into isopropanol ($1\,\mathrm{ml\,h^{-1}}$) to wash the obtained nanogel-containing chitosan membrane. After the oil phase is completely removed, the isopropanol is changed into pure water for washing for another 30 min. The views from the top of the blank membrane and nanogel-containing membranes in microchannels are observed by digital microscope and CLSM with excitation at ~488 and ~543 nm. The results show that the chitosan membrane is successfully *in situ* fabricated in the microchannel by interfacial cross-linking of the chitosan in the aqueous phase with terephthalaldehyde in the oil phase (Figure 12.2d). When PNIPAM nanogels are added in the aqueous chitosan solution, such cross-linking reaction simply traps the nanogels inside the cross-linked chitosan matrix and produces a nanogel-containing membrane in the microchip (Figure 12.2e). Both membranes have a uniform thickness of ~80 μm, with each edge closely connected to the wall of the microchannel. The incorporation of the fluorescent PNIPAM nanogels in the membranes is confirmed by CLSM. As shown in Figure 12.2f,g, both blank membrane (Figure 12.2f) and nanogel-containing membrane (Figure 12.2g) show green fluorescence (Figure 12.2f2,g2) due to the autofluorescence of the cross-linked chitosan matrix. However, compared with the blank membrane (Figure 12.2f3), the nanogel-containing membrane shows red fluorescence in Figure 12.2g3 due to the presence of fluorescent PNIPAM nanogels in the cross-linked membrane. The results indicate the successful *in situ* formation of nanogel-containing chitosan membrane in the microchannel by *in situ* trapping of nanogels inside the cross-linked membrane matrix.

12.3.2 Temperature-Responsive Self-Regulation of the Membrane Permeability

The embedded PNIPAM nanogels that allow temperature- and ethanol-responsive volume transitions can serve as excellent nanovalves to control the permeability of the chitosan membrane. The permeability self-regulation of the membrane in response to changes in temperature and ethanol concentration is investigated by using FITC molecules for tracing and using the CLSM for observation. The membranes integrated in microchips are placed on a thermostatic heating stage for temperature control. PBS (phosphate-buffered saline) with pH 7.4 is used as the aqueous solution. To investigate the temperature-responsive permeability control, the diffusion flux of FITC through the membrane at 25 and 40 °C are studied. Briefly, aqueous solution containing FITC is flowed in microchannel-1 at $0.5\,\mathrm{ml\,h^{-1}}$, and aqueous solution without FITC is co-flowed in microchannel-2 at $0.5\,\mathrm{ml\,h^{-1}}$ for 30 min to reach a stable flow state and a desired temperature. After that, the aqueous solution without FITC is stopped, and the diffusional behavior of FITC through the membrane, from microchannel-1

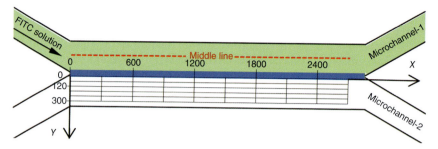

Figure 12.3 Schematic illustration showing the coordinate in microchannel-2 for monitoring the diffusion flux. (Sun et al. 2014 [20]. Reproduced with permission of The Royal Society of Chemistry.)

to microchannel-2, is recorded by the CLSM. For measuring the diffusional flux of FITC through the membrane, the relative fluorescence intensity (I_r) of FITC at different locations in the region of microchannel-2 that divided by X–Y coordinate (Figure 12.3) is monitored. For each location (X_i, Y_i), $I_r = I_{(X_i,Y_i)}/I_{X_i}$, where $I_{(X_i,Y_i)}$ is the fluorescence intensity at location (X_i, Y_i), and I_{X_i} is the fluorescence intensity at location (Y_m, X_i) in the middle line of microchannel-1 (Figure 12.3).

The temperature-responsive control of the membrane permeability is studied by measuring the FITC flux through the membrane upon increasing temperature. As shown in Figure 12.4, at the same time, the I_r value of the blank membrane only slightly increases upon increasing the temperature from 25 (Figure 12.4a–c) to 40 °C (Figure 12.4d–f), due to the effect of increased temperature on diffusion coefficient. Incorporation of thermo- and ethanol-responsive PNIPAM nanogels in the chitosan membrane enables the control of the membrane permeability based on the volume-phase transition of the PNIPAM nanogels. The nanogel contents in the membrane can be adjusted by varying the nanogel concentration in the aqueous phase. The effect of the nanogel content (Φ) on the permeability control is studied by measuring the ratio of the fluorescence intensity of FITC through membranes with different nanogel contents at location (0, 120) in microchannel-2 at 40 °C to that at 25 °C (I_{40}/I_{25}). On increasing the nanogel content from 0 to 10, 20, 40 wt% in the chitosan membrane, the I_{40}/I_{25} value increases from 1.11 to 1.13, 1.18, and 1.79, respectively. The more the PNIPAM nanogels in the membrane, the more significant the temperature-responsive permeability control. However, it is worth noting that, even if the nanogel content is increased to higher than 40 wt% (e.g., 60 wt%), it is difficult to form stable interface between aqueous phase and organic phase for preparing stable membranes anymore. Thus, chitosan membrane containing 40 wt% PNIPAM nanogels is selected for subsequent investigations.

As shown in Figure 12.5, the nanogel-containing membrane ($\Phi = 40$ wt%) shows a significant increase of I_r value upon increasing temperature from 25 (Figure 12.5a–c) to 40 °C (Figure 12.5d–f), due to the temperature-responsive volume transitions of PNIPAM nanogels across the VPTT (volume phase transition temperature). At temperatures lower than the VPTT, the swollen

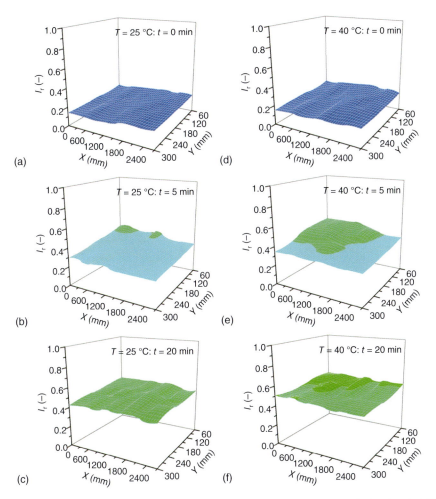

Figure 12.4 Diffusion flux of FITC through blank chitosan membrane at 25 °C (a–c) and 40 °C (d–f). (Sun et al. 2014 [20]. Reproduced with permission of The Royal Society of Chemistry.)

PNIPAM nanogels in the membrane result in large resistance for FITC diffusion. When the temperature is increased higher than the VPTT, the PNIPAM nanogels dramatically shrink and create voids in the membrane, which lead to less membrane resistance and increased transmembrane diffusion flux. At 25 °C, the nanogel-containing membrane shows lower I_r value than that of the blank membrane (Figure 12.5a–c). This is because the blank membrane shows a much larger swelling degree ($Q = 142.2$) than that of the nanogel-containing membrane ($Q = 45.6$), indicating a smaller cross-linking degree than that of nanogel-containing membrane. Thus, the nanogel-containing membrane with denser cross-linking matrix results in a larger membrane resistance and a less FITC diffusion flux as compared to the blank membrane at 25 °C.

More quantitative studies on the I_r value changes of both blank membrane and nanogel-containing membrane at typical locations in microchannel-2 upon

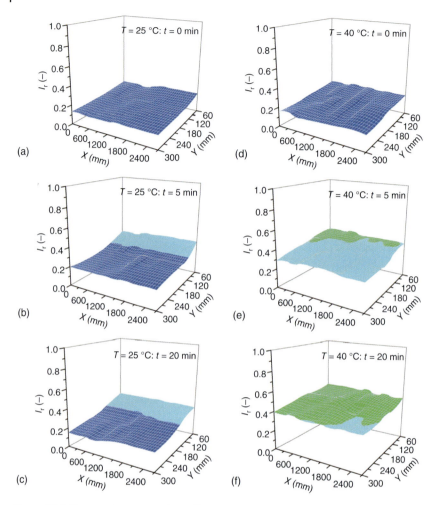

Figure 12.5 Temperature-responsive self-regulation of the membrane permeability. Diffusion flux of FITC through the nanogel-containing chitosan membrane ($\Phi = 40$ wt%) at 25 °C (a–c) and 40 °C (d–f). (Sun et al. 2014 [20]. Reproduced with permission of The Royal Society of Chemistry.)

the temperature change are shown in Figure 12.6. Compared with the blank membrane (Figure 12.6a,c), the nanogel-containing membrane (Figure 12.6b,d) exhibits an obviously increased I_r value for all the typical locations. The results indicate the excellent temperature-responsive permeability self-regulation of the nanogel-containing membrane.

12.3.3 Ethanol-Responsive Self-Regulation of the Membrane Permeability

The volume transitions of PNIPAM nanogels in response to ethanol concentration changes also enable similar self-regulation of the membrane permeability. The ethanol-responsive permeability control is investigated by measuring the diffusion flux of FITC through the membrane at different ethanol concentrations

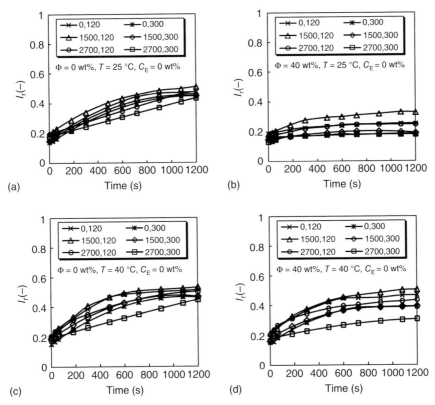

Figure 12.6 Diffusion flux of FITC through blank membrane ($\Phi = 0$ wt%) (a, c) and nanogel-containing membrane ($\Phi = 40$ wt%) (b, d) at typical locations under conditions of different temperatures. (Sun et al. 2014 [20]. Reproduced with permission of The Royal Society of Chemistry.)

(0 and 30 wt%). In this case, aqueous solution containing ethanol and FITC is flowed in microchannel-1, while aqueous solution without ethanol and FITC is flowed in microchannel-2.

For the blank membrane (Figure 12.4a–c), addition of 30 wt% ethanol shows no significant impact on I_r value as well as the FITC flux (Figure 12.7a–c). By contrast, the nanogel-containing membrane exhibits significant increase in I_r value upon addition of 30 wt% ethanol due to the ethanol-responsive volume shrinking of PNIPAM nanogels (Figure 12.7d–f). At 25 °C, when the ethanol concentration is lower than C_C, the PNIPAM nanogels show swollen state and lead to lower membrane permeability. When the ethanol concentration is increased higher than the C_C value, the competition of hydrophobic hydration between ethanol and PNIPAM polymers results in dramatic shrinking of the PNIPAM nanogels for increased membrane permeability.

The I_r value changes of the blank membrane (Figures 12.8a and 12.9) and nanogel-containing membrane (Figures 12.8b and 12.9) at typical locations in microchannel-2 upon ethanol addition also demonstrate significant ethanol-responsive permeability changes for the nanogel-containing membrane.

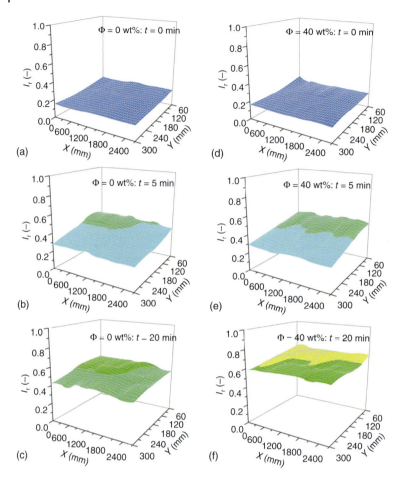

Figure 12.7 Ethanol-responsive self-regulation of membrane permeability. Diffusion flux of FITC through the blank chitosan membrane (a–c) and nanogel-containing chitosan membrane ($\Phi = 40$ wt%) (d–f) at 25 °C and 30 wt% ethanol concentration. (Sun et al. 2014 [20]. Reproduced with permission of The Royal Society of Chemistry.)

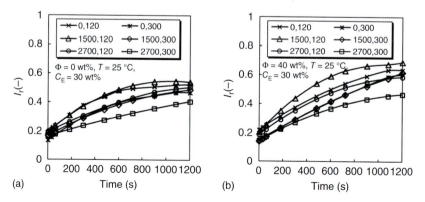

Figure 12.8 Diffusion flux of FITC through blank membrane ($\Phi = 0$ wt%) (a) and nanogel-containing membrane ($\Phi = 40$ wt%) (b) at typical locations under $C_E = 30$ wt%. (Sun et al. 2014 [20]. Reproduced with permission of The Royal Society of Chemistry.)

Figure 12.9 Diffusion flux of FITC through the blank and nanogel-containing membranes at location (0, 120) under conditions of different temperatures and ethanol concentrations (C_E). (Sun et al. 2014 [20]. Reproduced with permission of The Royal Society of Chemistry.)

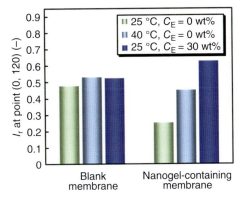

The results confirm the excellent performance of the nanogel-containing membrane for ethanol-responsive permeability control. As compared with the blank membranes, such nanogel-containing membranes exhibit outstanding properties in intelligently controlling the membrane permeability in response to changes in temperature and ethanol concentration in the environment.

12.3.4 Reversible and Repeated Thermo/Ethanol-Responsive Self-Regulation of the Membrane Permeability

The reversible and repeatable volume transitions of PNIPAM nanogels upon changes in temperature and ethanol concentration enable reversible and repeatable self-regulation of the membrane permeability. As shown in Figure 12.10, the I_r value of the nanogel-containing membranes at typical location (0, 120) increases with increasing temperature from 25 to 40 °C. Further increase of the I_r value is achieved by addition of 30 wt% ethanol at 25 °C. Such cycle of I_r value changes can be reversed and repeated by again changing the temperature and ethanol concentration in the abovementioned order. The results indicate excellent reversible and repeated thermo/ethanol-responsive characteristics of the membrane for permeability regulation.

Figure 12.10 Repeated temperature- and ethanol-responsive self-regulation of permeability at location (0, 120) of nanogel-containing membrane ($\Phi = 0$ wt%). (Sun et al. 2014 [20]. Reproduced with permission of The Royal Society of Chemistry.)

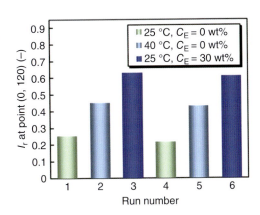

12.4 Summary

In this chapter, a facile and flexible microfluidic approach is introduced for *in situ* formation of nanogel-containing smart membranes in microchips for permeability control. The nanogel-containing smart membrane is fabricated by trapping PNIPAM nanogels as smart nanovalves in the chitosan membranes via *in situ* interfacial cross-linking at the interface between two laminar flows. The nanogel-containing membranes allow self-regulation of the membrane permeability in response to changes in temperature and ethanol concentration, due to the reversible and repeatable swelling/shrinking of the embedded PNIPAM nanogels. Such smart-membrane-in-chips show great potential as detectors for monitoring temperature and C_E changes in bioreactors by simply detecting flux change. For example, for biomass fermentation in which the yeast would become inactive at C_E higher than ~8 wt% [22], the C_E changes could be monitored by incorporating such membrane-in-chips in the bioreactor for indicating and adjusting the fermentation performance. Moreover, the facile approach presented here also allows development of versatile stimuli-responsive smart membranes by simply changing the functional nanogels such as pH-responsive nanogels [23] and molecular-responsive nanogels [24] in the membrane. Therefore, this microfluidic approach provides a powerful strategy for *in situ* fabrication of nanogel-containing smart membranes within microchips for biomedical applications such as on-line detection, separation, and controlled release.

References

1 Surmeian, M., Slyadnev, M.N., Hisamoto, H., Hibara, A., Uchiyama, K., and Kitamori, T. (2002) Three-layer flow membrane system on a microchip for investigation of molecular transport. *Anal. Chem.*, **74**, 2014–2020.

2 Kuo, T.C., Cannon, D.M., Chen, Y., Tulock, J.J., Shannon, M.A., Sweedler, J.V., and Bohn, P.W. (2003) Gateable nanofluidic interconnects for multilayered microfluidic separation systems. *Anal. Chem.*, **75**, 1861–1867.

3 Noblitt, S.D., Kraly, J.R., VanBuren, J.M., Hering, S.V., Collett, J.L., and Henry, C.S. (2007) Integrated membrane filters for minimizing hydrodynamic flow and filtering in microfluidic devices. *Anal. Chem.*, **79**, 6249–6254.

4 Wei, H., Chueh, B.H., Wu, H., Hall, E.W., Li, C.W., Schirhagl, R., Lin, J.M., and Zare, R.N. (2011) Particle sorting using a porous membrane in a microfluidic device. *Lab Chip*, **11**, 238–245.

5 Hisamoto, H., Shimizu, Y., Uchiyama, K., Tokeshi, M., Kikutani, Y., Hibara, A., and Kitamori, T. (2003) Chemicofunctional membrane for integrated chemical processes on a microchip. *Anal. Chem.*, **75**, 350–354.

6 Kiwi-Minsker, L. and Renken, A. (2005) Microstructured reactors for catalytic reactions. *Catal. Today*, **110**, 2–14.

7 Uozumi, Y., Yamada, Y.M.A., Beppu, T., Fukuyama, N., Ueno, M., and Kitamori, T. (2006) Instantaneous carbon–carbon bond formation using a

microchannel reactor with a catalytic membrane. *J. Am. Chem. Soc.*, **128**, 15994–15995.

8 Choi, E., Jun, I., Chang, H.K., Park, K.M., Shin, H., Park, K.D., and Park, J. (2012) Quantitatively controlled *in situ* formation of hydrogel membranes in microchannels for generation of stable chemical gradients. *Lab Chip*, **12**, 302–308.

9 Braschler, T., Johann, R., Heule, M., Metref, L., and Renaud, P. (2005) Gentle cell trapping and release on a microfluidic chip by *in situ* alginate hydrogel formation. *Lab Chip*, **5**, 553–559.

10 De Jong, J., Lammertink, R.G.H., and Wessling, M. (2006) Membranes and microfluidics: a review. *Lab Chip*, **6**, 1125–1139.

11 Chu, M.K.L., Chen, J., Gordijo, C.R., Chiang, S., Ivovic, A., Koulajian, K., Giacca, A., Wu, X.Y., and Sun, Y. (2012) *In vitro* and *in vivo* testing of glucose-responsive insulin-delivery microdevices in diabetic rats. *Lab Chip*, **12**, 2533–2539.

12 Long, Z., Shen, Z., Wu, D., Qin, J., and Lin, B. (2007) Integrated multilayer microfluidic device with a nanoporous membrane interconnect for online coupling of solid-phase extraction to microchip electrophoresis. *Lab Chip*, **7**, 1819–1824.

13 Liu, Z.B., Zhang, Y., Yu, J.J., Mak, A.F.T., Li, Y., and Yang, M. (2010) A microfluidic chip with poly(ethylene glycol) hydrogel microarray on nanoporous alumina membrane for cell patterning and drug testing. *Sens. Actuators, B*, **143**, 776–783.

14 Randall, G.C. and Doyle, P.S. (2005) Permeation-driven flow in poly(dimethylsiloxane) microfluidic devices. *Proc. Natl. Acad. Sci. U. S. A.*, **102**, 10813–10818.

15 Cabodi, M., Choi, N.W., Gleghorn, J.P., Lee, C.S., Bonassar, L.J., and Stroock, A.D. (2005) A microfluidic biomaterial. *J. Am. Chem. Soc.*, **127**, 13788–13789.

16 Kenis, P.J.A., Ismagilov, R.F., and Whitesides, G.M. (1999) Microfabrication inside capillaries using multiphase laminar flow patterning. *Science*, **285**, 83–85.

17 Kim, D. and Beebe, D.J. (2008) Interfacial formation of porous membranes with poly(ethylene glycol) in a microfluidic environment. *J. Appl. Polym. Sci.*, **110**, 1581–1589.

18 Orhan, J.B., Knaack, R., Parashar, V.K., and Gijs, M.A.M. (2008) In situ fabrication of a poly-acrylamide membrane in a microfluidic channel. *Microelectron. Eng.*, **85**, 1083–1085.

19 Luo, X., Berlin, D.L., Betz, J., Payne, G.F., Bentley, W.E., and Rubloff, G.W. (2010) *In situ* generation of pH gradients in microfluidic devices for biofabrication of freestanding, semi-permeable chitosan membranes. *Lab Chip*, **10**, 59–65.

20 Sun, Y.M., Wang, W., Wei, Y.Y., Deng, N.N., Liu, Z., Ju, X.J., Xie, R., and Chu, L.Y. (2014) *In situ* fabrication of temperature- and ethanol-responsive smart membrane in microchip. *Lab Chip*, **14**, 2418–2427.

21 Deng, N.N., Meng, Z.J., Xie, R., Ju, X.J., Mou, C.L., Wang, W., and Chu, L.Y. (2011) Simple and cheap microfluidic devices for preparation of monodisperse emulsions. *Lab Chip*, **11**, 3963–3969.

22 Ito, Y., Ito, T., Takaba, H., and Nakao, S. (2005) Development of gating membranes that are sensitive to the concentration of ethanol. *J. Membr. Sci.*, **261**, 145–151.

23 Hu, L., Chu, L.Y., Yang, M., Wang, H.D., and Niu, C.H. (2007) Preparation and characterization of novel cationic pH-responsive poly(N,N'-dimethylamino ethyl methacrylate) microgels. *J. Colloid Interface Sci.*, **311**, 110–117.

24 Ju, X.J., Liu, L., Xie, R., Niu, C.H., and Chu, L.Y. (2009) Dual thermo-responsive and ion-recognizable monodisperse microspheres. *Polymer*, **50**, 922–929.

13

Microfluidic Fabrication of Microvalve-in-a-Chip

13.1 Introduction

Microchips with highly integrated functional units for fluid manipulation are widely used in various fields, such as biological/chemical analysis [1], disease diagnosis [2], high-throughput screening [3], and cell culture [4], due to their short analysis time, low cost, and reduced reagent consumption [5]. Integration of smart-hydrogel-based microvalves in microchips allows controlling flow direction and flux via their stimuli-induced shape changes [6, 7]. The swelling/shrinking changes of such hydrogel microvalves in response to various stimuli, such as temperature [8–15], pH [16–19], magnetic field [20], electric field [21], and glucose [22], allow controlling the "on/off" switch of microchannels for flow manipulation. Thus, such smart-microvalve-integrated microchips create new opportunities for developing smart microsystems for myriad applications.

For biomedical applications, self-regulation of temperature in these microsystems is crucial for their use in cell culture, bio/chemical reactions, and those cases that require constant temperature conditions. For example, constant temperature is crucial to ensure efficient gene amplification and cell growth [23–26], and temperature maintenance can also benefit the reaction rate and safety of exothermic reactions. Usually, thermostatic control is achieved by electronic systems; however, these systems require external power supply and complex control algorithms [27]. Alternatively, the smart hydrogel microvalves integrate both sensing and actuating functions and require no external power supply; thus, they show great potential for developing microsystems for thermostatic biomedical applications. However, although the abovementioned microvalves allow "on–off" switch in response to a wide range of stimuli for flow manipulation, utilization of such microvalves in microsystems for thermostatic control still remains difficult to be achieved. Moreover, current methods for fabricating microvalves usually require expensive instruments and troublesome multistep fabrication processes. Therefore, simple techniques for design and fabrication of responsive hydrogel microvalves within microchips for thermostatic control are still essentially required.

For detection applications, timely detection of trace threat analytes that are harmful to environment and human health is critical for environmental protection [28], disease treatment [29, 30], and epidemic prevention [31]. The key challenge is how to efficiently convert and amplify the analyte signal into simple readouts for real-time detection. Based on the stimuli-responsive volume changes of smart hydrogels [32, 33], current techniques allow converting the stimulus signals into electrical or optical signals for detection [34, 35]. However, these methods inefficiently utilize the three-dimensional hydrogel deformation, thus possessing poor detection limit. Improved sensitivity can be achieved by converting the target signals into optical signals [36–41]. However, all these techniques require sophisticated equipment and professionals for detecting and analyzing. To sum up, the platforms with electrical signals provide easy use and low cost, but the detection limit is poor, whereas the platforms with optical signals offer improved sensitivity, but require sophisticated analyzing protocols, which restrict the applications for real-time online detection. Up to now, development of simple and ultrasensitive detection platforms for real-time online detection of trace analytes still remains a challenge.

In this chapter, focused on the abovementioned two problems, two microchips, incorporated with smart hydrogel microvalves, have been developed for self-regulation of temperature and timely detection of trace threat analytes. The microchip with a thermo-responsive hydrogel microvalve enables "on–off" switch by sensing temperature fluctuations to control the fluid flux as well as the fluid heat exchange for self-regulating the temperature at a constant range. Such a microvalve-incorporated microchip can be incorporated into the flow circulation loop of a micro-heat-exchanging system for thermostatic control and cell culture. The other microchip with Pb^{2+}-responsive hydrogel microvalve can be used for real-time detection of trace threat analytes. The microgel swells in response to Pb^{2+} concentration ($[Pb^{2+}]$) changes, thus converting trace $[Pb^{2+}]$ into significantly amplified signal of flow rate change for highly sensitive, fast and selective detection, which can be monitored on cell phone for timely warning and terminating of pollution.

13.2 Microfluidic Strategy for Fabrication of Microvalve-in-a-Chip

Fabrication of hydrogel microvalve in the microchannel of microfluidic chip can be achieved by *in situ* polymerization of the selective part of the monomer solution in microchannel. This can be realized by using sacrificed materials for blocking the microchannel for confining and shaping the injected monomer solution for polymerization [42], or using mask-based lithography for exposing the selective part of the monomer solution for polymerization [43]. In this section, both strategies for fabrication of microvalve-in-a-chip are introduced.

13.2.1 Fabrication of Thermo-Responsive Hydrogel Microvalve within Microchip for Thermostatic Control

A microchip with star-shaped microchannels is used as the platform for integrating the thermo-responsive hydrogel microvalve for thermostatic control. The microchip with star-shaped microchannels is constructed by assembling coverslips with thickness of ~150 μm on glass slides according to the literature [44].

13.2 Microfluidic Strategy for Fabrication of Microvalve-in-a-Chip

The gaps between the assembled coverslips create star-shaped microchannels for flowing fluid. For *in situ* fabrication of the thermo-responsive hydrogel microvalve in the desired site within the microchip, as shown in Figure 13.1a, an L-shaped chamber is created in the star-shaped microchannels for hydrogel polymerization. The L-shaped chamber is constructed by using melt paraffin

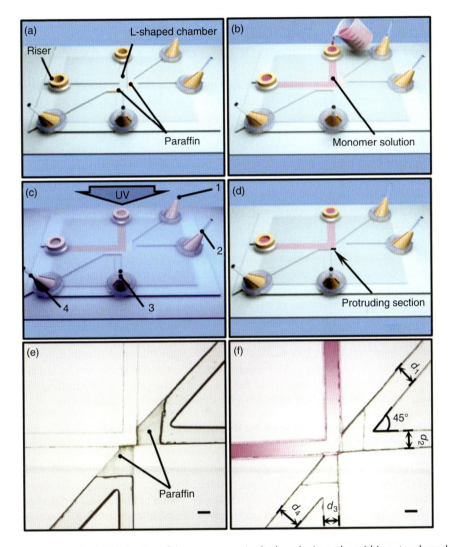

Figure 13.1 *In situ* fabrication of thermo-responsive hydrogel microvalve within a star-shaped microchip. (a–d) Schematic illustration showing the fabrication process: first, construct an L-shaped chamber in the star-shaped microchannel for hydrogel synthesis (a), then, inject monomer solution in the L-shaped chamber (b), followed by UV polymerization (c) and paraffin removal for producing the hydrogel microvalve (d). (e, f) Optical micrographs showing the L-shaped chamber (depth: 150 µm, width: 350 µm) confined by paraffin (e) and the L-shaped hydrogel microvalve labeled with rhodamine B (f). The widths of the microchannels are as follows: $d_1 = \sim 460$ µm, $d_2 = \sim 380$ µm, $d_3 = \sim 370$ µm, $d_4 = \sim 550$ µm, and all the depths are ~150 µm. Scale bars are 300 µm. (Lin *et al.* 2014 [42]. Reproduced with permission of The Royal Society of Chemistry.)

to selectively block the microchannel. Then, to seal the gaps for flowing fluid, another coverslip is fixed on the top of the gaps to create confined microchannels. After that, UV-curable adhesive is used to ensure tight sealing between the coverslips and the glass slide. Next, needles are set at the inlet and outlet of the microchannel and sealed by epoxy resin for further connecting polyethylene (PE) pipes (Figure 13.1a).

For synthesis of the thermo-responsive hydrogel microvalve within the microchip, monomer solution containing deionized water (10 ml), monomer N-isopropylacrylamide (NIPAM) (1.13 g), crosslinker N,N-methylene-bis-acrylamide (MBA) (0.0308 g), and photoinitiators 1-hydroxy-cyclohexyl-phenylketone (0.00452 g) and benzophenone (0.00452 g) is injected into the L-shaped chamber (Figure 13.1b). Two risers are set separately at the inlet and outlet of the L-shaped chamber for injecting the monomer solution and preserving the excess monomer solution from the L-shaped chamber. Then, the monomer-solution-containing microchip is placed in an ice bath and exposed to UV light for 8 min to polymerize the monomer solution for synthesizing the hydrogel microvalve (Figure 13.1c). After sealing the risers with epoxy resin, the blocking paraffin remained in the microchannel is washed off with hot water. Thus, a microchip incorporated with thermo-responsive hydrogel microvalve is finally obtained (Figure 13.1d).

Figure 13.1e shows the optical micrograph of the star-shaped microchannel with an L-shaped chamber that is confined by the microchannel and paraffin for hydrogel synthesis. Injection of the monomer solution in the L-shaped chamber followed by UV polymerization can *in situ* produce a thermo-responsive poly(N-isopropylacrylamide) (PNIPAM) hydrogel microvalve with L-shape in the microchip. The PNIPAM hydrogel microvalve is dyed with rhodamine B (red color) to clearly show the L-shape (Figure 13.1f). This strategy provides a versatile approach for *in situ* fabrication of hydrogel microvalve with different compositions in microchips. Moreover, since the chamber shape can be simply changed by adjusting the microchannel shape and the position of paraffin, this approach allows the creation of hydrogel microvalves with flexible shapes.

13.2.2 Fabrication of Pb^{2+}-Responsive Hydrogel Microvalve within Microchip for Pb^{2+} Detection

A microfluidic chip containing glass-capillary microchannel integrated with cylinder-shaped smart microgel as microvalve allows highly sensitive, fast, and selective detection of trace threat analytes. For detection of Pb^{2+}, cylinder-shaped microgel consisting of poly(N-isopropylacrylamide-*co*-benzo-18-crown-6-acrylamide) (P(NIPAM-*co*-B18C6Am)) microgel, with NIPAM units as actuators and B18C6Am units as ion signal sensing receptors to selectively recognize trace Pb^{2+}, is incorporated in the microchip (Figure 13.2). The uniform cylinder-shaped P(NIPAM-*co*-B18C6Am) microgel is *in situ* synthesized within a glass-capillary microchannel by an advanced rotation-based UV-irradiation method (Figure 13.2a,b). Typically, aqueous solution (2 ml) containing monomer B18C6Am (0.3 mmol) and NIPAM (2.0 mmol), crosslinker MBA (0.04 mmol), and photoinitiator 2,2'-azobis(2-amidinopropane dihydrochloride) (0.037 mmol) is injected into a glass capillary of 250 μm inner diameter and 960 μm outer

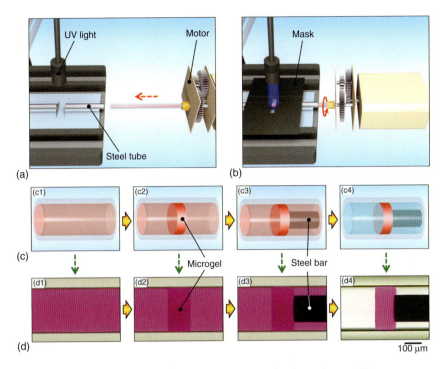

Figure 13.2 *In situ* synthesis of Pb^{2+}-responsive microgel within a glass-capillary microchannel. (a, b) An advanced rotation-based method is developed for fabricating uniform cylinder-shaped microgel in glass-capillary microchannel. The monomer-solution-containing glass capillary that is fixed on a motor is inserted into two steel tubes (a), and then covered by a mask for 360° uniform UV irradiation under rotation for polymerization (b). (c, d) Schematics (c) and optical micrographs (d) showing the fabrication and fixation of microgel. In the monomer-solution-containing glass capillary (c1, d1), uniform cylinder-shaped microgel is synthesized via rotation-assisted UV polymerization (c2, d2), and then stably supported by skillfully introducing a coaxially placed stainless steel bar (c3, d3), followed by removing the unpolymerized solution (c4, d4). (Lin *et al*. 2016 [43]. Reproduced with permission of Proceedings of the National Academy of Sciences.)

diameter. One end of the monomer-solution-loaded capillary is fixed on a rotating motor, while the other one end is inserted into two stainless steel tubes with inner diameter of 1000 μm. The stainless steel tubes are placed on a thermostatic stage, which is used to keep the synthesis temperature fixed at 0 °C. The exposed part of the glass capillary between the two steel tubes is covered by a patterned mask, with a transparent rectangular area (size: 130 μm × 1 cm) for UV irradiation. Then, the monomer-solution-loaded glass capillary is rotated at a speed of 60 rpm and treated with UV irradiation ($\lambda = 365$ nm) for 2 min. This allows *in situ* synthesis of a uniform cylinder-shaped microgel inside the glass capillary. Such a rotation-based UV-irradiation method ensures 360° uniform UV irradiation for efficiently converting the loaded monomer solution (Figure 13.2c1,d1) into uniform cylinder-shaped microgel (Figure 13.2c2,d2).

Next, the microgel-incorporated glass capillary is used for assembling the Pb^{2+} detection microfluidic chip. First, the microgel-incorporated glass capillary is

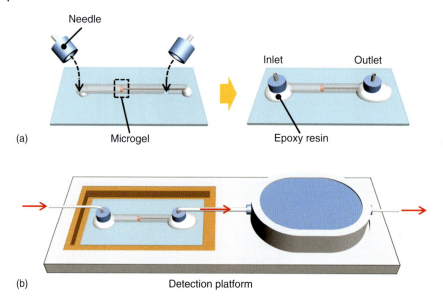

Figure 13.3 Schematic illustration for fabrication of the detection platform incorporated with Pb^{2+}-responsive hydrogel microvalve. (a) Two needles are connected to the two ends of the microgel-incorporated glass capillary on the microfluidic chip and then sealed with epoxy resin as the inlet and outlet. (b) The outlet of the microfluidic chip is connected to an online flowmeter for constructing the detection platform. (Lin et al. 2016 [43]. Reproduced with permission of Proceedings of the National Academy of Sciences.)

fixed on a glass slide by epoxy resin and then clamped by two fixtures for fixation. Then, another glass capillary, with inner diameter of 170 μm and outer diameter of 960 μm, is placed into the interstice between the two fixtures and fixed on the glass slide for coaxial alignment with the first glass capillary. After that, a cylinder bar of stainless steel with diameter of 165 μm is coaxially inserted into the capillaries to support the microgel (Figure 13.2c3,d3), followed by fixation of the cylinder bar on the glass slide and seal of the second capillary by epoxy resin, and finally removal of the fixtures. After the assembly, the unpolymerized solution in the glass capillary is removed (Figure 13.2c4,d4), and thus the microgel-based Pb^{2+} sensor is obtained. Next, two needles are connected to the two ends of the microgel-incorporated glass capillary and sealed with epoxy resin as the inlet and outlet for fabricating the microfluidic detection chip (Figure 13.3a). Finally, the outlet of the detection chip is connected to an online flowmeter of microfluidic control system by PE pipes for constructing the Pb^{2+} detection platform (Figure 13.3b).

13.3 Smart Microvalve-in-a-Chip with Thermostatic Control for Cell Culture

In this section, the microchip incorporated with thermo-responsive hydrogel microvalve, which was introduced in Section 13.2.1, is integrated into the flow circulation loop of a micro-heat-exchanging system for thermostatic control. Moreover, such a microvalve-incorporated microchip is further employed for culturing cells under thermostatic conditions.

13.3.1 Setup of Microvalve-Integrated Micro-heat-Exchanging System

The microvalve-incorporated microchip is integrated into the flow circulation loop of a micro-heat-exchanging system for thermostatic control. A micro-heat-exchanger is fabricated by a method similar to that used for fabricating the microchip (Figure 13.4a–c). Briefly, two reservoirs, each with size of 0.5 cm × 0.5 cm × 150 μm, are separated by a coverslip and constructed in the micro-heat-exchanger for flowing hot (lower reservoir) and cold (upper reservoir) fluids (Figure 13.4a–c). To create the micro-heat-exchanging system with circulation loop for flowing the cold fluid, the microvalve-incorporated microchip for controlling flux and a peristaltic pump for powering flow circulation are connected with the lower reservoir of the micro-heat-exchanger by PE pipes. Throughout the experiment, a microfluidic flow control system (MFCS) is used to inject cold water with a constant pressure of 40 kPa and measure the flux (Figure 13.4d–f).

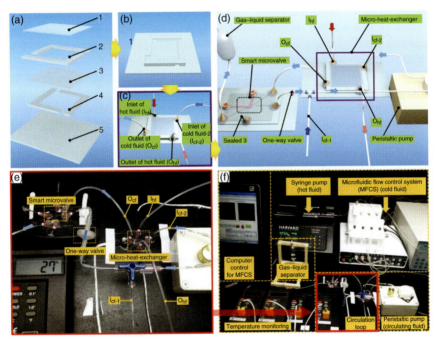

Figure 13.4 Construction of hydrogel-microvalve-integrated micro-heat-exchanging system. (a–c) Schematic illustration showing the structure of the micro-heat-exchanger (b, c) made of four coverslips (1–4) and a glass slide (5) (a). (d) Schematic illustration showing the flow circulation loop that contains the smart microvalve, the micro-heat-exchanger, the microvalve-integrated microchip, and a peristaltic pump. (e) Magnified photograph of the flow circulation loop in the micro-heat-exchanging system. (f) Photograph of the hydrogel-microvalve-integrated micro-heat-exchanging system (Lin *et al.* 2014 [42]. Reproduced with permission of The Royal Society of Chemistry.)

In the micro-heat-exchanging system, cold water is injected into the circulation loop via the inlet (I_{cf-1}) that is connected to the peristaltic pump. In the micro-heat-exchanger, hot water with a constant flow rate is injected via the inlet (I_{hf}) by a pump and flowed out via the outlet (O_{hf}). In the microchip (Figure 13.1c), microchannel-1 is used as the entrance for the circulating cold water, and microchannel-2 and microchannel-4 are used as the exits. To avoid entrance of air into the circulation loop, a gas–liquid separator is set in microchannel-4. Meanwhile, a one-way valve is used downstream the microchannel-2 to avoid backflow. To monitor the fluid temperature, probes of temperature sensors are incorporated in each of the inlets (I_{cf-2} and I_{hf}) and outlets (O_{cf} and O_{hf}) of the micro-heat-exchanger (Figure 13.4d–f).

13.3.2 Thermo-Responsive Switch Performance of Hydrogel Microvalve

The hydrogel microvalve consists of PNIPAM networks, which allow reversible volume changes when changing temperature across the volume-phase transition temperature (VPTT) (~32 °C). Such a thermo-responsive volume transition behavior enables changing the microvalve shape in response to temperature fluctuations for controlling the "on–off" of the microchannel for flux manipulation (Figure 13.5a–c). Such a thermo-responsive switch behavior is systematically studied by flowing the microchannel with deionized water at different temperatures for flow rate measurement (Figure 13.5d). Briefly, microchannel-1 is sealed by blocking the needle with a plug to ensure the fluid flowing through the hydrogel microvalve. Then, water with temperatures ranging from 28 to 40 °C is injected through microchannel-2 by the MFCS (Figure 13.5d). The flux of microchannel-2, which equals the flux sum of microchannel-3 and microchannel-4, is measured as the flux through the microvalve to indicate the microvalve switch performance.

As shown in Figure 13.5e, when water with a temperature of 28 °C is injected, the hydrogel microvalve swells and blocks the microchannel. Thus, the flow rate through the microvalve is ~0 μl min^{-1}, indicating the complete "off" state of the microvalve. When the water temperature increases, the hydrogel microvalve gradually shrinks and opens the microchannel for the flow of water, resulting in an increased flow rate. Due to the dramatic volume change of PNIPAM near its VPTT, a significant increase of the flow rate can be observed near 32 °C. Especially, the critical temperature for opening the hydrogel microvalve is observed at about 29–30 °C. This critical temperature mainly depends on the VPTT of the PNIPAM hydrogel, because the VPTT affects the shrinking/swelling behaviors. When the water temperature further increases to higher than 35 °C, a maximum flow rate can be obtained. This indicates that the shrunken hydrogel microvalve reaches its equilibrium state for complete opening of the microchannel.

Next, the response time for switching the microvalve from "off" state to "on" state is studied by suddenly changing the flowing cold water (25 °C) that flows in microchannel-2 to hot water (45 °C) (Figure 13.5f). In the microchip, the microchannel-2 is used for injecting water, while all other microchannels are unsealed for draining the water. To indicate the response time for microvalve switch, the flux change of microchannel-2 is continuously monitored during the

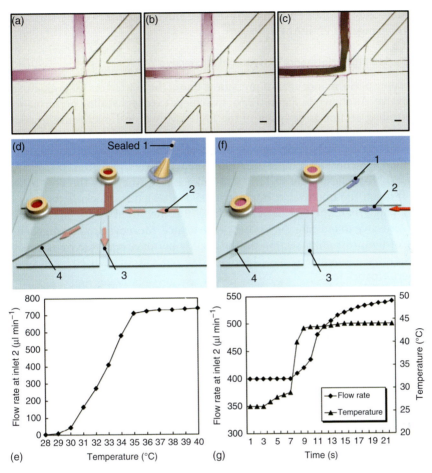

Figure 13.5 Thermo-responsive switch performance of the hydrogel microvalve. (a–c) Optical micrographs showing the volume transitions of the hydrogel microvalve at temperatures of 29 °C (a), 32 °C (b), and 35 °C (c). Scale bars are 300 μm. (d, e) Schematic illustration showing the switch of the hydrogel microvalve when flowing water with different temperatures (d) for controlling the water flux through the microvalve (e). (f, g) Schematic illustration showing the switch of the hydrogel microvalve when suddenly changing the cold water (25 °C) into hot water (45 °C) (f) for controlling the water flux through the microvalve (g). The lengths of the protruding sections in (d, f) are both 230 μm. (Lin et al. 2014 [42]. Reproduced with permission of The Royal Society of Chemistry.)

process. After suddenly switching the cold water into hot water, the water temperature increases. This leads to volume transition of the hydrogel microvalve from a swollen state to a shrunken state, thus opening the microchannel for an increased water flux (Figure 13.5g). Especially, a rapid increase in the flow rate is observed after 6 s, because of the significant increase of the water temperature. Then, after another ~6 s, the flow rate reaches ~80% of the maximum flow rate, indicating a fast response. These results show the excellent "on–off" switch behavior of the hydrogel microvalve. Moreover, such an "on–off" switch can be

13.3.3 Sealing Performance of Hydrogel Microvalve

The sealing performance against pressure is one of the key factors for evaluating a microvalve. As shown in Figure 13.6a, to achieve excellent sealing performance, a protruding section is designed in the PNIPAM hydrogel microvalve to resist the pressure impact. The sealing performance of the hydrogel microvalve is studied by measuring the flux through the microvalve when injecting microchannel-2 with fluid at a constant temperature (25 °C) but different pressures (0–100 kPa) (Figure 13.6a). During the sealing performance test, to ensure the flow of the fluid through the microvalve, microchannel-1 is sealed by blocking the needle with a plug. Then, the flux of microchannel-2 is measured as the flux through the microvalve to indicate the sealing performance. Moreover, to investigate the effect of the protruding section on the sealing performance, microvalves with protruding sections at lengths of 140, 170, 200, and 230 μm are used for the sealing performance test.

When the microchannel in "off" state is closed by the swollen microvalve, this protruding section is first deformed by the flowing water due to the pressure impact. Thus, this protruding section protects the microvalve from detaching from the microchannel wall, ensuring the closure of the microchannel. Moreover, the protruding section with longer length can provide better protection for the microvalve to achieve excellent sealing performance. As shown in

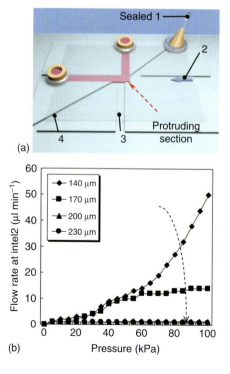

Figure 13.6 Sealing performance of the hydrogel microvalve. (a) Schematic illustration showing the experiment setup for testing the sealing performance of the hydrogel microvalve, in which the water flux through the hydrogel microvalve under different flow pressures is measured. (b) Effect of flow pressure on the water flux through the hydrogel microvalve. Hydrogel microvalves containing protruding sections with lengths of 140, 170, 200, and 230 μm are used. (Lin et al. 2014 [42]. Reproduced with permission of The Royal Society of Chemistry.)

Figure 13.6b, for hydrogel microvalves with protruding sections of 140 and 170 μm, a significant increase of water flux that leaks through the microvalve is observed when increasing the water pressure from 0 to 100 kPa. Contrarily, for hydrogel microvalves with protruding sections of 200 and 230 μm, the water flux that leaks through the microvalve remains very low (~1 μl min^{-1}) under the pressure of 0–100 kPa, indicating excellent sealing performance.

13.3.4 Temperature Self-Regulation of Hydrogel Microvalve for Thermostatic Control

The hydrogel microvalve with excellent thermo-responsive switch performance for flow control provides a novel platform based on fluid heat exchange for thermostatic control. Such a thermostatic control is studied by cooling hot fluid that flows in the lower reservoir with cold fluid that flows in the upper reservoir of the microvalve-integrated micro-heat-exchanging system (Figure 13.4e,f) for temperature self-regulation.

To highlight the advantage of the hydrogel microvalve for thermostatic control, three different cooling methods are compared, as shown in Figure 13.7. Typically, air at room temperature (25 °C), water (22 °C) without microvalve control, and microvalve-controlled circulating water are used to cool the hot water at different temperatures. During the cooling process, water without microvalve control is injected via the inlet (I_{cf-2}) in the micro-heat-exchanger; thus the water can flow out of the circulation loop via microchannel-4 due to the absence of the hydrogel microvalve. For both water-cooling methods, cold water at a constant pressure of 100 kPa and a constant temperature of 22 °C is supplied into the circulation loop via I_{cf-1} by the MFCS. Meanwhile, hot water with temperatures ranging from 33 to 45 °C is supplied with a constant flow rate of 1000 μl min^{-1} by the syringe pump.

As shown in Figure 13.7a, for the air-cooling method, the water temperature at the outlet (T_{O-hf}) remains almost unchanged as compared to that at the inlet (T_{I-hf}), showing low cooling efficiency. Meanwhile, the variation of T_{O-hf} from 33.5 to 42.7 °C with increasing T_{I-hf} indicates the poor thermostatic control of the air-cooling method (Figure 13.7a). For the water-cooling without microvalve control, the T_{O-hf} is largely reduced due to the more efficient heat exchange between hot and cold water as compared to that between hot water and air (Figure 13.7a). However, the results still exhibit limited thermostatic control performance. Because when the temperature of hot water changes ~10 °C, T_{O-hf} shows a temperature variation of ~4 °C. As compared to the abovementioned two cooling methods, the microvalve-controlled water-cooling method shows excellent temperature self-regulation for thermostatic control. As shown in Figure 13.7b, when T_{I-hf} is increased from ~34 to ~44 °C, the T_{O-hf} remains nearly unchanged at ~30 °C after cooling with the microvalve-controlled cold water. This excellent thermostatic control performance results from the thermo-responsive switch of the hydrogel microvalve for intelligently controlling the flux of cold water. In the micro-heat-exchanger, the T_{O-cf} increases with increasing T_{I-hf} due to heat exchange. Such an increase of T_{O-cf} makes the hydrogel microvalve shrink and leads to part of the water draining out of the microchip through microchannel-4. To compensate the water loss for

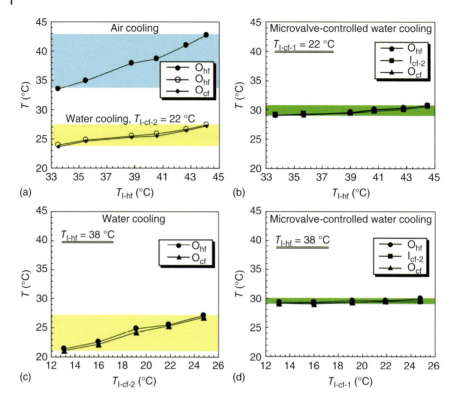

Figure 13.7 Hydrogel microvalve for self-regulating the temperature of hot water for thermostatic control. (a, b) Hot water at different temperatures are cooled with air (25 °C) (a, upper), water without microvalve control ($T_{\text{l-cf-2}} = 22$ °C) (a, lower) and microvalve-controlled circulating water ($T_{\text{l-cf-1}} = 22$ °C) (b). (c, d) Hot water with constant temperature of 38 °C is cooled with water without microvalve control (c) and microvalve-controlled circulating water (d). The hydrogel microvalve contains protruding sections with length of 230 μm. (Lin *et al.* 2014 [42]. Reproduced with permission of The Royal Society of Chemistry.)

maintaining the pressure, another flow of cold water (22 °C) enters into the circulation loop via $I_{\text{cf-1}}$ and mixes with the circulating water for cooling the water temperature. Because the flux of $I_{\text{cf-1}}$ is in direct proportion to the water loss through microchannel-4, the thermo-responsive switch of the hydrogel microvalve allows adjusting the temperature of circulating cold water, which then controls the heat exchange between the hot water and cold water in the micro-heat-exchanger for thermostatic control.

Next, the effect of the temperature of cold water on the thermostatic control is investigated by using cold water with a constant pressure of 100 kPa but different temperatures for cooling the hot fluid (38 °C) (Figure 13.7c,d). For water-cooling method without hydrogel microvalve control, both $T_{\text{O-hf}}$ and $T_{\text{O-cf}}$ decrease with decreasing $T_{\text{l-cf}}$, indicating poor thermostatic control with a temperature variation of ~6 °C (Figure 13.7c). By contrast, based on the thermo-responsive switch of microvalve for flow manipulation, the microvalve-controlled water-cooling method exhibits excellent thermostatic control with small temperature variation within ~0.7 °C (Figure 13.7d). All the results show the excellent performance of the hydrogel microvalve in the micro-heat-exchanging system for thermostatic control.

13.3.5 Temperature Self-Regulation with Hydrogel Microvalve for Cell Culture

The potential of the hydrogel-microvalve-integrated microchip for thermostatic biomedical applications is demonstrated by taking advantage of the temperature self-regulation performance of the hydrogel microvalve for culturing *Chlorella pyrenoidosa* cells (Figure 13.9a). The micro-heat-exchanger in the micro-heat-exchanging system is integrated with a reservoir (size: 5 mm × 5 mm × 3 mm) to construct a culturing microdevice for cell growth (Figure 13.8). The cell growth in the reservoir of the culturing microdevices with and without the smart hydrogel microvalve for thermostatic control is monitored. Briefly, two sets of *C. pyrenoidosa* cells are separately cultured in 75 µl Bold's Basal medium in the reservoirs of two culturing microdevices. For one of the culturing microdevices with hydrogel microvalve, hot fluid (44 °C), which is cooled with microvalve-controlled circulating water (22 °C) in the micro-heat-exchanging system, is flowed under the cell-containing reservoir for controlling the culture temperature. For the other one without hydrogel microvalve, hot fluid (44 °C), which is cooled with air at room temperature, is flowed in the cell-containing reservoir. Due to the fluorescent chloroplasts in *C. pyrenoidosa* cells, the cell growth can be monitored by confocal laser scanning microscope (CLSM).

As shown in Figure 13.9b,d, the number of the *C. pyrenoidosa* cells remains nearly unchanged after 48 h, when conventional air cooling is used for

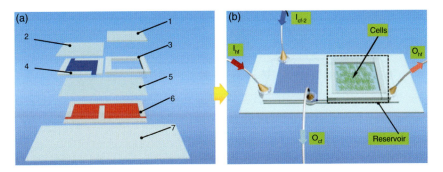

Figure 13.8 Fabrication of the culturing microdevice for cell growth. (a) Six coverslips (1–6) are assembled on a glass slide (7), with five coverslips (2–6) bonded on the glass slide by UV-curable adhesive. The coverslip (3), covered with coverslip (1), is used as a reservoir for cell culture. The reservoir spaces in coverslip (4) (blue area) and coverslip (6) (red area) are, respectively, used for flowing cold fluid and hot fluid. (b) Schematic illustration of the culturing microdevice for cell growth. In the micro-heat-exchanger (left) of this microdevice, hot fluid flows in the lower channel from I_{hf} to O_{hf}, and cold fluid flows in the upper channel from I_{cf-2} to O_{cf}. Hot fluid that is cooled by the cold fluid flows under the reservoir for controlling the temperature for cell culture. (Lin et al. 2014 [42]. Reproduced with permission of The Royal Society of Chemistry.)

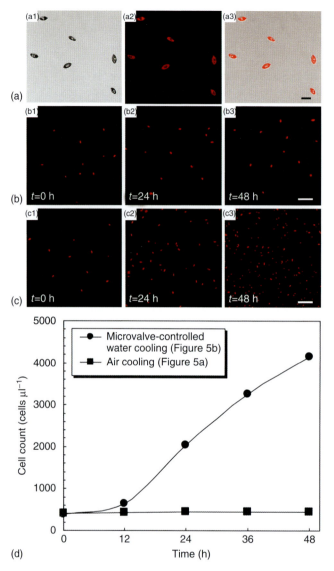

Figure 13.9 Temperature self-regulation with hydrogel microvalve for culture of *Chlorella pyrenoidosa* cells. (a) Optical (a1), fluorescent (a2), and overlapped (a3) CLSM images of the *Chlorella pyrenoidosa* cells. (b, c) Fluorescent images showing the growth of *Chlorella pyrenoidosa* cells at 0, 24, and 48 h, with culturing temperature controlled by air cooling (b) and microvalve-controlled water cooling (c). (d) The growth curves of *Chlorella pyrenoidosa* cells at temperature controlled by air cooling (Figure 13.7a) and microvalve-controlled water cooling (Figure 13.7b). Scale bars are 10 μm in (a) and 100 μm in (b, c). (Lin *et al.* 2014 [42]. Reproduced with permission of The Royal Society of Chemistry.)

temperature control. By contrast, when microvalve-controlled water cooling is used for temperature control, the *C. pyrenoidosa* cells grow a lot within 48 h (Figure 13.9c,d). Thus, the microvalve-controlled water-cooling method can provide better temperature conditions than the one with air cooling for the growth of *C. pyrenoidosa* cells. The results show the excellent performance of the hydrogel-microvalve-integrated microchip for self-regulating temperature for cell culture.

13.4 Smart Microvalve-in-a-Chip with Ultrasensitivity for Real-Time Detection

In this section, the microchip incorporated with Pb^{2+}-responsive hydrogel microvalve, which was introduced in Section 13.2.2, is integrated into a Pb^{2+} detection platform for real-time Pb^{2+} detection, with fast response, highly sensitivity, and selectivity.

13.4.1 Concept of the Microchip Incorporated with Pb^{2+}-Responsive Microgel for Real-Time Online Detection of Trace Pb^{2+}

As schematically illustrated in Figure 13.10a, at the beginning, the Pb^{2+}-responsive P(NIPAM-*co*-B18C6Am) microgel is in a shrunken state at operation temperature (T_o) (Figure 13.10b) higher than the VPTT of the initial microgel ($VPTT_1$). When trace Pb^{2+} appears, the B18C6Am units of the P(NIPAM-*co*-B18C6Am) microgel capture Pb^{2+}; this forms stable B18C6Am/Pb^{2+} host–guest complexes via ionic molecular recognition [33, 45] (Figure 13.10b,c). Thus, the VPTT of P(NIPAM-*co*-B18C6Am) microgel shifts from $VPTT_1$ to a higher $VPTT_2$ due to the electrostatic repulsion among the charged B18C6Am/Pb^{2+} complex groups [45–47]. So, the P(NIPAM-*co*-B18C6Am) microgel isothermally swells at T_o because of the shift of $VPTT_1$ to a higher value than T_o and the enhanced osmotic pressure within the microgel networks based on Donnan potential [45–47]. Such a Pb^{2+}-responsive volume transition is employed to convert the trace Pb^{2+} into flow rate change that can be simply measured for trace Pb^{2+} detection. This is achieved by constructing a P(NIPAM-*co*-B18C6Am) microgel with cylinder shape inside a glass-capillary microchannel. The interstice between the microgel and the inner microchannel of the glass capillary creates a crescent-moon-shaped microspace for flowing fluids (Figure 13.10d,e). In response to Pb^{2+} in the flowing solution, the P(NIPAM-*co*-B18C6Am) microgel isothermally swells to a certain degree, with the swelling volume depending on the Pb^{2+} concentration ($[Pb^{2+}]$) [45, 46]. Thus, the flowing area of the crescent-moon-shaped microspace in the microchannel decreases, and the flow rate decreases correspondingly from Q to Q' ($Q > Q'$). Based on the Hagen–Poiseuille's law, the flow rate through a microchannel is governed by the fourth power of the hydraulic equivalent diameter of flowing space [48]. So, the Pb^{2+}-induced swelling of the P(NIPAM-*co*-B18C6Am) microgel in the glass-capillary microchannel can greatly affect the flow rate. As a result, with the proposed microchip that is incorporated with Pb^{2+}-responsive hydrogel

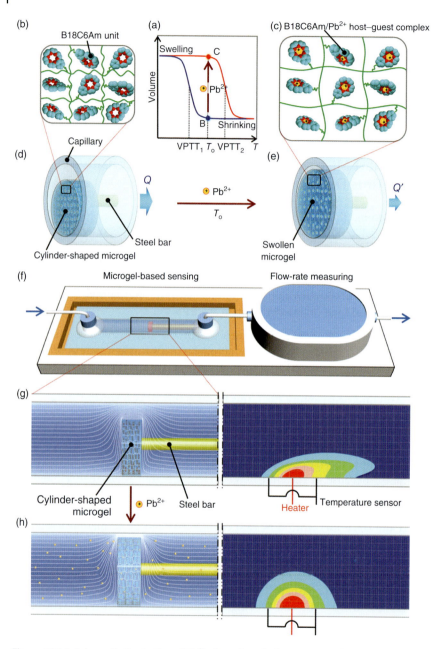

Figure 13.10 Schematic illustration of Pb^{2+} detection platform equipped with a microchip containing Pb^{2+}-responsive P(NIPAM-co-B18C6Am) microgel. (a–c) P(NIPAM-co-B18C6Am) microgel can isothermally swell after recognizing Pb^{2+} via forming stable host–guest complexes. (d, e) By incorporating cylinder-shaped P(NIPAM-co-B18C6Am) microgel inside the glass capillary as Pb^{2+} sensor, the trace Pb^{2+} signal can be efficiently converted into significantly amplified signal of flow rate change ($Q \rightarrow Q'$, in which $Q > Q'$). (f–h) Pb^{2+} detection platform (f), equipped with a microfluidic chip for Pb^{2+} sensing based on the abovementioned principle and an on-line flowmeter for flow rate measurement based on the flow-rate-dependent temperature distribution (g, h), enables real-time online quantitative detection of trace Pb^{2+}. (Lin et al. 2016 [43]. Reproduced with permission of Proceedings of the National Academy of Sciences.)

microvalve, the trace Pb^{2+} signals can be efficiently converted into significantly amplified signals of flow rate change. Then, use of a simple online flowmeter downstream the microchip for measuring the flow rates enables quantitative detection of the trace Pb^{2+} (Figure 13.10f–h). Meanwhile, since the characteristic time of hydrogel swelling is proportional to the square of a linear dimension of the hydrogel [49], the micron-scale size of the P(NIPAM-co-B18C6Am) microgel allows rapid swelling in response to Pb^{2+}. Moreover, the mass transfer of Pb^{2+} into the networks of P(NIPAM-co-B18C6Am) microgel can be enhanced by the continuous flowing of solution around the microgel, which also benefits the rapid microgel swelling. Therefore, based on the abovementioned advantages, ultrasensitive and real-time detection of Pb^{2+} can be realized.

13.4.2 Sensitivity of the Pb^{2+} Detection Platform

The Pb^{2+} detection platform enables highly sensitive and fast detection of trace Pb^{2+} solution. First, to determine the optimal operation temperature for the Pb^{2+} detection, effects of temperature and $[Pb^{2+}]$ on the flow rates are studied. Briefly, pure water with different temperatures is supplied into the Pb^{2+} detection platform. Next, aqueous solution containing Pb^{2+} with a certain concentration is supplied into the Pb^{2+} detection platform. The temperature of the Pb^{2+}-containing solution is kept the same as that of the pure water. Both the pure water and Pb^{2+}-containing solution are supplied by the MFCS under a constant pressure of 30 kPa. After being supplied for 15 min, the equilibrated flow rates of the pure water and Pb^{2+}-containing solution are measured by the online flowmeter for evaluating the equilibrated flow rate changes ($\Delta J = Q - Q'$). During the experiments, the temperature of pure water and Pb^{2+}-containing solution is changed from 30 to 40 °C, and $[Pb^{2+}]$ is varied from 10^{-9} to 10^{-4} M. The temperature at which the Pb^{2+} detection platform exhibits the most significant ΔJ value is defined as the optimal operation temperature for Pb^{2+} detection. As shown in Figure 13.11a, the optimal operation temperature of the Pb^{2+} detection platform is 34 °C, because most significant ΔJ in response to each $[Pb^{2+}]$ is observed at 34 °C.

Next, the response time and sensitivity of the Pb^{2+} detection platform are investigated by monitoring the dynamic change of flow rates after switching pure water to solutions containing Pb^{2+} with concentrations varied from 10^{-10} to 10^{-2} M. At the optimal operation temperature of 34 °C, after recognizing Pb^{2+} with concentrations varied from 10^{-10} to 10^{-5} M within 5 min, significantly decreased flow rate can be detected (Figure 13.11b). It is worth noting that obviously decreased flow rate can also be detected even in the $[Pb^{2+}]$ range from 10^{-10} to 10^{-8} M. Such a $[Pb^{2+}]$ range is much lower than the guideline value of World Health Organization (WHO) for drinking water (4.83×10^{-8} M). These results show the ultrasensitivity of the Pb^{2+} detection platform for detecting trace Pb^{2+}. Meanwhile, the slope of $\Delta J_t/\Delta J_{max}$ curves at $\Delta J_t/\Delta J_{max} = 50\%$ (S_{50}) increases with increasing $[Pb^{2+}]$, where ΔJ_t is the ΔJ at t seconds and ΔJ_{max} is the maximum value of ΔJ (Figure 13.11c). Moreover, the time required for change of $\Delta J_t/\Delta J_{max}$ from 5% to 50% ($t_{50}-t_5$) and for change of $\Delta J_t/\Delta J_{max}$ from 5% to 90% ($t_{90}-t_5$) decreases with increasing $[Pb^{2+}]$ (Figure 13.11c, inset). All the results indicate that, due to

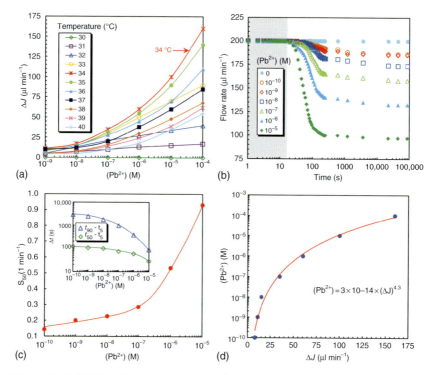

Figure 13.11 (a) Effect of temperature and [Pb^{2+}] on the ΔJ after switching pure water to Pb^{2+}-containing solution for 15 min. (b) Time-dependent changes of flow rate in response to different [Pb^{2+}] at 34 °C. (c) Effect of [Pb^{2+}] on the slope (S_{50}) of $\Delta J_t/\Delta J_{max}$ curves at $\Delta J_t/\Delta J_{max} = 50\%$. Inset shows the effect of [Pb^{2+}] on the time of ($t_{50}-t_5$) and ($t_{90}-t_5$). (d) Quantitative relationship between [Pb^{2+}] and ΔJ after switching pure water to Pb^{2+}-containing solution for 15 min. (Lin et al. 2016 [43]. Reproduced with permission of Proceedings of the National Academy of Sciences.)

the faster formation of B18C6Am/Pb^{2+} complex groups at higher [Pb^{2+}], a faster dynamic swelling rate can be obtained. To precisely detect [Pb^{2+}], the quantitative relationship between [Pb^{2+}] and ΔJ is obtained from Figure 13.11d, which can be expressed as $[Pb^{2+}] = 3 \times 10^{-14} \times (\Delta J)^{4.3}$.

13.4.3 Selectivity and Repeatability of the Pb^{2+} Detection Platform

The Pb^{2+} detection platform also shows excellent selectivity and repeatability for Pb^{2+} detection. First, the selectivity of the Pb^{2+} detection platform is investigated by monitoring the dynamic change of flow rates after switching pure water to solution containing Pb^{2+} and to those containing other interfering ions with concentrations varied from 10^{-10} to 10^{-3} M. Then, the ΔJ after switching the pure water to aqueous solution that concurrently contains Pb^{2+} and interfering ions, including Ba^{2+}, Sr^{2+}, K$^+$, and Na$^+$, for 15 min is investigated. The concentration for each of the interfering ions is 1–1000 times as large as the [Pb^{2+}]. All the solutions are flowed at 34 °C under a constant pressure of 30 kPa. As shown in Figure 13.12a, the decrease of flow rate that is caused by the interferences of other

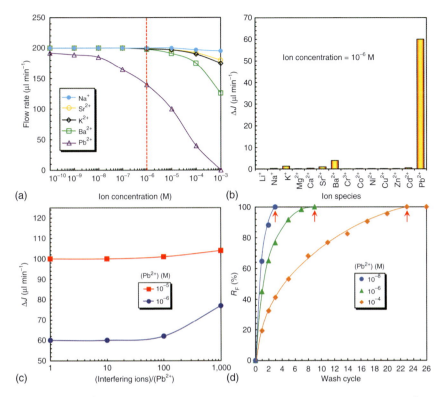

Figure 13.12 Pb^{2+} detection platform for highly selective and repeated detection of Pb^{2+}. (a) Equilibrated flow rates at different concentrations of Na^+, Sr^{2+}, K^{2+}, Ba^{2+}, and Pb^{2+} at 34 °C. (b) Effect of ion species on ΔJ after switching pure water to the solution containing each ion with concentration of 10^{-6} M for 15 min at 34 °C. (c) Effects of interfering ions including Ba^{2+}, Sr^{2+}, K^+, and Na^+ on ΔJ. (d) R_F of Pb^{2+} detection platforms after different Pb^{2+} wash cycles. The Pb^{2+} detection platforms are used for detecting different $[Pb^{2+}]$. (Lin et al. 2016 [43]. Reproduced with permission of Proceedings of the National Academy of Sciences.)

ions can only be observed when the concentrations of interfering ion are larger than 10^{-6} M. Under a concentration of 10^{-6} M, Pb^{2+} can cause a significant flow rate decrease of ~60 µl min^{-1}, while the flow rate decrease caused by interferences from other ions are less than 2 µl min^{-1}, which is negligible (Figure 13.12b). Such interferences from Ba^{2+}, Sr^{2+}, K^+, and Na^+ are negligible even when increasing their concentrations 100 times that of $[Pb^{2+}]$ at $[Pb^{2+}] = 10^{-6}$ M or even 1000 times that of $[Pb^{2+}]$ at $[Pb^{2+}] = 10^{-5}$ M (Figure 13.12c). All the results show the excellent selectivity of the detection platform for Pb^{2+} detection.

Moreover, the Pb^{2+} detection platform enables repeated use after removal of the captured Pb^{2+}. Such a removal process can be conducted by alternatively and repeatedly injecting pure water of 55 and 25 °C into the Pb^{2+} detection platform, each for 3 min, for Pb^{2+} removal. The flux recovery ratio ($R_F = (\Delta J_{max} - \Delta J_t)/\Delta J_{max}$) is measured after different wash cycles to estimate Pb^{2+} removal performance. After complete removal of Pb^{2+} ($R_F = 100\%$), the Pb^{2+} detection platform is employed for repeated Pb^{2+} detection.

On washing with hot water (55 °C, >VPTT$_2$), the P(NIPAM-co-B18C6Am) microgel shrinks and the B18C6Am moieties within the microgel networks become closer to each other. This produces electrostatic repulsions among the ions against the formation of stable B18C6Am/Pb^{2+} complexes, resulting in decomplexation of Pb^{2+} from B18C6Am units [47]. Meanwhile, upon heating, inclusion constant between Pb^{2+} and B18C6Am decreases, facilitating the Pb^{2+} decomplexation [46]. On washing with cold water (25 °C), the P(NIPAM-co-B18C6Am) microgel swells again; this absorbs fresh water into the microgel networks for removing Pb^{2+}. Repeated shrinking/swelling cycles of the P(NIPAM-co-B18C6Am) microgel upon heating and cooling can enhance the transportation of water into and out of the microgel networks for Pb^{2+} removal. After different wash cycles, the Pb^{2+} detection platform is used for Pb^{2+} detection to estimate the recovery of the detecting performance. As shown in Figure 13.12d, for detection of Pb^{2+} with different concentrations, the R_F of Pb^{2+} detection platform can reach 100% after repeated wash cycles, indicating the complete removal of Pb^{2+} from the P(NIPAM-co-B18C6Am) microgel. Meanwhile, the cycle times required for 100% recovery of the detecting performance increase with increasing [Pb^{2+}] (Figure 13.12d). Therefore, the Pb^{2+} detection platform shows excellent recovery performance for repeated and accurate detection of Pb^{2+} with different concentrations.

The detection mechanism and portability of the Pb^{2+} detection platform enable its flexible and facile utilization as an online system for real-time detection of Pb^{2+}. This is demonstrated by using the Pb^{2+} detection platform for real-time online detection of Pb^{2+} in tap water and in wastewater from a model industrial factory for pollution warning and terminating. To construct a detection system for such detections, as shown in Figure 13.13, the Pb^{2+} detection platform is connected with a peristaltic pump, a buffer tank, a solenoid

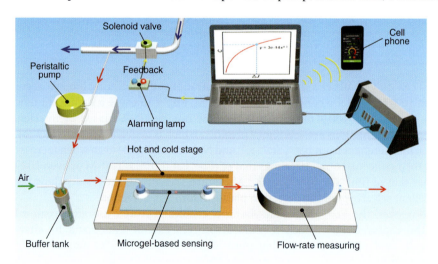

Figure 13.13 Schematic illustration showing the setup of Pb^{2+} detection system for real-time online detection of Pb^{2+} in liquid. (Lin et al. 2016 [43]. Reproduced with permission of Proceedings of the National Academy of Sciences.)

valve, and a computer combined with a cell phone and an alarm lamp. The testing solution is sucked by the peristaltic pump from the water pipeline into the buffer tank and then supplied into the microchip with constant pressure. Once the [Pb^{2+}] value exceeds the preset critical value, which can be simply set via the APP software on cell phone, both the cell phone and lamp alarm for pollution warning. Meanwhile, the solenoid valve is automatically closed for terminating the discharge of Pb^{2+}-containing solution for preventing pollution.

13.4.4 Setup of Pb^{2+} Detection System for Real-Time Online Detection of Pb^{2+} in Tap Water for Pollution Warning

For the real-time online detection of Pb^{2+} in tap water for pollution warning, the Pb^{2+} detection platform is incorporated with a peristaltic pump, a buffer tank, a computer, and a cell phone to construct the Pb^{2+} detection system (Figure 13.14a). The peristaltic pump is used to suck water from the tap into the buffer tank, while the constant pressure pump of MFCS is used to supply air (30 kPa) into the buffer tank for driving water into the Pb^{2+} detection platform. After microgel-based sensing and flow rate measurement, the measured [Pb^{2+}] value in the Pb^{2+} detection platform is analyzed by the computer. A cell phone

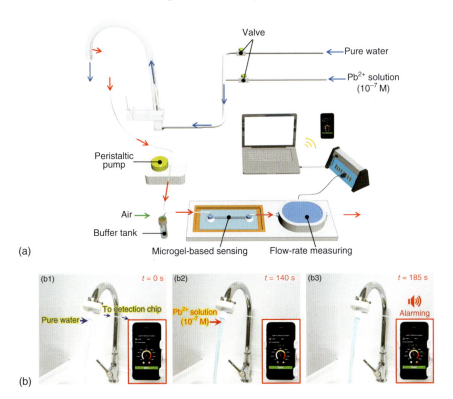

Figure 13.14 Pb^{2+} detection system for real-time online detection of Pb^{2+} in tap water for pollution warning. (a) Schematic illustration showing the setup of the Pb^{2+} detection system. (b) Detection of 10^{-7} M Pb^{2+} in tap water for Pb^{2+}-pollution warning. (Lin et al. 2016 [43]). Reproduced with permission of Proceedings of the National Academy of Sciences.)

is connected with the computer via Wi-Fi for facile [Pb^{2+}] monitoring and pollution warning. The cell phone is amounted with a custom-made APP software for displaying the measured [Pb^{2+}] value and alarming pollution. Once the [Pb^{2+}] value exceeds the preset critical value, which can be simply and flexibly set with the APP software, the cell phone alarms for pollution warning. For Pb^{2+} detection, the Pb^{2+} detection system is simply connected to the tap water pipeline as a branch. The preset critical value of [Pb^{2+}] for pollution warning is set at 4.83×10^{-8} M, which is the guideline value of WHO for drinking water. As shown in Figure 13.14b, when pure water flows in the pipeline as well as the Pb^{2+} detection platform, the displayed [Pb^{2+}] value remains 0 M. After switching the pure water to aqueous solution containing Pb^{2+} with concentration of 10^{-7} M via valve switch, 185 s later the displayed [Pb^{2+}] value increases beyond the preset value of 4.83×10^{-8} M, meanwhile the phone alarms for pollution warning.

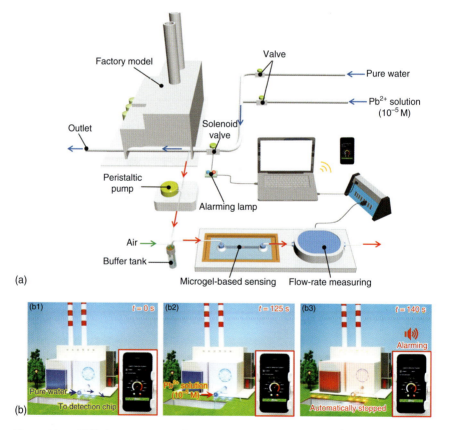

Figure 13.15 Pb^{2+} detection system for real-time online detection of Pb^{2+} in wastewater from a model industrial factory for pollution warning and terminating. (a) Schematic illustration of the setup of the Pb^{2+} detection system. (b) Detection of 10^{-5} M Pb^{2+} in wastewater emitted from a model industrial factory for Pb^{2+}-pollution warning and terminating. (Lin et al. 2016 [43]. Reproduced with permission of Proceedings of the National Academy of Sciences.)

13.4.5 Setup of Pb^{2+} Detection System for Real-Time Online Detection of Pb^{2+} in Wastewater from a Model Industrial Factory for Pollution Warning and Terminating

The setup of Pb^{2+} detection system for real-time online detection of Pb^{2+} in wastewater is similar to that of the Pb^{2+} detection system for the detection of Pb^{2+} in tap water. Especially, an alarming lamp and a solenoid valve are added in this system for warning the pollution and terminating the Pb^{2+} pollution discharge (Figure 13.15a). Thus, once the detected $[Pb^{2+}]$ value exceeds the preset critical value, both the cell phone and the lamp can alarm for pollution warning; meanwhile, the solenoid valve can be automatically closed for terminating the Pb^{2+} pollution discharge. In this experiment, the preset critical value of $[Pb^{2+}]$ for pollution warning and terminating is set at 2.42×10^{-6} M, which is the guideline value of China for wastewater. As shown in Figure 13.15b, after switching pure water to Pb^{2+}-containing solution (10^{-5} M), 140 s later the displayed $[Pb^{2+}]$ value exceeds the preset value of 2.42×10^{-6} M. Meanwhile, both the cell phone and lamp alarm for warning the Pb^{2+} pollution, and the solenoid valve is automatically closed to stop the discharge of Pb^{2+}-containing solution for pollution terminating (Figure 13.15b).

13.5 Summary

In summary, integration of smart hydrogel microvalve in microchips provides advanced function for flow manipulation for myriad applications. As we introduced in this chapter, the microchips with PNIPAM hydrogel microvalve enable thermo-responsive "on–off" switch for flow control and thus manipulate the heat exchange for self-regulating fluid temperature in microsystems at a constant temperature. Such smart microvalves integrated in microchips are promising as temperature regulators for applications that require thermostatic conditions such as cell culture and biomolecule synthesis. Meanwhile, the microchip with P(NIPAM-co-B18C6Am) hydrogel microvalve exhibits highly sensitive, fast, and selective detection performance, and possesses flexible and facile utility as an online unit for real-time Pb^{2+} detection. Such a combination of highly sensitive and selective detection, real-time online operation, and simple readouts, along with the easy construction, makes the proposed microchip platforms ideal candidates for further investigations and applications.

References

1 Auroux, P.-A., Koc, Y., Manz, A., and Day, P. (2004) Miniaturised nucleic acid analysis. *Lab Chip*, **4**, 534–546.
2 Nagrath, S., Sequist, L.V., Maheswaran, S., Bell, D.W., Irimia, D., Ulkus, L., Smith, M.R., Kwak, E.L., Digumarthy, S., and Muzikansky, A. (2007) Isolation of rare circulating tumour cells in cancer patients by microchip technology. *Nature*, **450**, 1235–1239.

3 Dittrich, P.S. and Manz, A. (2006) Lab-on-a-chip: microfluidics in drug discovery. *Nat. Rev. Drug Discovery*, **5**, 210–218.

4 Baker, S.C., Rohman, G., Southgate, J., and Cameron, N.R. (2009) The relationship between the mechanical properties and cell behaviour on PLGA and PCL scaffolds for bladder tissue engineering. *Biomaterials*, **30**, 1321–1328.

5 Whitesides, G.M. (2006) The origins and the future of microfluidics. *Nature*, **442**, 368–373.

6 Ionov, L. (2013) Biomimetic hydrogel-based actuating systems. *Adv. Funct. Mater.*, **23**, 4555–4570.

7 Beebe, D.J., Moore, J.S., Bauer, J.M., Yu, Q., Liu, R.H., Devadoss, C., and Jo, B.-H. (2000) Functional hydrogel structures for autonomous flow control inside microfluidic channels. *Nature*, **404**, 588–590.

8 Zhang, Y., Kato, S., and Anazawa, T. (2008) A microchannel concentrator controlled by integral thermoresponsive valves. *Sens. Actuators, B*, **129**, 481–486.

9 Richter, A., Wenzel, J., and Kretschmer, K. (2007) Mechanically adjustable chemostats based on stimuli-responsive polymers. *Sens. Actuators, B*, **125**, 569–573.

10 Richter, A., Türke, A., and Pich, A. (2007) Controlled double-sensitivity of microgels applied to electronically adjustable chemostats. *Adv. Mater.*, **19**, 1109–1112.

11 Wang, J., Chen, Z., Mauk, M., Hong, K.-S., Li, M., Yang, S., and Bau, H.H. (2005) Self-actuated, thermo-responsive hydrogel valves for lab on a chip. *Biomed. Microdevices*, **7**, 313–322.

12 Richter, A., Howitz, S., Kuckling, D., and Arndt, K.-F. (2004) Influence of volume phase transition phenomena on the behavior of hydrogel-based valves. *Sens. Actuators, B*, **99**, 451–458.

13 Yu, C., Mutlu, S., Selvaganapathy, P., Mastrangelo, C.H., Svec, F., and Fréchet, J.M. (2003) Flow control valves for analytical microfluidic chips without mechanical parts based on thermally responsive monolithic polymers. *Anal. Chem.*, **75**, 1958–1961.

14 Richter, A., Kuckling, D., Howitz, S., Gehring, T., and Arndt, K.-F. (2003) Electronically controllable microvalves based on smart hydrogels: magnitudes and potential applications. *J. Microelectromech. Syst.*, **12**, 748–753.

15 Arndt, K.-F., Kuckling, D., and Richter, A. (2000) Application of sensitive hydrogels in flow control. *Polym. Adv. Technol.*, **11**, 496–505.

16 Liu, C., Park, J.Y., Xu, Y., and Lee, S. (2007) Arrayed pH-responsive microvalves controlled by multiphase laminar flow. *J. Micromech. Microeng.*, **17**, 1985.

17 Kim, D., Kim, S., Park, J., Baek, J., Kim, S., Sun, K., Lee, T., and Lee, S. (2007) Hydrodynamic fabrication and characterization of a pH-responsive microscale spherical actuating element. *Sens. Actuators, A*, **134**, 321–328.

18 Park, J.Y., Oh, H.J., Kim, D.J., Baek, J.Y., and Lee, S.H. (2006) A polymeric microfluidic valve employing a pH-responsive hydrogel microsphere as an actuating source. *J. Micromech. Microeng.*, **16**, 656.

19 Kuckling, D., Richter, A., and Arndt, K.F. (2003) Temperature and pH-dependent swelling behavior of poly (*N*-isopropylacrylamide) copolymer hydrogels and their use in flow control. *Macromol. Mater. Eng.*, **288**, 144–151.

20 Satarkar, N.S., Zhang, W., Eitel, R.E., and Hilt, J.Z. (2009) Magnetic hydrogel nanocomposites as remote controlled microfluidic valves. *Lab Chip*, **9**, 1773–1779.

21 Kwon, G.H., Choi, Y.Y., Park, J.Y., Woo, D.H., Lee, K.B., Kim, J.H., and Lee, S.-H. (2010) Electrically-driven hydrogel actuators in microfluidic channels: fabrication, characterization, and biological application. *Lab Chip*, **10**, 1604–1610.

22 Baldi, A., Gu, Y., Loftness, P.E., Siegel, R.A., and Ziaie, B. (2003) A hydrogel-actuated environmentally sensitive microvalve for active flow control. *J. Microelectromech. Syst.*, **12**, 613–621.

23 Sahiner, N., Ozay, H., Ozay, O., and Aktas, N. (2010) A soft hydrogel reactor for cobalt nanoparticle preparation and use in the reduction of nitrophenols. *Appl. Catal., B*, **101**, 137–143.

24 Godin, M., Delgado, F.F., Son, S., Grover, W.H., Bryan, A.K., Tzur, A., Jorgensen, P., Payer, K., Grossman, A.D., and Kirschner, M.W. (2010) Using buoyant mass to measure the growth of single cells. *Nat. Methods*, **7**, 387–390.

25 Connolly, A.R. and Trau, M. (2010) Isothermal detection of dNA by beacon-assisted detection amplification. *Angew. Chem. Int. Ed.*, **49**, 2720–2723.

26 Spits, C., Le Caignec, C., De Rycke, M., Van Haute, L., Van Steirteghem, A., Liebaers, I., and Sermon, K. (2006) Whole-genome multiple displacement amplification from single cells. *Nat. Protoc.*, **1**, 1965–1970.

27 Dong, L. and Jiang, H. (2007) Autonomous microfluidics with stimuli-responsive hydrogels. *Soft Matter*, **3**, 1223–1230.

28 Quang, D.T. and Kim, J.S. (2010) Fluoro- and chromogenic chemodosimeters for heavy metal ion detection in solution and biospecimens. *Chem. Rev.*, **110**, 6280–6301.

29 Kosaka, P.M., Pini, V., Ruz, J., da Silva, R., González, M., Ramos, D., Calleja, M., and Tamayo, J. (2014) Detection of cancer biomarkers in serum using a hybrid mechanical and optoplasmonic nanosensor. *Nat. Nanotechnol.*, **9**, 1047–1053.

30 Canfield, R.L., Henderson, C.R. Jr.,, Cory-Slechta, D.A., Cox, C., Jusko, T.A., and Lanphear, B.P. (2003) Intellectual impairment in children with blood lead concentrations below 10 μg per deciliter. *N. Engl. J. Med.*, **348**, 1517–1526.

31 Park, J.-S., Cho, M.K., Lee, E.J., Ahn, K.-Y., Lee, K.E., Jung, J.H., Cho, Y., Han, S.-S., Kim, Y.K., and Lee, J. (2009) A highly sensitive and selective diagnostic assay based on virus nanoparticles. *Nat. Nanotechnol.*, **4**, 259–264.

32 Döring, A., Birnbaum, W., and Kuckling, D. (2013) Responsive hydrogels-structurally and dimensionally optimized smart frameworks for

applications in catalysis, micro-system technology and material science. *Chem. Soc. Rev.*, **42**, 7391–7420.

33 Chu, L.-Y., Xie, R., Ju, X.-J., and Wang, W. (2013) *Smart Hydrogel Functional Materials*, Springer, New York.

34 Matsumoto, A., Sato, N., Sakata, T., Yoshida, R., Kataoka, K., and Miyahara, Y. (2009) Chemical-to-electrical-signal transduction synchronized with smart gel volume phase transition. *Adv. Mater.*, **21**, 4372–4378.

35 Guenther, M., Kuckling, D., Corten, C., Gerlach, G., Sorber, J., Suchaneck, G., and Arndt, K.F. (2007) Chemical sensors based on multiresponsive block copolymer hydrogels. *Sens. Actuators, B*, **126**, 97–106.

36 Yuan, Y., Li, Z., Liu, Y., Gao, J., Pan, Z., and Liu, Y. (2012) Hydrogel photonic sensor for the detection of 3-pyridinecarboxamide. *Chem. – Eur. J.*, **18**, 303–309.

37 Jacobi, Z.E., Li, L., and Liu, J. (2012) Visual detection of lead (II) using a label-free DNA-based sensor and its immobilization within a monolithic hydrogel. *Analyst*, **137**, 704–709.

38 Hong, W., Li, W., Hu, X., Zhao, B., Zhang, F., and Zhang, D. (2011) Highly sensitive colorimetric sensing for heavy metal ions by strong polyelectrolyte photonic hydrogels. *J. Mater. Chem.*, **21**, 17193–17201.

39 Zhao, Y., Zhao, X., Tang, B., Xu, W., Li, J., Hu, J., and Gu, Z. (2010) Quantum-dot-tagged bioresponsive hydrogel suspension array for multiplex label-free DNA detection. *Adv. Funct. Mater.*, **20**, 976–982.

40 Ye, G., Yang, C., and Wang, X. (2010) Sensing diffraction gratings of antigen-responsive hydrogel for human immunoglobulin-G detection. *Macromol. Rapid Commun.*, **31**, 1332–1336.

41 Ye, G. and Wang, X. (2010) Glucose sensing through diffraction grating of hydrogel bearing phenylboronic acid groups. *Biosens. Bioelectron.*, **26**, 772–777.

42 Lin, S., Wang, W., Ju, X.-J., Xie, R., and Chu, L.-Y. (2014) A simple strategy for in situ fabrication of a smart hydrogel microvalve within microchannels for thermostatic control. *Lab Chip*, **14**, 2626–2634.

43 Lin, S., Wang, W., Ju, X.-J., Xie, R., Liu, Z., Yu, H.-R., Zhang, C., and Chu, L.-Y. (2016) Ultrasensitive microchip based on smart microgel for real-time online detection of trace threat analytes. *Proc. Natl. Acad. Sci. U. S. A.*, **113**, 2023–2028.

44 Deng, N.-N., Meng, Z.-J., Xie, R., Ju, X.-J., Mou, C.-L., Wang, W., and Chu, L.-Y. (2011) Simple and cheap microfluidic devices for the preparation of monodisperse emulsions. *Lab Chip*, **11**, 3963–3969.

45 Zhang, B., Ju, X.-J., Xie, R., Liu, Z., Pi, S.-W., and Chu, L.-Y. (2012) Comprehensive effects of metal ions on responsive characteristics of P(NIPAM-co-B18C6Am). *J. Phys. Chem. B*, **116**, 5527–5536.

46 Liu, Z., Luo, F., Ju, X.-J., Xie, R., Sun, Y.-M., Wang, W., and Chu, L.-Y. (2013) Gating membranes for water treatment: detection and removal of trace Pb^{2+} ions based on molecular recognition and polymer phase transition. *J. Mater. Chem. A*, **1**, 9659–9671.

47 Ju, X.-J., Zhang, S.-B., Zhou, M.-Y., Xie, R., Yang, L., and Chu, L.-Y. (2009) Novel heavy-metal adsorption material: ion-recognition P(NIPAM-*co*-BCAm) hydrogels for removal of lead (II) ions. *J. Hazard. Mater.*, **167**, 114–118.

48 Bird, R., Stewart, W., and Lightfoot, E. (2006) *Transport Phenomena*, 2nd edn, John Wiley & Sons, New York.

49 Tanaka, T., Sato, E., Hirokawa, Y., Hirotsu, S., and Peetermans, J. (1985) Critical kinetics of volume phase transition of gels. *Phys. Rev. Lett.*, **55**, 2455.

14

Summary and Perspective

14.1 Summary

New materials, especially new functional materials, are the material basis prerequisite for the sustainable development of society as well as the modern scientific development and technological innovation and also the prerequisite and support for all high-tech development. After entering the twenty-first century, people's concern and attention to materials science and materials industry have increased to an unprecedented height. At present, the new materials industry, in which functional materials are the mainstream, has been recognized as one of the most important and the fastest developing high-tech industries in the world. The realization of structural functionalization, functional diversification and intelligentization of materials, is the frontier and hot spot in the field of new materials research. By changing the preparation processes, structures, and functions of materials, such as by introducing miniaturization, fiber, film and membrane, composite, multifunctionalization, intelligentization, and integration of materials and elements, it is very promising to improve the performances of existing materials, to obtain new functions, or even to design new features of functional materials.

Due to its excellent ability to control fluid interfaces as well as excellent heat and mass transfer performances, the microfluidic technology provides a novel and promising technology platform for the design and controllable fabrication of new functional materials and brings new opportunities for the realization of new structures and new functions of functional materials. As mentioned in the earlier chapters, microfluidic technology has emerged in the construction of precisely controllable microstructured new functional materials with high performances, such as microcapsules and microspheres (Figure 14.1), membranes in microchannels and microfiber materials, and especially shows incomparable creativity and superiority compared with traditional technology in the design and preparation of some novel functional materials with high added values [1–15].

14.2 Perspective

To date, the investigations of fabrications of new functional materials by using microfluidic technology to build microscale liquid–liquid interfaces as templates

Figure 14.1 Functional polymeric microparticles engineered from controllable microfluidic emulsions. (Wang *et al.* 2014 [1]. Reproduced with permission of American Chemical Society.)

are still in the initial stage. There are still many problems in the construction of functional materials with new microstructures, for example, the basic laws and mechanisms for designing and regulating stable liquid–liquid interfaces and then fabricating functional materials with controllable structures and desirable performances still need further understanding. To sum up, the problems mainly exist in the following aspects: (i) Currently constructed microscale liquid–liquid interfaces and prepared functional materials are still relatively simple in the microstructures, while the design and preparation of materials with complex structures and multifunctions are still lacking. (ii) The in-depth study on the design and construction of a series of microstructural liquid–liquid interfaces as well as their stabilities is still lacking. (iii) The investigations on the relationships between the microstructures of liquid–liquid interfaces and the interfacial mass transfer and reaction in the processes of preparation of materials and their further effects and regulations on the formation processes of the material microstructures are still lacking.

From the experience of the development history of mobile phones (Figure 14.2a), multifunctional, smart or intelligent, and easy-to-use properties will also be the desired features of advanced functional materials, and microfluidic technology just has its unique advantages in achieving these features of functional materials (Figure 14.2b). Therefore, the abovementioned existing problems are not only challenges but also opportunities.

It is foreseeable that the investigations in the field of preparation of advanced functional materials through microfluidic method will be focused on the following issues in the next step: (i) more rational design and more controllable preparation of new functional materials with the help of theoretical knowledge and research methods of fluid flow, interfacial phenomena, mass transfer and reaction, and so on and (ii) systematic basic researches on the microfluidic

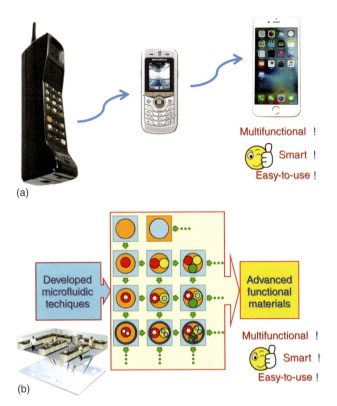

Figure 14.2 (a) History of development of mobile phones. (b) Perspective of microfluidic technology platform for the fabrication of advanced functional materials.

fabrication processes for novel functional materials, and then the methodologies for design and preparation of important functional materials with high added values, thereby laying the theoretical and technical base for the formation of new-generation materials industry.

References

1 Wang, W., Zhang, M.J., and Chu, L.Y. (2014) Functional polymeric microparticles engineered from controllable microfluidic emulsions. *Acc. Chem. Res.*, **47**, 373–384.
2 Atencia, J. and Beebe, D.J. (2005) Controlled microfluidic interfaces. *Nature*, **437**, 648–655.
3 Joanicot, M. and Ajdari, A. (2005) Applied physics – droplet control for microfluidics. *Science*, **309**, 887–888.
4 de Jong, J., Lammertink, R.G.H., and Wessling, M. (2006) Membranes and microfluidics: a review. *Lab Chip*, **6**, 1125–1139.
5 Chu, L.Y., Utada, A.S., Shah, R.K., Kim, J.W., and Weitz, D.A. (2007) Controllable monodisperse multiple emulsions. *Angew. Chem. Int. Ed.*, **46**, 8970–8974.

6 Utada, A.S., Chu, L.Y., Fernandez-Nieves, A., Link, D.R., Holtze, C., and Weitz, D.A. (2007) Dripping, jetting, drops, and wetting: the magic of microfluidics. *MRS Bull.*, **32**, 702–708.

7 Shah, R.K., Shum, H.C., Rowat, A.C., Lee, D., Agresti, J.J., Utada, A.S., Chu, L.Y., Kim, J.W., Fernandez-Nieves, A., Martinez, C.J., and Weitz, D.A. (2008) Designer emulsions using microfluidics. *Mater. Today*, **11**, 18–27.

8 Kenis, P.J.A., Ismagilov, R.F., and Whitesides, G.M. (1999) Microfabrication inside capillaries using multiphase laminar flow patterning. *Science*, **285**, 83–85.

9 Tumarkin, E. and Kumacheva, E. (2009) Microfluidic generation of microgels from synthetic and natural polymers. *Chem. Soc. Rev.*, **38**, 2161–2168.

10 Marre, S. and Jensen, K.F. (2010) Synthesis of micro and nanostructures in microfluidic systems. *Chem. Soc. Rev.*, **39**, 1183–1202.

11 Wang, W., Zhang, M.J., Xie, R., Ju, X.J., Yang, C., Mou, C.L., Weitz, D.A., and Chu, L.Y. (2013) Hole–shell microparticles from controllably evolved double emulsions. *Angew. Chem. Int. Ed.*, **52**, 8084–8087.

12 Liu, Y.M., Wang, W., Zheng, W.C., Ju, X.J., Xie, R., Zerrouki, D., Deng, N.N., and Chu, L.Y. (2013) Hydrogel-based micro-actuators with remote-controlled locomotion and fast Pb^{2+}-response for micromanipulation. *ACS Appl. Mater. Interfaces*, **5**, 7219–7226.

13 Sun, Y.M., Wang, W., Wei, Y.Y., Deng, N.N., Liu, Z., Ju, X.J., Xie, R., and Chu, L.Y. (2014) *In situ* fabrication of temperature- and ethanol-responsive smart membrane in microchip. *Lab Chip*, **14**, 2418–2427.

14 He, X.H., Wang, W., Liu, Y.M., Jiang, M., Wu, F., Deng, K., Liu, Z., Ju, X.J., Xie, R., and Chu, L.Y. (2015) Microfluidic fabrication of bio-inspired microfibers with controllable magnetic spindle-knots for 3D assembly and water collection. *ACS Appl. Mater. Interfaces*, **7**, 17471–17481.

15 Lin, S., Wang, W., Ju, X.J., Xie, R., Liu, Z., Yu, H.R., Zhang, C., and Chu, L.Y. (2016) Ultrasensitive microchip based on smart microgel for real-time on-line detection of trace threat analytes. *Proc. Natl. Acad. Sci. USA*, **113**, 2023–2028.

Index

a

acaleph-like microcrawler 191
acetic acid (HAc) 125, 127, 131, 236
acid-induced swelling 151
acid-triggered burst release 176, 177, 178, 179, 181, 182
acorn-shaped configuration 203
acrylamide (AAm) 56, 68, 138, 149, 163, 174–177, 190, 195, 270
3-acrylamidophenylboronic acid (AAPBA) 136, 137, 138, 140, 141, 158
acrylic acid (AAc) 137, 138, 141, 142, 143, 144, 158
acrylonitrile butadiene styrene (ABS) 231, 233
actuators 55, 68, 79, 85, 103, 136, 137, 174, 207, 216, 270
adhesion energy 203, 204
adjustable controlled-release rate 145
adsorption capacity 73, 74, 75, 117
air-cooling 277, 279, 280, 281
alcohol-responsive burst release 171–174
algae cells 205
alginate(s) 6, 123, 131–136, 158, 211, 241, 242, 243, 244–248, 253
alginate hydrogel materials 135
alginate microcapsules 132
alginate microparticles 211
amide groups 55, 60, 61, 65, 171
ammonium persulfate (APS) 56, 57, 62, 86, 163, 190, 213, 217, 218
artificial spider web 248
asymmetry swelling/shrinking volume changes 189
2,2'-azobis(2-amidi-nopropane dihydrochloride) 168, 195
2,2'-azobis(2-amidi-nopropanedihydro-chloride) 168
2,2'-azobis(2-methylpropionamidine) dihydrochloride (V-50) 91

b

B/A1/C double emulsions 39
B15C5Am. *See* benzo-15-crown-5-acrylamide (B15C5Am)
B18C6Am 68, 71, 73, 75, 195–200, 270, 281, 282, 283, 284, 286, 289
B18C6Am/Pb2+ complexes 68, 73, 75, 286
B18C6Am/Pb2+ host-guest complexes 68, 281
benzo-18-crown-6 73
benzo-15-crown-5-acrylamide (B15C5Am) 174, 175, 176, 177
benzophenone 270
benzyl benzoate (BB) 48, 91, 132, 147, 178, 179, 204, 236, 256
biomimetic soft microrobots 194
Bold's Basal medium 279
bovine serum albumin (BSA) 117, 118, 135, 136, 195, 196, 197, 200, 201
bowl shape 188, 203, 204, 206
bromoeosin 236, 240, 241
BSA adsorption 117, 118

BSA-FITC   135, 136
  soaked microcapsules   135
burst release   162–182
butyl acetate   168, 169, 179

C

Ca-alginate   134, 135, 136, 241, 242, 243, 244–248
  microcapsules   134, 135
  microfibers   244
CaCO$_3$ nanoparticles   48, 132, 138
capillary microfluidic device   93
capillary number   15
carboxymethylcellulose sodium (CMC)   225
cationic pH-responsive microcapsules   151
cell capture   205
cell culture   253, 267, 268, 272–281, 289
cell growth   223, 267, 279
chemical co-precipitation   145
chitosan (CS)   5, 6, 123, 125, 145, 146, 147, 148, 149, 150, 151–154, 156–158, 176, 177–182, 235, 236–241, 253, 254, 255, 256, 257, 258, 259, 260, 262, 264
chitosan microcapsules (CS)   147
Chlorella pyrenoidosa cells   279, 280
chlorotrimethylsilane   47
coaxial three-phase jets   224
co-delivery   217–218, 240
coefficient of variation (CV)   15, 40, 57, 82, 125, 163
co-encapsulation   11, 12, 24, 211, 212–217, 219
co-flow   4, 12, 13, 14, 15, 16, 86, 124, 132, 256, 257
  microchannel   4
collection tube   16, 20, 24, 62, 91, 108, 124, 132, 133, 138, 201, 224, 226, 242
colloidal-scale hole-shell microparticles   187
complete engulfing configuration   36, 37
complexes' stability constant (logK)   73

confined microreaction   161, 187, 201–207, 211
confocal laser scanning microscope (CLSM)   40, 41, 65, 66, 94, 95, 96, 111, 134, 135, 142, 148, 149, 150, 167, 169, 170, 177, 179, 181, 182, 203, 205, 240, 255, 256, 257, 258, 279, 280
constant-flow pumps   14
constant-pressure pumps   14, 287
contact angle   203, 204
continuous fluid   13, 14, 15, 25, 37
controllable double emulsions   16
controllable emulsion droplets   12
controllable monodisperse single emulsions   14
controllably deformed emulsions   110
controllably deformed W/O/W emulsions   108, 118
controlled capture   187, 201–207
controlled release   4, 136, 145, 146, 147, 149, 151, 154–158, 161, 168, 182, 183, 187, 211, 216–219, 253, 255, 264
copper   47, 48, 49
core compartment   161, 162
core flow   223, 224, 236
core-sheath flow jet   224, 225, 226
  templates   226
core-sheath microfibers   224–235
  fabrication   224
  morphological characterization   227
  temperature regulation   230
  thermal property   227
core-sheath poly(vinyl butyral) microfibers   224, 226
core-shell composite microfibers   5
core-shell configuration   203
core-shell microcapsules   4, 62, 64, 66, 75, 136, 161–183
core-shell microspheres   55, 68, 69, 70, 71, 72, 73, 74, 75
core-shell PNIPAM microcapsules   55, 63, 64, 65, 67, 75, 95, 169
coverslips   13, 14, 47, 255, 268, 269, 270, 273, 279

critical ethanol concentration value
    (CC)   255
cross linked polymer network   85
cross-linker   270
  glutaraldehyde (GA)   146
crosslinking   3, 5, 92, 96, 99, 101, 132,
    146, 147, 149, 152, 177, 178, 181,
    225, 236, 239, 241, 242, 243, 255,
    257, 259, 264
cross linking degree   92, 96, 99, 101,
    132, 259
18-crown-6   68
crown ether   73, 198
crystallization enthalpy   229
CS-M-T microcapsules   152
cyclohexane   168, 169
cylinder glass capillaries   12

d

DC749   44, 48
dehydrated Ca-alginate microfibers
    244, 246
dehydrated spider-silk-like Ca-alginate
    microfibers   243, 244, 246
dehydration   69, 97, 244
deionized water   68, 69, 70, 71, 73, 74,
    75, 81, 83, 108, 135, 138, 139,
    163, 168, 169, 170, 173, 174, 175,
    179, 190, 257, 270, 274
dense skin layer   98, 172, 227
density mismatch   188, 201, 207
de-swelling   59, 61, 85, 86, 145, 149,
    151, 172, 198
detection   1, 4, 5, 67, 68, 255–264, 267,
    268, 270–272, 281–289
dewetting   109, 110, 188, 207
differential scanning calorimeter (DSC)
    227, 229, 230
diffusional permeability   145
diffusion-driven release
    systems   165
dimethoxy-2-phenylacetophenone
    (BDK)   63, 80, 81, 91, 138, 163,
    168, 190, 191, 195, 213
2,2-dimethoxy-2-phenylacetophenone
    (BDK)   138, 163, 213

dimethyl sulfoxide
    (DMSO)   225, 226
diphenyliodonium nitrate
    (PAG)   132
direction-specific burst release   170,
    171
dispersed fluid   14, 15
Disperse Red   242
Donnan potential   137, 281
double emulsion droplets   125
double emulsions, 3, 12, 35, 62, 123, 161,
    188, 212
Dow Corning 749 (DC749)   38, 41, 44,
    45, 47, 48, 50, 217
dripping   15, 18, 24
droplet coalescence   37, 47, 48, 50
droplet maker   22, 24, 25, 27
droplet-making units   12, 13, 14, 21
droplet-pairing   44, 45–47
droplet-triggered droplet
    formation   44
    pairing   44
drug delivery   3, 11, 55, 79, 80, 85, 103,
    105, 123, 136, 142, 144, 145, 156,
    158, 163, 166, 219, 224
drug delivery systems   79
drug loading levels   145
drug release kinetics   79, 105, 123
dry film photoresist (DFR)   13
dynamic swelling rate   284
dynamic thermo-responsive swelling
    ratio   101

e

eccentric core-shell structures   188
eccentric oil core   168
EC microcapsule preparation   125
EGDMA dyed with LR300   116
emulsification   4, 11, 16, 17, 18, 19, 20,
    21, 35, 91, 92, 93, 123, 124, 127,
    132, 133, 158, 214
emulsion droplet   1, 2, 3, 4, 12–16, 19,
    22, 36, 37, 38, 39, 41, 42, 43, 44,
    46, 47, 48, 49, 50, 51, 57, 58, 62,
    63, 64, 81, 93, 94, 125, 127, 136,
    140, 187
emulsion droplet system   1

emulsion templates   57, 63, 69, 70, 71, 91, 92, 93, 94, 95, 96, 109, 110, 113, 114, 116, 133, 135, 136, 139, 140, 149, 150, 166, 169, 181, 187, 188, 195, 196, 197, 203, 213, 217, 218
encapsulation ratio   230, 231
energy storage   123, 223
  systems   223
enhanced mass transfer   106, 118
epoxy resin   14, 80, 270, 272
equilibrium deswelling ratio   149
equilibrium state   58, 59, 99, 102, 274
ethanol-responsive permeability control   260, 263
ethanol-responsive volume change   255
ethoxylated trimethylolpropane triacrylate (ETPTA)   201, 202, 203, 204, 207
ethyl cellulose (EC)   123, 124–131
ethylene glycol dimethacrylate (EGDMA)   81, 108, 111, 112, 113, 114, 115, 116, 117, 118
ethyl gallate (EG)   55, 62–67
evolved double emulsions   188, 207
expanded microchamber   38, 41, 45, 46

f

fast-responsive PNIPAM microgels   206, 207
Fe_3O_4 magnetic nanoparticles (MNPs)   68, 195
Fe_3O_4 nanoparticles   163, 164
ferrofluid   163, 168, 195, 241
fishbowl-shaped hole-shell microparticles   201, 204
FITC-dextran   156, 157
FITC-insulin   142, 143, 144
FITC-labeled BSA (BSA-FITC)   135, 136
FITC labeled insulin (FITC-insulin)   142, 143, 144
FITC-PNIPAM nanogels   201, 203, 205
flow circulation loop   268, 273
flow-focusing   4, 12, 13, 14, 21, 38, 41, 44, 46, 47, 50, 80

cross-junction geometry   14
  geometries   38, 41, 44, 46, 47, 50
flow rate   14, 15, 17–19, 24, 25, 26, 38, 44, 45, 47, 48, 50, 63, 80, 91, 92, 93, 94, 109, 124, 126, 128, 129, 130, 131, 133, 139, 204, 205, 214, 215, 223, 226, 236–239, 242, 243, 248, 257, 268, 274, 275, 277, 281, 282, 283, 284, 285, 287
flow rate change   268, 281, 282, 283
fluid heat exchange   268, 277
fluorescein isothiocyanate (FITC)   135, 136, 142, 143, 144, 156, 157, 201, 203, 205, 255, 257, 258, 259, 260, 261, 262, 263
fluorescence intensity   111, 143, 144, 156, 167, 258
fluorescent dye   62, 91, 180, 203
fluorescent dye Lumogen Red 300 (LR300)   62
fluorescent nanoparticles   167
fluorinated oil   37
Fluorochrome LR300   170
free radical polymerization   56, 80
freeze-dried PNIPAM microgels   96
functional microfibers   223–248
  core-sheath microfibers (see core-sheath microfibers)
  cylinder core flow   223
  electrospinning and wet spinning   223
  fabrication   224
  peapod-like microfibers   235
  spider-silk-like microfibers   241
functional nanoporous poly(MMA-co-EGDMA-co-GMA) microparticles   118

g

glass-capillary microfluidic device   12, 14, 15, 16, 17, 18, 20, 22, 108, 201, 224
glass plates   12, 21, 44, 46
glass slide   13, 14, 47, 80, 124, 132, 167, 255, 268, 270, 272, 273, 279
glucose-induced shrinking   138

glucose-induced swelling   137, 140, 141, 142, 143
glucose regulation   136
glucose-responsive microcapsules   123, 137, 138, 139, 140, 158
glucose-responsive release   142, 143, 144
glutaraldehyde (GA)   146, 147, 181, 236
glycerin   108, 138, 195
glycerol   62, 81, 91, 201, 212, 217, 218
glycidyl methacrylate (GMA)   114–116, 117, 118
gold nanoparticles   205

h
Hagen-Poiseuille's law   281
heavy metal   67, 71, 72
   ions   71, 72
hierarchically engineered poly(methyl methacrylate-*co*-ethylene glycol dimethacrylate) (poly(MMA-*co*-EGDMA)) microparticles   108
hierarchical porous microparticles   105–118
hierarchical porous poly(MMA-*co*-EGDMA-*co*-GMA) microparticles   108, 113, 114, 115, 116, 117, 118
hierarchical porous structures   105, 106, 107, 108, 116, 118
higher order multiple emulsions   11, 12, 19, 41–44, 46, 212, 219
high interconnectivity   105
highly-interconnected hierarchical porous structures   105, 106–108, 118
highly monodisperse size   15, 18
hole-shell microparticles   187–207
hole-shell structures   188, 203, 205
hollow calcium alginate microcapsules   123, 134, 158
hollow fiber membranes   5
hollow microcapsules   123–158, 162
hollow tubular microfibers   5
homogenizer-produced W/O emulsions   166

hydrated Ca-alginate microfibers   243
hydrogel microparticles   55–75
hydrogel microvalve   267, 268, 269, 270, 272, 273, 274, 275, 276, 277, 278, 279, 280, 281, 289
hydrogel-microvalve-integrated micro-heat-exchanging system   273, 277
hydrogen-bonding interaction   55
hydrophilic microlancets   48
hydrophilic spider-silk-like Ca-alginate microfiber   247
hydrophilic-swelling/hydrophobic-shrinking phase changes   162
hydrophobic interaction   60, 62, 65
1-hydroxy-cyclohexyl-phenylketone   270
hydroxyethyl cellulose (HEC)   47, 147, 151, 152, 178, 241
2-hydroxyethyl methacrylate (HEMA)   79–84, 103
2-hydroxy-2-methyl-1-phenyl-1-propanone (HMPP)   108, 201, 204

i
imitated solar irradiation   234, 235
in-chip membranes   253, 254, 264
inclusion constant   68, 286
independent single emulsion droplets   39
initiator   56, 57, 62, 81, 86, 87, 91, 138, 163, 213, 218
injection tube   14, 16, 57, 91, 108, 226, 236, 241
inner fluid   14, 22, 93, 94, 109, 125, 127, 138, 140, 146, 147, 178, 179, 201, 212, 214, 218, 226, 227
in situ polymerization   202, 268
insulin   136, 142, 143, 144, 253
interfaces   1, 2, 3–6, 37, 59, 93, 95, 105, 106, 117, 125, 163, 178, 191, 202, 204, 237, 253, 254, 255, 257, 258, 264, 295, 296
interfacial energy   36, 188
interfacial reactions   253, 254

interfacial tension   15, 16, 36, 37, 38, 39, 45, 83, 204
internal gelation   123, 132, 134, 158
ionic crosslinking   132
isolated co-encapsulation   211, 212, 213, 215, 217, 219
isopentyl acetate   195
isopropanol   86, 87, 91, 92, 93, 94, 95, 133, 139, 140, 148, 178, 257
isothermal volume phase transition   58, 59, 60

j
Janus microspheres   4
jetting   15

k
K$^+$-recognition   174, 175
K$^+$-responsive
  burst release   174–176
    core-shell microcapsules   174, 175
K$^+$-triggered volume shrinking   175

l
lab-on-a-chip   1, 2
laminar flow   1, 2, 4, 5, 6, 253, 254, 257, 264
  interfaces   1, 6, 253, 264
lead (Pb^{2+})   67
linear solid microfibers   5
liquid droplets   11
liquid extractors   22, 23, 25, 26, 213
liquid-liquid interfaces   1, 2, 3–6, 295, 296
lithographically fabricated devices   35
lithography   12, 268
lock-key   206
log K   73
lower critical solution temperature (LCST)   55–58, 61, 62
LR300. *See* Lumogen Red 300 (LR300)
Lumogen Red 300 (LR300)   39, 41, 43, 46, 50, 62, 91, 92, 116, 168, 169, 170, 179, 180, 181, 182, 203, 226, 227, 228, 239, 240

m
magnetic-guided assembly   245, 246
magnetic-guided patterning   244–246
magnetic-guided targeting delivery   152, 163, 164
magnetic hierarchical porous poly(MMA-*co*-EGDMA-*co*-GMA) microparticles   117
magnetic knots   242, 243, 244, 245
magnetic minipillars   244, 245, 246
magnetic nanoporous poly(MMA-*co*-EGDMA-*co*-GMA) microparticles   117
magnetic PNB core-shell microspheres   74, 75
magnetic spindle-knots   224, 241, 242, 243, 244, 246–248
magnetic targeting   145
manually assembled glass-capillary devices   35
mass transfer   1, 2, 11, 35, 79, 105, 107, 111, 112, 118, 207, 283, 295, 296
mass transport   187
MC-W-W EC
  microcapsules   127
mechanical strength   83, 112, 165, 217, 231
melting enthalpy   229, 230
membrane-in-a-chip
  in biomedical fields   253
  ethanol-responsive self-regulation   260
  fabrication of   254
  nanogel-containing smart membrane   253, 255
  reversible and repeated thermo/ethanol-responsive self-regulation   263
  in situ formation   253
  smart membranes   253
  temperature-responsive self-regulation of the membrane permeability   257

membrane permeability   254, 255, 256, 257–264
methanol   172, 173, 174
methyl methacrylate (MMA)   79–84, 103, 108, 111, 113, 114, 115, 116, 117, 118, 201
micro-actuator   195, 196, 197, 198, 199–201
micro-analysis   1, 4, 5
microcapsules, 1, 11, 35, 55, 91, 123, 161, 211, 295
microchannel, 1, 11, 37, 170, 195, 213, 223, 253, 267, 295
microfibers   2, 3–6, 223–248, 295
microfluidic-constructed stable phase interface structure systems   2
microfluidic device   1, 11–31, 35–51, 56, 62, 63, 68, 69, 70, 75, 80, 81, 86, 87, 91, 92, 93, 94, 108, 109, 124, 125, 132, 138, 139, 146, 147, 163, 169, 178, 190, 195, 196, 197, 201, 202, 212, 213, 217, 223, 224, 225, 226, 236, 241, 242
microfluidic emulsification   132, 133
microfluidic flow control system (MFCS)   273, 274, 277, 283, 287
microfluidic laminar flow technology   5
microfluidics   1–6, 11–35, 35–51, 55–75, 79–103, 105–118, 123–158, 161–183, 187–207, 211–219, 223–248, 253–264, 267–289
  controllable fabrication of functional materials   1
  microscale closed liquid-liquid interfaces   3
  microscale nonclosed annular laminar interfaces   5
  microscale nonclosed layered laminar interfaces   4
  technology   1, 2, 4, 6, 83, 136, 295, 296, 297
microgel-based
    Pb^{2+} sensor   272
microgel-incorporated glass capillary   271, 272
micro-heat-exchanger   273, 274, 277, 278, 279
micro-heat-exchanging system   268, 273–274, 277, 279
micro-lancet   37, 38, 47, 48, 49, 50, 51
microlancets   48
micrometer-sized pores   105, 106, 107, 108–118
micro-reaction   1, 4, 5, 50, 51, 161, 187, 201–207, 211, 219, 253
microscale closed liquid-liquid interfaces   3
microscale emulsion interfaces   4
microscale nonclosed layered laminar interfaces   4
micro-scale phase interfaces   3
microscope glass slides and coverslips   14
micro-separation   1, 4, 5
microsphere(s), 1, 11, 55, 79, 123, 161, 192, 295
microvalve-controlled water cooling   277, 278, 280, 281
microvalve-in-a-chip   267–289
  mask-based lithography   268
  Pb^{2+}-responsive hydrogel microvalve   270
  in situ polymerization   268
  thermo-responsive hydrogel microvalve   268
microvalve-integrated
    micro-heat-exchanging system   273
microvalves   123, 145, 147, 155, 157, 158, 267–289
middle fluid   16, 108, 109, 125, 138, 140, 146, 147, 163, 178, 190, 201, 204, 212, 213, 217, 218, 226, 227

monodisperse calcium alginate hollow
    microcapsules
  droplet generation and ionic
    cross-linking   132
  microfluidic fabrication strategy   132
  morphologies and structures of   133
monodisperse controllable double
    emulsions   38
monodisperse core-shell hydrogel
    microparticles, Pb^{2+}
  adsorption behaviors   71
  core-shell microspheres   68
  industrial wastewater   68
  microfluidic fabrication   68
  thermo-responsive swelling/shrinking
    configuration change   68
monodisperse core-shell microcapsules
  alcohol-responsive burst release   171
  chitosan microcapsules   179
  controllable fabrication   161
  direction-specific thermo-responsive
    burst release   168
  double emulsions   161
  fabrication   177
  K^+-responsive burst release   174
  microfluidic strategy   162
  nanoparticles   166
  oil-soluble substances   162
  pH-responsive burst release   176
monodisperse core-shell PNIPAM
    hydrogel microparticles, ethyl
    gallate
  antioxidant activity   62
  intact-to-broken transformation
    behaviors   65
  microfluidic fabrication   62
  thermo-responsive phase transition
    behaviors   65
  volume phase transition temperature
    62
monodisperse emulsion droplets   12,
    13
monodisperse ethyl cellulose hollow
    microcapsules
  microfluidic fabrication
    strategy   124
  morphologies and structures of   125

monodisperse glucose-responsive
    hollow microcapsules
  glucose-responsive behaviors of
    microcapsules   140
  glucose-responsive drug release
    behaviors   142
  microfluidic fabrication strategy   136
  sugar-responsive systems   136
  tumor cells   136
monodisperse higher-order multiple
    emulsions   41
monodisperse hole-shell microparticles
  core droplet   188
  effect of inner cavity   191
  functionality   205
  interfacial properties   188
  microfluidic fabrication   201
  microfluidic strategy   188
  particle-template/emulsion-template
    methods   187
  Pb^{2+} sensing and actuating   195
  poly(NIPAM-co-B18C6Am)   195,
    196, 199, 200
  shell droplet   188
  structure control   203
  thermo-driven crawling movement
    188, 190, 193
  versatility   187
monodisperse multi-stimuli-responsive
    hollow microcapsules
  controlled-release characteristics
    154
  environmental changes   144
  intelligent drug delivery systems   144
  microfluidic fabrication strategy   145
  microvalves   145
  "on-off" mechanism   144
  pH   145
  site-specific targeting   158
  stimuli-responsive behaviors   150
monodisperse oil droplets   239
monodisperse oil-in-water (O/W)
    emulsions   80
monodisperse oil-in-water-in-oil
    (O/W/O) emulsions   41, 68, 91
monodisperse PNIPAM hydrogel
    microparticles, tannic acid

microfluidic fabrication   56
volume phase transition behaviors   57
monodisperse porous microparticles   79
monodisperse porous poly(HEMA-MMA) microparticles
  biodegradability   79
  microfluidic fabrication strategy   80
  structures   82
monodisperse quadruple-component O/W/O double emulsions   214
monodispersity   11, 14, 24, 40, 55, 57, 71, 79, 80, 82, 83, 93, 131, 134, 140, 149, 150, 187, 204
monomer   62, 80, 81, 86, 87, 91, 92, 96, 103, 106, 108, 138–140, 149, 163, 174, 187, 201, 212, 224, 268, 269, 270, 271
multicompartmental microparticles   211–219
  compound-fluidic electrospray technique   211
  controllable co-encapsulation   213
  encapsulation systems   211
  fabrication   212
  immunoprotection   211
  isolated co-encapsulation   211
  microbioreactions   211
  multi-core/shell microparticles   212
  synergistic release   216
  Trojan-horse-like microparticles   217
  troublesome multi-step process   211
multicomponent multiple emulsions   11, 12, 14, 22–31, 27, 213
multi-core microspheres   4
multi-core/shell microparticles   212–217
multiple emulsions   3, 4, 11–31, 35–51, 158, 211, 212, 213, 219
multi-stimuli-responsive microcapsules   123, 144–157

n

Na-alginate   132, 133, 241
nanogel(s)   201, 203, 253, 254, 255–264
nanogel-containing chitosan membrane   255, 256, 257, 260, 262
nanogel-containing membrane   254, 257, 258, 259, 260, 261, 262, 263, 264
nanogel-containing smart membranes   253, 254, 255–257, 264
nanometer-sized pores   105, 106, 107, 108, 111, 112, 114, 118
nanoparticles   166
nanoparticles-in-microcapsule system   166
nanoporous PEGDMA microparticles   113, 117
nano-porous poly(MMA-co-EGDMA-co-GMA) microparticles   117, 118
nano-porous structure   107, 108, 111, 112, 113, 114, 116, 117, 118
nanovalves   255, 256, 257, 264
N-butyl acetate   241, 242, 244
N-isopropylacrylamide (NIPAM)   55, 56, 62, 68, 79, 86, 91, 92, 93, 96, 97, 99, 100, 102, 136, 138, 149, 158, 162, 163, 168, 174, 175, 176, 177, 189, 190, 191, 195, 196, 199–201, 212, 213, 216, 217, 254, 270, 281, 282, 283, 286, 289
N-2 microgel   98
N,N'-methylene-bis-acryamide (MBA)   86, 190
N,N'-methylene-bis-acryamide (MBA)   56, 62, 86, 91, 92, 93, 96, 97, 99, 100, 102, 138, 163, 168, 190, 195, 212, 217, 270
N,N,N',N'-tetramethylethylene-diamine (TEMED)   56, 57, 86, 218
n-octane   227
non-engulfing configuration   36
nonporous PEGDMA microparticles   113, 117
non-spherical particles   4, 11, 35

O

oil-core/hydrogel-shell microcapsules 163
oil-filled compartments 162
oil-in-water-in-oil (O/W/O) emulsion droplets 62
oil-in-water-in-oil-in-oil (O/W/O/O) triple emulsions 43
oil-in-water-in-oil-in-water-in-oil (O/W/O/W/O) quadruple emulsions 41
oil-in-water (O/W) primary emulsions 91
oil jets 44, 45, 46
oil-soluble 2,2-dimethoxy-2-phenyl-acetophenone (BDK) 80
oil-soluble substances 162
oil/water interface 95
oleic-acid-modified magnetic nano-particles (OA-MNPs) 195, 197
on-demand release 211, 212
OP-10 181
open-celled porous microgels 96
open-celled porous PNIPAM microgels 91, 92, 93, 102
open-celled porous structure 90–103
OPH-2 and OPI-2 microgels 99
osmotic pressure 73, 127, 131, 198, 281
outer fluid 14, 69, 108, 109, 124, 125, 126, 127, 128, 129, 130, 131, 133, 138, 139, 146, 147, 163, 178, 190, 201, 213, 225, 226
O/W emulsions 3, 80
O/W/O double emulsion droplets 64
O/W/O double emulsions 3, 4, 18, 24, 27, 39, 41, 42, 43, 44, 45, 46, 50, 51, 63, 64, 68, 70, 71, 132, 133, 134, 138, 139, 146, 147, 149, 162, 163, 164, 177, 178, 181, 190, 212, 213, 214
O/W/O/O triple emulsions 43, 46
O/W/O/W triple emulsions 28

p

P(NIPAM-co-AAm) 149
paraffin wax 225, 226, 227, 228, 229, 230, 231, 232, 233, 234, 235
parallel laminar flows 253, 254
partial engulfing configuration 36, 37
partially dewetting 110, 188
pathologically acidic conditions 154
Pb^{2+}-adsorption 73, 74, 75
Pb^{2+} detection platform 272, 281, 282, 283–288
Pb^{2+} pollution discharge 289
Pb^{2+}-recognition 68
Pb^{2+}-responsive hydrogel microvalve 268, 270, 272, 281
Pb^{2+}-responsive microgel
  pollution warning 287
  real-time online detection of trace $Pb2+$ 281
  selectivity and repeatability of the $Pb2+$ detection platform 284
  sensitivity of the $Pb2+$ detection platform 283
  wastewater 289
Pb^{2+}-responsive P(NIPAM-co-B18C6Am) microgel 281, 282
Pb^{2+} sensing 195–201, 282
Pb^{2+} sensing and actuating 195
PDMS microfluidic device 12, 14, 15
peapod-like chitosan microfibers 235, 236, 237, 239, 240
peapod-like jet 236–239
peapod-like jet containing discrete oil droplets 237
peapod-like microfibers 235–241
  fabrication 236
  flow rates 236
  synergistic encapsulation 240
peristaltic pump 273, 274, 286, 287
permeability coefficient 154, 155, 156, 157
PGPR90 48
phase change materials 223, 224–235, 248
phase transition 55, 56–62, 65, 66, 86, 89, 91, 137, 138, 140, 149, 162, 174, 175, 176, 189, 191–193, 205, 216, 227, 254, 258, 274
phenols 60, 62, 65
photoacid generator 132, 133

photoacid generator diphenyliodonium nitrate (PAG)   132
photo-initiator   63, 80, 108, 138, 163, 190, 191, 213, 270
photoinitiator dimethoxy-2-phenyl-acetophenone (BDK)   63
pH-responsive burst release   176–182
pH-responsive capsule membrane   123, 158
pH-responsive controlled-release behaviors   154
pH-responsive core-shell microcapsules   177
pH-responsive swelling   147, 151, 152
physiological temperature   136, 137, 141, 142
pK_a 137, 140, 146, 147, 151, 154, 516
plug-n-play microfluidic devices   21
Pluronic F127   62, 91, 108, 125, 132, 138, 147, 163, 168, 178, 190, 195, 201, 212
P(NIPAM-co-B18C6Am) microgel   270, 281, 282, 283, 286
PNB microactuators   195, 196, 197, 198, 199, 200, 201
PNIPAM hydrogel   55–61, 95, 166, 172, 189, 191, 192, 193, 195, 216, 270, 274, 276, 289
  microvalve   270
  shell   166, 216
PNIPAM microactuators   198
PNIPAM microgels   55, 56, 57, 58, 59, 60, 61, 75, 79, 85, 86, 88, 89, 90, 91, 92, 93, 94, 95, 96, 97, 98, 99, 100, 101, 102, 103, 206, 207
PNIPAM microparticles with open-celled porous structure
  microfluidic fabrication strategy   91
  morphologies and microstructures   93
  thermo-responsive volume change behaviors   98
PNIPAM microspheres   57, 58, 62, 65, 192, 193
PNIPAM polymeric networks   59, 61, 73
pollution terminating   289

pollution warning   286, 287–289
poly(EGDMA) (PEGDMA)   111
poly(HEMA-MMA)   79–84, 103
poly(hydroxyethyl methacrylate-methyl methacrylate) (poly(HEMA-MMA))   79
poly(methyl methacrylate) (PMMA)   79
poly(MMA-co-EGDMA)   108, 111, 113, 114, 115, 116
poly(N-isopropylacrylamide) (PNIPAM)   55, 79, 136, 162, 189, 254, 270
poly(N-isopropylacrylamide-co-acrylamide) (P(NIPAM-co-AAm))   149
poly(N-isopropylacrylamide-co-benzo-15-crown-5-acrylamide) (PNB)   175
poly(N-isopropylacrylamide-co-benzo-18-crown-6-acrylamide) (PNB)   68, 270
poly(NIPAM-co-AAm-co-B15C5Am) (PNAB)   175, 176, 177
poly(NIPAM-co-AAPBA) (PNA)   138, 140, 141
poly(NIPAM-co-AAPBA-co-AAc) (PNAA)   138, 141, 142, 143, 144, 158
poly(NIPAM-co-B15C5Am)   176
poly(vinyl alcohol) (PVA)   81, 217, 218
poly(vinyl butyral) (PVB)   224, 225, 226, 227, 228–235
poly(vinyl pyrrolidone) (PVP)   79, 80, 81, 82, 83, 103
poly(HEMA-MMA)
  copolymers   80
polydimethylsiloxane (PDMS)   12, 14, 15, 217, 244, 253
  plates   12
polyelectrolyte microparticles   38, 211
polyethylene (PE)   80, 132, 270
Polyfluor   570, 255, 256
polyglycerol polyricinoleate (PGPR 90 or PGPR)   38, 41, 43, 44, 45, 46, 50, 56, 57, 62, 63, 81, 86, 91, 108, 110, 111, 112, 113, 116, 132, 138, 147, 163, 168, 178, 190, 195, 204, 212

poly(NIPAM-*co*-B18C6Am) hole-shell microparticles
  effect of hollow cavity   199
  effect of Pb^{2+} 196
  magnetic-guided targeting behavior   195
  micromanipulation   200
poly(*N*-isopropylacrylamide) (PNIPAM) hydrogel   189
poly(PNIPAM-*co*-B18C6Am) (PNB) hydrogel   195
polymeric microparticles   187
polymeric networks   55, 59, 60, 61, 73, 80, 86, 88, 100, 102
polymerization   3, 5, 56, 57, 63, 68, 69, 71, 80, 81, 83, 86, 92, 94, 108, 110, 123, 138, 139, 140, 145, 158, 168, 169, 187, 191, 195, 196, 201, 202, 204, 213, 255, 268, 269, 270, 271
poly(NIPAM-*co*-AAPBA-*co*-AAc) (PNAA) microcapsule   138
poly(NIPAM-*co*-AAm-*co*-B15C5Am) microcapsules   177
poly(*N*-isopropylacrylamide-*co*-benzo-18-crown-6-acrylamide) (P(NIPAM-*co*-B18C6Am)) microgel   68, 270, 281, 282, 283, 286
poly(*N*-isopropylacrylamide) (PNIPAM) microgels   85
poly(methyl methacrylate-*co*-ethylene glycol dimethacrylate) (poly(MMA-*co*-EGDMA)) microparticles   108, 111, 113, 114, 115, 1164
poly(MMA-*co*-EGDMA) microparticles   108, 111
poly(MMA-*co*-EGDMA-*co*-GMA) microparticles   114, 115, 116, 117, 118
poly(*N*-isopropylacrylamide) (PNIPAM) nanogels   254

poly(N-isopropylacrylamide-*co*-methyl methacrylate-*co*-allylamine) nanogels   201
poly(dimethylsiloxane) oil (PDMS)   217
poly(methyl methacrylate) (PMMA) particles   79
poly(HEMA-MMA) porous microspheres   80, 82
poly(vinyl butyral) (PVB) resin   225
polystyrene (PS)   86, 87, 88, 195, 196
polystyrene beads   86, 87, 88
poly(*N*-isopropylacrylamide-*co*-acrylamide) (P(NIPAM-*co*-AAm)) sub-microspheres   149
pore size   5, 105, 106, 108, 113, 118, 135
porogens   79, 80, 82, 83, 86, 91, 103
porosity   83, 103, 105, 106, 108, 109, 112, 113, 118
porous microparticles   79–103, 105–118
porous PNIPAM microparticles with tunable response behaviors
  dramatic response and stimuli-specific behavior   85
  gel swelling   85
  heterogeneous internal microstructures   85
  linear side chains   85
  microfluidic fabrication strategy   86
  temperature-dependent equilibrium volume-deswelling ratio   85
  tunable response behaviors   87
  void structures   86
porous polymeric microparticles   105, 106
porous structures   79, 80, 83, 90–103, 105, 106, 107, 108, 109, 111, 112, 113, 114, 116, 117, 118
post-array-containing microfluidic device   16
precipitation polymerization   145, 255
precursor droplets   40, 41, 42, 44
programmed synergistic release

P(NIPAM-*co*-AAm) sub-microspheres 149
pulsed release 218
PVB/8P microfibers 229
PVB/24P microfibers 230, 233, 235

q
quadruple-component double emulsions 22–28, 169, 214
quadruple-component O/W/O double emulsions 213
quadruple emulsions 21, 212, 219
quintuple-component double emulsions 22, 27, 29
quintuple-component multiple emulsions 29
quintuple-component triple emulsions 22, 27, 29

r
real-time detection 268, 281–289
real-time on-line detection of trace Pb^{2+} 281–283
real-time Pb^{2+} detection 281, 289
response rate 59, 88, 90, 91, 99, 100, 103, 144, 192, 193, 200
reversible glucose-induced swelling/shrinking behaviors 137
reversible glucose-responsive swelling/shrinking 138
reversible swelling/shrinking volume transitions 254, 255
rhodamine B 50, 111, 142, 143, 269, 270
route-specific targeting drug delivery 163

s
scale-up 14, 16, 21, 22, 27
Scanning Electron Microscope (SEM) 83, 84, 85, 96, 97, 111, 112, 113, 114, 126, 127, 128, 129, 130, 131, 205, 206, 227, 228, 239, 243, 255, 256
Schiff base 177
  bonding 177
sealing performance 253, 276–277
selective adsorption 55, 68, 75
selective detection 268, 270, 289
selective permeability 5
self-regulated drug delivery 136
self-regulated permeability 253–264
sensors 55, 62, 67, 75, 79, 85, 103, 136, 138, 140, 174, 207, 230, 253, 255, 272, 274, 282
sequential shear-induced emulsifications 35
sextuple-component triple emulsions 22, 27, 28, 30, 31
shape angle 203, 204
shear-induced generation of controllable multiple emulsions
  controllable double emulsions 16
  controllable emulsion droplets 12
  controllable monodisperse single emulsions 14
  controllable quadruple-component double emulsions 22
  controllable triple emulsions 19
  inner droplet number 11
  microchannel wettability 12
  multicomponent multiple emulsions 27
  sequential bulk emulsification 11
  T-junction geometries 12
  volume fractions 11
sheath flow 224, 225, 226, 236
shell permeability 143
shell thickness 40, 41, 110, 158, 175, 237, 238, 239
shrunken state 99, 137, 141, 146, 174, 189, 205, 275, 281
silicone oil (SiO) 37, 38, 40, 41, 44, 45, 46, 47, 48, 50
single emulsions 4, 12, 13, 14–16, 31, 39, 41, 42, 43, 46, 50, 51, 63, 79, 81, 103, 111, 124, 132, 133
site-specific targeting drug delivery 145
size-classification 206

size distributions 4, 14, 15, 57, 58, 63, 64, 70, 71, 82, 93, 94, 96, 112, 123, 125, 131, 133, 134, 139, 140, 145, 163
size-match 206
size monodispersity 11, 40, 71, 134, 187
smart core-shell microcapsules
 alcohol-responsive burst release 171
 core-shell chitosan microcapsules 179
 direction-specific thermo-responsive burst release 168
 fabrication 177
 K^+-responsive burst release 174
 nanoparticles 166
 oil-soluble substances 162
 pH-responsive burst release 176
smart hydrogel microvalves 267, 268, 279, 289
smart-membrane-in-chip 253, 264
smart microvalve-in-a-chip
 cell culture 279
 microvalve-integrated micro-heat-exchanging system 273
 pollution warning 287
 real-time online detection of trace Pb^{2+} 281
 sealing performance 276
 selectivity and repeatability of the Pb^{2+} detection platform 284
 sensitivity of the Pb^{2+} detection platform 283
 temperature self-regulation 277
 thermo-responsive switch performance 274
 wastewater 289
smart-microvalve-integrated microchips 267
sodium acrylate 217
sodium alginate (Na-alginate) 132, 241
sodium dodecyl sulfate (SDS) 38, 41, 44, 45, 46, 47, 50, 191, 192, 193
solvent diffusion 123, 125, 158
solvent evaporation 3, 68, 69, 169, 195, 196
solvent quality 204
soybean oil (SO) 38, 56, 57, 62, 63, 91, 132, 138, 147, 163, 168, 169, 170, 178, 179, 190, 195, 196, 212, 256
soybean oil containing emulsifier PGPR 90, 62
specific surface area 5, 79, 83, 108, 117, 118, 192
spider-silk-like Ca-alginate microfibers 241, 242, 243, 244, 246, 247
spider-silk-like microfibers 241–248
 fabrication 241
 magnetic-guided patterning and assembling 244
 morphological characterization 242
 water collection ability 246
spider-web-like structures 224, 247, 248
spreading coefficient 36, 37, 38, 39, 41, 42, 43, 45, 46
square glass tubes 12
squirting cucumbers 166
stable microscale phase interfaces 3
star-shaped microchannels 268, 269, 270
stimuli-responsive controlled-release systems 145
stimuli-responsive hydrogel microparticles 55, 75
stimuli-responsive hydrogel shell
stimuli-responsive microcapsules 144
stimuli-responsive microparticles 55
stimuli-responsive smart membranes 5, 264
stimuli-sensitive hydrogel microparticles/microgels 55
stress-strain curves 232
Sudan Black 236, 240, 241
Sudan III 163, 164, 195, 212
Sudan Red 139, 190
superparamagnetic nanoparticles 152
supramolecular host-guest complexes 68

surface wettability 13, 37, 47, 48, 51, 205
surfactant bilayers 109, 110
sustained drug release 143, 151
swelling ratios 101, 151, 152
swollen state 57, 99, 137, 140, 141, 143, 144, 146, 174, 182, 205, 261, 275
symmetry breaking 187, 191, 194
synergistic delivery systems 11
synergistic drug delivery 224
synergistic encapsulation 235–241, 248
synergistic release 212–218

t

tannic acid (TA) 55–61, 56
tap water pipeline 288
temperature- and ethanol-responsive smart membrane
    ethanol-responsive self-regulation 260
    nanogel-containing smart membrane 255
    reversible and repeated thermo/ethanol-responsive self-regulation 263
    temperature-responsive self-regulation of the membrane permeability 257
temperature-dependent volume change 140, 147, 155, 198, 206
temperature-dependent volume phase transition 149
temperature regulation 145, 224, 230–235, 248
temperature-responsive permeability self-regulation 260
temperature-responsive sub-microspheres 123, 145, 146, 147, 148, 149, 150, 152, 154, 155, 156, 157, 158
temperature-responsive volume change 73, 155
temperature self-regulation 277–281
template-directed synthesis 105

tensile strength 231
terephthalaldehyde 178
terephthalaldehyde-crosslinked chitosan hydrogel shell 176
tetramethylammonium hydroxide 164
thermal energy 233, 234
thermal stability 227, 230, 231
thermo-driven crawling movement 188–194
thermo-driven locomotion 189, 193, 194
thermo-driven soft microcrawlers 188
thermo-induced shrinking process 100
thermo-regulation 235
thermo-responsive burst release 168–171
thermo-responsive hydrogel microvalve 268, 269, 270, 272
thermo-responsive phase transition 58, 60, 61, 65
thermo-responsive PNIPAM shell 165
thermo-responsive poly(N-isopropylacrylamide) (PNIPAM) hydrogel microvalve 270
thermo-responsive polymer 55
thermo-responsive swelling behaviors 100
thermo-responsive swelling process 100, 101
thermo-responsive switch 274–278
thermo-responsive volume phase transitions 189
thermostatic control 267, 268–270, 272–281
thermo-triggered actuator 216
thermo-triggered burst release 162–171
thermo-triggered synergistic release 216
3D assembly 241–248
3D glass-capillary microfluidic device 15
3D microchannel 12
3D printing techniques 21

time-dependent deswelling ratio   172, 173
T-junction   4, 12, 13
  geometries   4, 12
trace analytes detection   268
trace threat analytes   267, 268, 270
transition tube   16, 19, 20, 21, 62, 108, 124, 133, 201, 224, 226, 236, 241, 242
trans-membrane diffusional permeation   145
trans-membrane diffusion flux   259
triggered release   161, 191, 211, 216
3,4,5-trihydroxybenzoic acid ethyl ester   62
triple emulsions   12, 14, 16, 19–22, 27, 28, 29, 30, 31, 41, 43, 46, 212, 217–218
Trojan-horse-like microparticles   217–218
Trojan-horse-like structures   212
truncated-sphere shape   203, 204
tubular flows   223
tubular microfibers   5, 225, 226, 248
tunable pore size   105, 106
2D microchannel   12, 13
2D PDMS microfluidic device   15
two-stage glass capillary microfluidic device   108, 109, 224
two-step sequential emulsification   16

u

ultra-thin shell   35, 39, 40, 41, 49, 50
ultra-thin-shelled double emulsions   40
uniform hierarchical porous microparticles
  enhanced protein adsorption   117
  highly interconnected hierarchical porous structures   106
  magnetic-guided oil removal   115
  microdrop interfaces   106
  microfluidic strategy   106
  monodisperse emulsion drops   106
  nanometer-sized pores   105
  nanometer-sized pores and micrometer-sized pores   111

oil removal   114
porosities   105
porous polymeric microparticles   105
protein adsorption   116
W/O/W emulsions   108
UV-curable adhesive   14, 47, 270, 279
UV-initiated polymerization   68, 69, 108, 138, 140, 169, 196
UV-initiator   91
UV irradiation   63, 80, 81, 92, 93, 94, 133, 139, 163, 201, 202, 270, 271

v

vitamin B12 (VB12)   154
voidless microgel   89
voidless PNIPAM microspheres   192, 193
volume phase transition   55, 57–62, 86, 89, 91, 137, 138, 140, 149, 162, 176, 189, 191–193, 205, 216, 254, 258, 274
  behaviors   57
  kinetics   86, 89
volume phase transition temperature (VPTT)   62, 65, 66, 67, 73, 98, 137, 138, 140, 141, 162, 163, 165, 166, 170, 174, 175, 189, 190, 191, 193, 195, 196, 198, 216, 217, 218, 254, 258, 259, 274, 281

w

wastewater   64, 68, 286, 288, 289
water collection   224, 241–248
water condensation   247
water-cooling   277, 278, 280, 281
water flux   275, 276, 277
water-in-oil-in-oil (W/O/O) double emulsion   38–39, 41, 43
water-in-oil-in-water-in-oil (W/O/W/O) droplets   19
water-soluble monomer   80
Weber number   15
wettability modification   12, 18, 25
wetting-induced droplet coalescence   37, 47

wetting-induced emulsion generation 44
  droplet-triggered droplet pairing 44
  monodisperse controllable double emulsions 38
  monodisperse higher-order multiple emulsions 41
  wetting-induced coalescing 37–38
  wetting-induced droplet coalescing 47
  wetting-induced spreading 36
wetting-induced spreading 36–47, 50
W/O emulsion droplets 57
W/O emulsions 44, 50, 92, 93, 95, 166
W/O/O double emulsions 39, 40, 43, 44, 45
(O1 + O2)/W/O quadruple-component double emulsions 169
W/O/W double emulsions 25, 124, 125, 127, 188, 201

x
X-shaped microchannels 256

y
Young's modulus 244